Process control in textile manufacturing

The Textile Institute and Woodhead Publishing

The Textile Institute is a unique organisation in textiles, clothing and footwear. Incorporated in England by a Royal Charter granted in 1925, the Institute has individual and corporate members in over 90 countries. The aim of the Institute is to facilitate learning, recognise achievement, reward excellence and disseminate information within the global textiles, clothing and footwear industries.

Historically, The Textile Institute has published books of interest to its members and the textile industry. To maintain this policy, the Institute has entered into partnership with Woodhead Publishing Limited to ensure that Institute members and the textile industry continue to have access to high calibre titles on textile science and technology.

Most Woodhead titles on textiles are now published in collaboration with The Textile Institute. Through this arrangement, the Institute provides an Editorial Board which advises Woodhead on appropriate titles for future publication and suggests possible editors and authors for these books. Each book published under this arrangement carries the Institute's logo.

Woodhead books published in collaboration with The Textile Institute are offered to Textile Institute members at a substantial discount. These books, together with those published by The Textile Institute that are still in print, are offered on the Woodhead web site at: www.woodheadpublishing.com. Textile Institute books still in print are also available directly from the Institute's web site at: www.textileinstitutebooks.com

A list of Woodhead books on textiles science and technology, most of which have been published in collaboration with the Textile Institute, can be found towards the end of the contents pages.

Woodhead Publishing Series in Textiles: Number 131

Process control in textile manufacturing

Edited by
Abhijit Majumdar, Apurba Das, R. Alagirusamy and V. K. Kothari

The Textile Institute

WOODHEAD PUBLISHING

Oxford Cambridge Philadelphia New Delhi

Published by Woodhead Publishing Limited in association with The Textile Institute
Woodhead Publishing Limited, 80 High Street, Sawston, Cambridge CB22 3HJ, UK
www.woodheadpublishing.com
www.woodheadpublishingonline.com

Woodhead Publishing, 1518 Walnut Street, Suite 1100, Philadelphia, PA 19102-3406, USA

Woodhead Publishing India Private Limited, G-2, Vardaan House, 7/28 Ansari Road, Daryaganj, New Delhi – 110002, India
www.woodheadpublishingindia.com

First published 2013, Woodhead Publishing Limited

British Library Cataloguing in Publication Data
A catalogue record for this book is available from the British Library.

Library of Congress Control Number: 2012949006

ISBN 978-0-85709-027-0 (print)
ISSN 2042-0803 Woodhead Publishing Series in Textiles (print)

The publisher's policy is to use permanent paper from mills that operate a sustainable forestry policy, and which has been manufactured from pulp which is processed using acid-free and elemental chlorine-free practices. Furthermore, the publisher ensures that the text paper and cover board used have met acceptable environmental accreditation standards.

Typeset by Newgen Knowledge Works Pvt Ltd, India
Printed and bound in the UK by the MPG Books Group

Contents

Contributor contact details

(* main contact)

Editors

Dr Abhijit Majumdar,* Dr Apurba Das, Prof. R. Alagirusamy and Prof. Vijay Kumar Kothari
Department of Textile Technology
Indian Institute of Technology Delhi
Hauz Khas
New Delhi - 110016
India

Email: majumdar@textile.iitd.ac.in; apurba@textile.iitd.ernet.in; alagiru@gmail.com; kothari@textile.iitd.ernet.in;

Chapter 1

Vedpal and Dr Vipul Jain*
Department of Mechanical Engineering
Indian Institute of Technology Delhi
Hauz Khas
New Delhi - 110016
India

Email: vjain@mech.iitd.ac.in

Chapter 2

Dr S. S. Saha
Department of Electrical Engineering
Government College of Engineering & Textile Technology
Berhampore
West Bengal – 742101
India

Email: sahashib@hotmail.com

Chapter 3

Dr Apurba Das
Department of Textile Technology
Indian Institute of Technology Delhi
Hauz Khas
New Delhi - 110016
India

Email: apurba@textile.iitd.ernet.in

Chapter 4

Dr M. Zimniewska* I. Frydrych and J. Mankowski
Department of Innovative Textile Technologies

Institute of Natural Fibres and Medicinal Plants
Wojska Plskiego 71b
60-630 Poznan
Poland

Email: gosiaz@inf.poznan.pl

Chapter 5

Prof. B. L. Deopura, A. Chatterjee and N. V. Padaki
Department of Textile Technology
Indian Institute of Technology Delhi
Hauz Khas
New Delhi - 110016
India

Email: bdeopura@yahoo.com

Chapter 6

Prof. R. Alagirusamy
Department of Textile Technology
Indian Institute of Technology Delhi
Hauz Khas
New Delhi - 110016
India

Email: alagiru@gmail.com

Chapter 7

Dr Anindya Ghosh*
Government College of Engineering and Textile Technology
Berhampore
West Bengal – 742101
India

Email: anindya.textile@gmail.com

and

Dr Abhijit Majumdar
Department of Textile Technology
Indian Institute of Technology Delhi
Hauz Khas
New Delhi - 110016
India

Email: majumdar@textile.iitd.ac.in

Chapter 8

Prof. Prabal Kumar Majumdar
Department of Textile Technology
Government College of Engineering and Textile Technology
Serampore - 712201 Dt. Hooghly (W.B.)
India

Email: pkm5@rediffmail.com

Chapter 9

Dr Arindam Basu
Central Silk Technological Research Institute
Central Silk Board
Madivala, BTM Layout
Bangalore - 560068
India

Email: arindambasu_dr@yahoo.co.in

Chapter 10

Prof. Sadhan Chandra Ray
Professor in Fabric Manufacture
Department of Jute and Fibre Technology
University of Calcutta
35 Ballygunge Circular Road
Kolkata - 700019
India

Email: sadhan53@yahoo.co.in

Chapter 11

Prof. Vijay Kumar Kothari
Department of Textile Technology
Indian Institute of Technology Delhi
Hauz Khas
New Delhi - 110016
India

Email: kothari@textile.iitd.ernet.in; iitkothari@gmail.com

Chapter 12

Doice Moyo, Dr Asis Patanaik* and Dr Rajesh D. Anandjiwala
CSIR Materials Science and Manufacturing
Polymers and Composites Competence Area
P.O. Box 1124
Port Elizabeth 6000
South Africa

and

Faculty of Science
Department of Textile Science
Nelson Mandela Metropolitan University
Port Elizabeth 6031
South Africa

Email: patnaik_asis@yahoo.com; APatnaik@csir.co.za

Chapter 13 and 14

Dr Songmin Shang
Institute of Textiles and Clothing
The Hong Kong Polytechnic University
Hung Hom
Kowloon
Hong Kong
China

Email: tcshang@inet.polyu.edu.hk

Chapter 15

Prof. Asim Kumar Roy Choudhury
Department of Textile Technology
Government College of Engineering and Textile Technology
Serampore - 712201 Dt. Hooghly (W.B.)
India

Email: akrc2008@yahoo.in

Chapter 16

G.Thilagavathi* and S.Viju
Department of Fashion Technology
PSG College of Technology
Peelamedu
Coimbatore - 641004
Tamilnadu
India

Email: thilagapsg@gmail.com

Woodhead Publishing Series in Textiles

23 **Coated and laminated textiles**
W. Fung
24 **Fancy yarns**
R. H. Gong and R. M. Wright
25 **Wool: Science and technology**
Edited by W. S. Simpson and G. Crawshaw
26 **Dictionary of textile finishing**
H-K. Rouette
27 **Environmental impact of textiles**
K. Slater
28 **Handbook of yarn production**
P. R. Lord
29 **Textile processing with enzymes**
Edited by A. Cavaco-Paulo and G. Gübitz
30 **The China and Hong Kong denim industry**
Y. Li, L. Yao and K. W. Yeung
31 **The World Trade Organization and international denim trading**
Y. Li, Y. Shen, L. Yao and E. Newton
32 **Chemical finishing of textiles**
W. D. Schindler and P. J. Hauser
33 **Clothing appearance and fit**
J. Fan, W. Yu and L. Hunter
34 **Handbook of fibre rope technology**
H. A. McKenna, J. W. S. Hearle and N. O'Hear
35 **Structure and mechanics of woven fabrics**
J. Hu
36 **Synthetic fibres: nylon, polyester, acrylic, polyolefin**
Edited by J. E. McIntyre
37 **Woollen and worsted woven fabric design**
E. G. Gilligan
38 **Analytical electrochemistry in textiles**
P. Westbroek, G. Priniotakis and P. Kiekens
39 **Bast and other plant fibres**
R. R. Franck
40 **Chemical testing of textiles**
Edited by Q. Fan
41 **Design and manufacture of textile composites**
Edited by A. C. Long
42 **Effect of mechanical and physical properties on fabric hand**
Edited by H. M. Behery
43 **New millennium fibers**
T. Hongu, M. Takigami and G. O. Phillips
44 **Textiles for protection**
Edited by R. A. Scott
45 **Textiles in sport**
Edited by R. Shishoo
46 **Wearable electronics and photonics**
Edited by X. M. Tao
47 **Biodegradable and sustainable fibres**
Edited by R. S. Blackburn
48 **Medical textiles and biomaterials for healthcare**
Edited by S. C. Anand, M. Miraftab, S. Rajendran and J. F. Kennedy
49 **Total colour management in textiles**
Edited by J. Xin
50 **Recycling in textiles**
Edited by Y. Wang
51 **Clothing biosensory engineering**
Y. Li and A. S. W. Wong

52 **Biomechanical engineering of textiles and clothing**
Edited by Y. Li and D. X-Q. Dai
53 **Digital printing of textiles**
Edited by H. Ujiie
54 **Intelligent textiles and clothing**
Edited by H. R. Mattila
55 **Innovation and technology of women's intimate apparel**
W. Yu, J. Fan, S. C. Harlock and S. P. Ng
56 **Thermal and moisture transport in fibrous materials**
Edited by N. Pan and P. Gibson
57 **Geosynthetics in civil engineering**
Edited by R. W. Sarsby
58 **Handbook of nonwovens**
Edited by S. Russell
59 **Cotton: Science and technology**
Edited by S. Gordon and Y-L. Hsieh
60 **Ecotextiles**
Edited by M. Miraftab and A. R. Horrocks
61 **Composite forming technologies**
Edited by A. C. Long
62 **Plasma technology for textiles**
Edited by R. Shishoo
63 **Smart textiles for medicine and healthcare**
Edited by L. Van Langenhove
64 **Sizing in clothing**
Edited by S. Ashdown
65 **Shape memory polymers and textiles**
J. Hu
66 **Environmental aspects of textile dyeing**
Edited by R. Christie
67 **Nanofibers and nanotechnology in textiles**
Edited by P. Brown and K. Stevens
68 **Physical properties of textile fibres Fourth edition**
W. E. Morton and J. W. S. Hearle
69 **Advances in apparel production**
Edited by C. Fairhurst
70 **Advances in fire retardant materials**
Edited by A. R. Horrocks and D. Price
71 **Polyesters and polyamides**
Edited by B. L. Deopura, R. Alagirusamy, M. Joshi and B. Gupta
72 **Advances in wool technology**
Edited by N. A. G. Johnson and I. Russell
73 **Military textiles**
Edited by E. Wilusz
74 **3D fibrous assemblies: Properties, applications and modelling of three-dimensional textile structures**
J. Hu
75 **Medical and healthcare textiles**
Edited by S. C. Anand, J. F. Kennedy, M. Miraftab and S. Rajendran
76 **Fabric testing**
Edited by J. Hu
77 **Biologically inspired textiles**
Edited by A. Abbott and M. Ellison
78 **Friction in textile materials**
Edited by B. S. Gupta
79 **Textile advances in the automotive industry**
Edited by R. Shishoo
80 **Structure and mechanics of textile fibre assemblies**
Ediited by P. Schwartz

81 **Engineering textiles: Integrating the design and manufacture of textile products**
Edited by Y. E. El-Mogahzy
82 **Polyolefin fibres: Industrial and medical applications**
Edited by S. C. O. Ugbolue
83 **Smart clothes and wearable technology**
Edited by J. McCann and D. Bryson
84 **Identification of textile fibres**
Edited by M. Houck
85 **Advanced textiles for wound care**
Edited by S. Rajendran
86 **Fatigue failure of textile fibres**
Edited by M. Miraftab
87 **Advances in carpet technology**
Edited by K. Goswami
88 **Handbook of textile fibre structure Volume 1 and Volume 2**
Edited by S. J. Eichhorn, J. W. S. Hearle, M. Jaffe and T. Kikutani
89 **Advances in knitting technology**
Edited by K-F. Au
90 **Smart textile coatings and laminates**
Edited by W. C. Smith
91 **Handbook of tensile properties of textile and technical fibres**
Edited by A. R. Bunsell
92 **Interior textiles: Design and developments**
Edited by T. Rowe
93 **Textiles for cold weather apparel**
Edited by J. T. Williams
94 **Modelling and predicting textile behaviour**
Edited by X. Chen
95 **Textiles, polymers and composites for buildings**
Edited by G. Pohl
96 **Engineering apparel fabrics and garments**
J. Fan and L. Hunter
97 **Surface modification of textiles**
Edited by Q. Wei
98 **Sustainable textiles**
Edited by R. S. Blackburn
99 **Advances in yarn spinning technology**
Edited by C. A. Lawrence
100 **Handbook of medical textiles**
Edited by V. T. Bartels
101 **Technical textile yarns**
Edited by R. Alagirusamy and A. Das
102 **Applications of nonwovens in technical textiles**
Edited by R. A. Chapman
103 **Colour measurement: Principles, advances and industrial applications**
Edited by M. L. Gulrajani
104 **Fibrous and composite materials for civil engineering applications**
Edited by R. Fangueiro
105 **New product development in textiles: Innovation and production**
Edited by L. Horne
106 **Improving comfort in clothing**
Edited by G. Song
107 **Advances in textile biotechnology**
Edited by V. A. Nierstrasz and A. Cavaco-Paulo
108 **Textiles for hygiene and infection control**
Edited by B. McCarthy
109 **Nanofunctional textiles**
Edited by Y. Li
110 **Joining textiles: Principles and applications**
Edited by I. Jones and G. Stylios

111 **Soft computing in textile engineering**
Edited by A. Majumdar
112 **Textile design**
Edited by A. Briggs-Goode and K. Townsend
113 **Biotextiles as medical implants**
Edited by M. King and B. Gupta
114 **Textile thermal bioengineering**
Edited by Y. Li
115 **Woven textile structure**
B. K. Behera and P. K. Hari
116 **Handbook of textile and industrial dyeing. Volume 1: Principles, processes and types of dyes**
Edited by M. Clark
117 **Handbook of textile and industrial dyeing. Volume 2: Applications of dyes**
Edited by M. Clark
118 **Handbook of natural fibres. Volume 1: Types, properties and factors affecting breeding and cultivation**
Edited by R. Kozłowski
119 **Handbook of natural fibres. Volume 2: Processing and applications**
Edited by R. Kozłowski
120 **Functional textiles for improved performance, protection and health**
Edited by N. Pan and G. Sun
121 **Computer technology for textiles and apparel**
Edited by J. Hu
122 **Advances in military textiles and personal equipment**
Edited by E. Sparks
123 **Specialist yarn and fabric structures**
Edited by R. H. Gong
124 **Handbook of sustainable textile production**
M. I. Tobler-Rohr
125 **Woven textiles: Principles, developments and applications**
Edited by K. Gandhi
126 **Textiles and fashion: Materials design and technology**
Edited by R. Sinclair
127 **Industrial cutting of textile materials**
I. Viļumsone-Nemes
128 **Colour design: Theories and applications**
Edited by J. Best
129 **False twist textured yarns**
C. Atkinson
130 **Modelling, simulation and control of the dyeing process**
R. Shamey and X. Zhao
131 **Process control in textile manufacturing**
Edited by A. Majumdar, A. Das, R. Alagirusamy and V. K. Kothari
132 **Understanding and improving the durability of textiles**
Edited by P. A. Annis
133 **Smart textiles for protection**
Edited by R. Chapman
134 **Functional nanofibers and applications**
Edited by Q. Wei
135 **The global textile and clothing industry: Technological advances and future challenges**
Edited by R. Shishoo
136 **Simulation in textile technology: Theory and applications**
Edited by D. Veit
137 **Pattern cutting for clothing using CAD: How to use Lectra Modaris pattern cutting software**
M. Stott

Foreword

I am pleased to write a Foreword for this unique and comprehensive volume on *Process control in textile manufacturing* edited by Abhijit Majumdar, Apurba Das, R. Alagirusamy and V. K. Kothari. The editors and authors have compiled urgently needed information, reflecting the highest standard of publication in this field, on a very important aspect of the conversion of fibers into the ultimate textile structures employed in various end-use applications. The subject matter discussed in this text is timely. The globalization of trade now demands that products marketed around the world meet certain basic quality and performance requirements. The current production of all types of fibers such as natural, manmade, and synthetics now exceeds 80 odd million tons annually. The use of suitable textile fibers in such highly sophisticated applications as medical devices, civil and construction engineering, automotive industry, filtration and aircraft and aerospace industries as well as for more mundane purposes, such as apparel, requires the maintenance of high quality products. 'Engineering With Fibers', a term that was coined and made popular by my mentor and professional colleague, Subhash Batra, is quite appropriate in the current environment of utilization of textiles in highly engineered textile structures. Consequently, the engineers and technologists engaged in machine design and processing of fibers must be knowledgeable regarding the availability and the performance of process control devices used in the industry. It has become mandatory on the part of textile manufacturers that they certify and guarantee the performance characteristics of highly engineered textile structures designed for composite applications.

The authors have done an excellent and noteworthy job of introducing the reader to the fundamentals of process and quality control in the cultivation of natural fibers and in the manufacture of synthetic fibers. Separate chapters on the process and control in the main systems of yarn production, nonwovens, knitting, weaving and subsequent handling of products in the dyeing and finishing and apparel manufacturing of textiles are important and extremely useful. In today's highly competitive environment the manufacturers must be diligent as well as expert in placing products in the market that are well-made and that meet the expectations of the industry and the customer. This volume will be be an important resource in the hands of textile engineers, technologists and managers engaged in the development and manufacture of textile products.

Bhuvenesh C. Goswami, Emeritus Alumni Distinguished Professor of Textiles, Clemson University, USA

Part I
General issues

1
Basics of process control in textile manufacturing

VEDPAL AND V. JAIN,
Indian Institute of Technology Delhi, India

DOI: 10.1533/9780857095633.1.3

Abstract: New technologies are constantly being developed and the service life of products is becoming shorter due to changes in customer taste. Statistical process control (SPC) methods for short production runs are therefore increasingly important. Multistage process surveillance and fault diagnosis have become a necessity. Process control incorporating SPC and hybrid approaches are discussed in this chapter. Overall the chapter helps in building a sound understanding about process control with a flavour of different approaches that are helpful in analysing a process which further helps in taking appropriate decisions.

Key words: statistical process control, control charts, manufacturing, quality, hybrid systems.

1.1 Introduction

In today's dynamic manufacturing environment there is a constant need for improvement in processes. A company cannot rest on past performance and expect to remain successful. Customers' expectations change rapidly and manufacturing processes have become very demand-driven. More precise specifications present a new challenge that may require the upgrading of existing processes and modern technologies. As a result, process control has become more challenging. Pressure on time and resources, the need for faster delivery, defect-free products, and the reliability and durability of parts and processes, are some of the critical factors facing manufacturing enterprises. Process control plays a key role in building recognition in a competitive market.

Modern technology is capable of the simultaneous control of many variables and the collection of data may appear easy. However, process control also involves costs, for example:

- expenditure on initial analysis of the process,
- expenditure on implementation and integration.

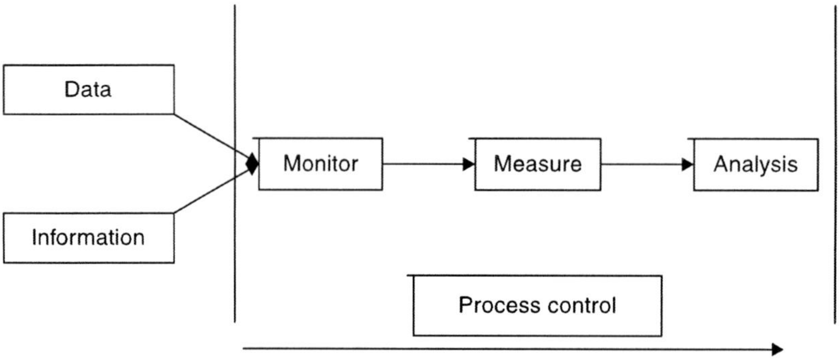

1.1 Some steps in process control.

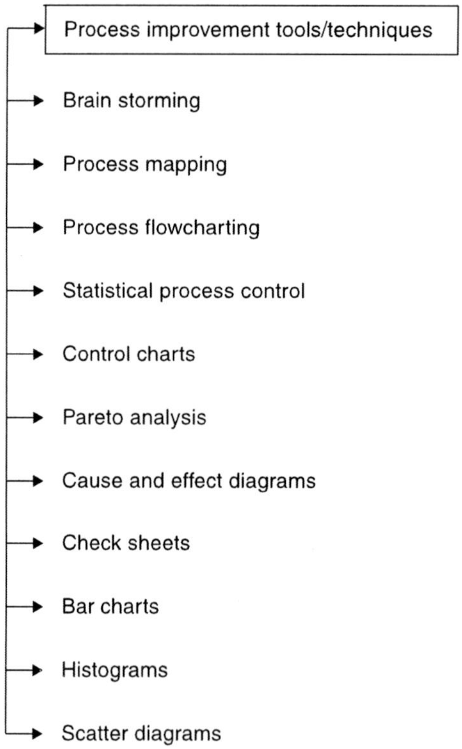

1.2 Process control tools and techniques.

It is important to be aware of the goals and benefits of process control at the outset. Process control can add value if used effectively in decision making, for example:

- If the process is under control, the desired product quality will be achieved.

- If process analysis shows a technology is not serving its purpose, it can be modified or replaced to improve product quality, save costs or improve productivity.

Some basic steps in process control are shown in Fig. 1.1. Some of the tools in process analysis, measurement and control are shown in Fig. 1.2. These techniques are discussed later in the chapter.

1.2 Process mapping, analysis and control

Product quality is directly dependent upon the process quality. To achieve an excellent quality and defect-free operations process mapping becomes very important. It starts with understanding the process, its approach and the application level and presenting the information with graphical representation. Once the process mapping is done, the control and analysis part comes into picture. This allows us to determine whether the process is in control or not and to analyse if the quality improvement efforts have the desired effect.

This is discussed in detail in the following subsections.

1.2.1 Process mapping

One of the fundamental steps in understanding or improving a process is process mapping. Information about the activities/steps that take place in a process is gathered and mapped so as to create a model representing the complete process. Complex interactions may be represented in a simplified manner. Activities can be grouped into several sub-processes. By the use of process maps, improvement teams are better able to understand the process and become more efficient in ensuring effective control and finding opportunities for improvement.

To ensure process control is effective, the correct product attributes must first be identified. Data or information about the product is collected. The specific characteristics of the product are referred to as the critical attributes and determine the type of measurement. Not all attributes will be critical for the customer. Some may be very important while others may add little value to the product. It is necessary to use appropriate tools to determine a ranking procedure. Among these may be failure mode and effect analysis (FMEA), which analyses the manner in which failure is observed, the causes of the failure, and the effect of failure on the product and other operations.

A process flowchart is an important tool in the construction of a process and provides a snapshot of the complete process. Standard symbols are used for drawing the flowchart. Through process flowcharting, a conclusion may be drawn as to why any redundant operations are being carried out, the final objective being to seek opportunities for process improvement.

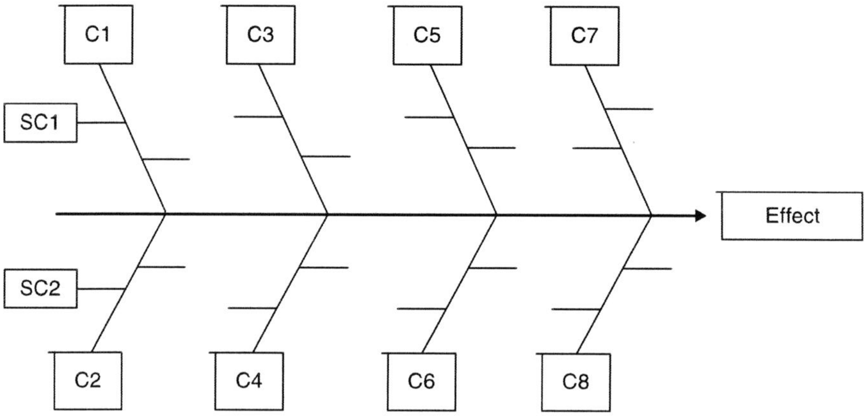

1.3 Cause–effect diagram.

A cause-and-effect diagram is also known as the Fishbone or Ishikawa diagram. It is an effective way of mapping the inputs that affect quality. It is also a very effective tool in problem solving.

In this approach, as shown in Fig. 1.3, the potential causes of the problem are written as C1 to C8. Then the sub-causes (SC1, SC2, etc.) that are relevant to the principal causes are identified. This method is very useful for identifying the causes of a problem. Brainstorming sessions may also be useful in understanding the possible causes of a problem.

1.2.2 Process analysis and control

Once a process has been mapped, its performance can be measured and assessed. It is important to be aware of the relationship between a manufacturing control system and statistical process control (SPC) system. In a manufacturing control system, the processes to be carried out are carefully planned and executed according to the initial plan. A processing run of the chosen products is processed in a manufacturing facility as defined by the manufacturing model. An SPC analysis is then carried out on the processed units to identify and analyse the process. The manufacturing facility is then modified according to the SPC analysis. These tools and techniques help a production manager to adjust a process and to prevent it from going out of control. An uncontrolled process may cause considerable wastage of resources and time. Organisations alert to this will invest in creating and upgrading the process control systems in their manufacturing facilities.

Various factors are important in controlling the different stages of a process. These are presented in Fig. 1.4. Defects are examined with the help of quality tools. Problems that are solvable online are taken care of by line

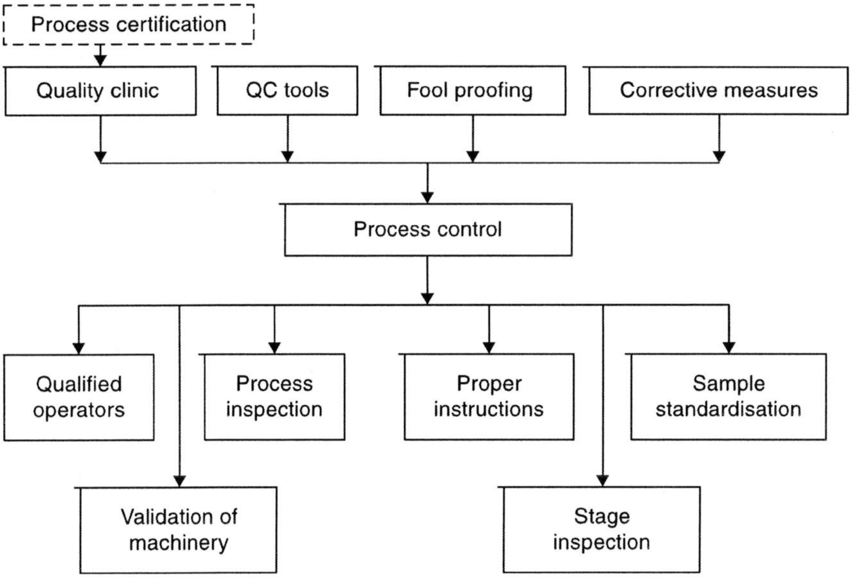

1.4 Manufacturing process control.

staff. More complex problems are dealt with in quality clinics. The aim of these is to ensure that durable, mistake-proof corrective actions are taken. Among the tools used in quality clinics is Root Cause Analysis. The processes that have the most impact on quality are identified and studied in detail to assess how they can be enhanced. Once implemented, the revised process is tested. The quality team will then carry out a final inspection by selecting samples according to a standard plan.

A check sheet is a tool to help in collecting the data in an efficient and organised way. Data is categorised to simplify the task of analysis. The data is collected and ordered by adding check marks. A simple example is given in Table 1.1. Control charts are used to monitor processes with the help of measures such as arithmetic means and other ranges. Control charts measure limits above and below the mean. These are called the upper and lower control limits. These control charts give real time information about the process and are very helpful in detecting and predicting variations when used with SPC. A typical control chart consists of a centre line and two control limits (upper and lower). The control limits are normally located at $\pm 3\sigma$ of this statistic and the centre line refers to the average level of the statistically controlled process. The X bar and R (Range) chart are the most commonly used charts for monitoring variations and measuring variable quality characteristics. If a particular variation is observed in the process, it is necessary to identify an assignable cause for this variation. If observations show fluctuations outside the control limits, the process is defined as going out

Table 1.1 Check sheet for data collection

Check sheet category	Sub-category	Check marks	Total
Failure	Run outs	IIII IIII III	11
	Wears	IIII IIII IIII	12
Appearance	Finish	IIII I	5

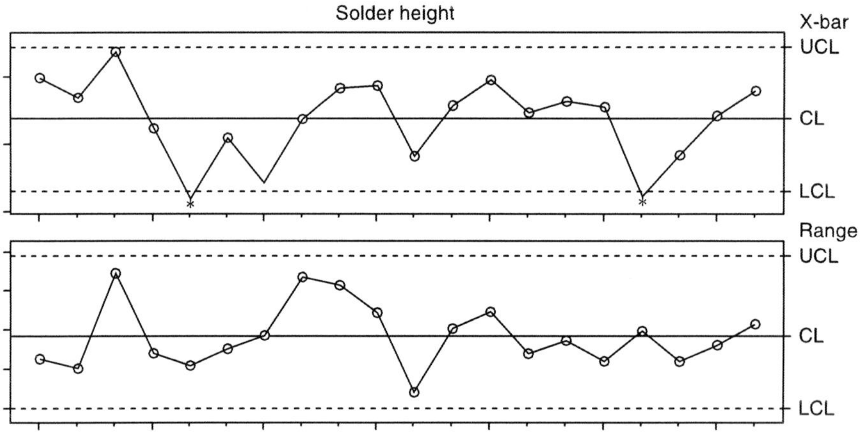

1.5 X bar and range (R) chart.

of control and the assignable causes must be eliminated in order to bring it back under control. However, a process may be defined as out of control even if all the observed points are within the control limits (Montgomery, 1996). This occurs when an unnatural pattern is displayed. Clues from this pattern of observation provide the user with information about the process and assist in eliminating the causes of variation.

In Fig. 1.5, the X bar and R chart show the observations recorded at different time intervals for the solder heights. The central limit (CL), the upper (UCL) and lower (LCL) control limits are set according to the process requirements. In the R chart, all observations are within the control limits, whereas in the X chart, the observations marked with (*) are below the LCL, representing an out-of-control process. This shows that the process average is influenced by some external factor and corrective actions are needed to eliminate the causes.

A bar chart is a visual display tool in which the relative sizes of the measured data are displayed by the height of the bars. The bars are separated to show where data are not continuous. The bar chart helps in comparing different types of data. A scatter diagram gives a graphical representation of the data. It shows the changes in a variable between one state and another. These variables are plotted at right angles to each other and the scatter marks made

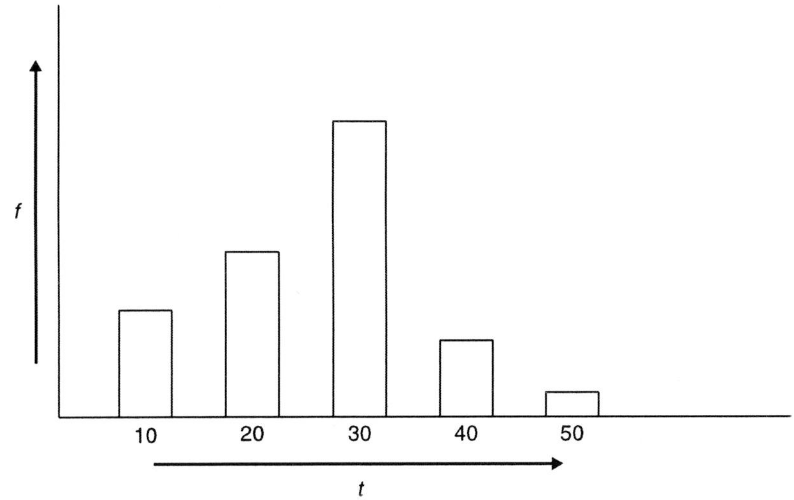

1.6 Histogram as a tool for process analysis.

accordingly. This tool shows how two variables are related, how one variable changes if another is changed, and if there is a relationship between them. Many predictions can be made depending on the range of data available.

A histogram shows the probability of a particular value occurring. Here, the data are grouped in cells and the relative frequency represented in bars. Figure 1.6 illustrates a histogram taking frequency (f) and time (t) (or any other parameter) into consideration on the vertical and horizontal axes respectively. It is used to assess large amounts of data with a wide range.

1.3 Statistical process control (SPC) and improving processes

Statistical techniques can be applied to data at diverse levels of organisations to identify when a process is behaving unusually. Further, each organisation has numerous processes and variables that can affect product and service outcomes such as quality, productivity etc. Hence, there is always a scope for improvement. Since the improvement is an ongoing process and requires substantial efforts to achieve better results.

The detailed discussion is as follows.

1.3.1 SPC

SPC is a toolkit for managing processes. It is a method of controlling manufacturing processes to ensure that the outputs conform to specifications.

A process may go out of control for various reasons, which may include variability in materials and machinery, variability in the process parameters, or may be due to some chance variations. A process running out of control is a cause of concern and immediate steps must be taken to identify the causes. SPC provides the means of analysing such a process. SPC was first developed in the USA during the 1940s. Initially, it was largely ignored until the Japanese (who were introduced to SPC by W. Edwards Deming in 1950) launched a national effort to improve the quality and productivity of manufacturing (Ishikawa, 1985). In SPC, decisions are taken on the basis of information for a particular process in a data recording system. SPC incorporates various tools such as control charts, histograms and check sheets, which are used for recording and analysing data.

SPC attempts to understand the critical variables in each sequence of a manufacturing process and their intercorrelation. Through statistical sampling and experimentation, these interrelations may be understood and thus rendered controllable (Bushe, 1988). In Japanese factories, workers sample parts as they move through the manufacturing processes using control charts. If any deviation from the specifications is found, it is readjusted. Therefore all finished products meet the specifications. Discovery of the location of deviations improves the tolerances and capabilities of the processes and enhances continuous improvement of product quality and efficiency. There can be various cultural issues in understanding and implementing SPC successfully in the workplace. As an example, factories tend to be highly segmented in their technical, structural, cultural and political dimensions. Work is also highly segmented and there is a tendency to deal with each problem in isolation. However, SPC takes a holistic approach and requires multivariate thinking and working.

Artificial intelligence (AI) has great potential in an automated SPC system (Medsker, 1995). In this system, control and diagnosis are done by neural networks and expert systems respectively. The neural networks are used for control-chart pattern recognition and the expert systems are used for monitoring variations in the process. The expert system contains the process-specific knowledge, which enables it to diagnose an out-of-control process and to suggest what corrective actions could be taken. By incorporating quality–cost simulation technology, expert systems such as IntelliSPC have been able to monitor and predict costs over time (Kuo and Huang, 1997; Guh *et al.*, 1999).

Consistency in quality and machine utilisation is very important in the Flexible Manufacturing System (FMS). FMS can be modelled with Petri nets. However, these models lack the functionality of SPC and therefore do not make possible a complete FMS. To overcome this, the Petri net-based SPC model is used where measured data from the inspection machines and sensor data from the devices is utilised. The cause/effect of the product's

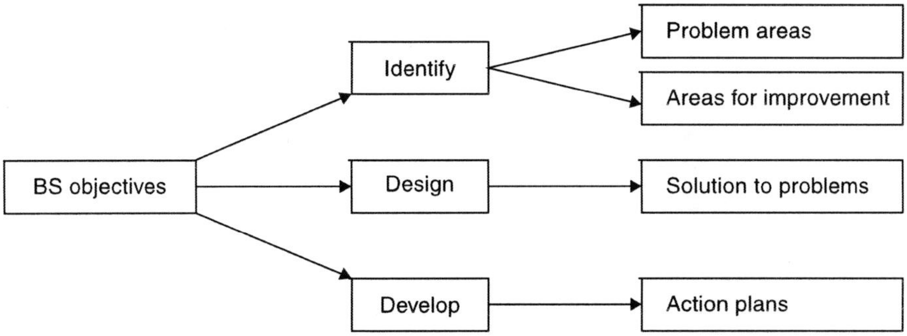

1.7 Brainstorming objectives and their further classifications.

defects and quality can be analysed by the Petri net-based SPC model and the diagnosis can easily be integrated into the model.

1.3.2 Improving processes

A brainstorming (BS) process may be called to discuss the improvement plans for a process. The problem brief is assisted by the process maps. The BS process is carried out without any criticism or comparison and all ideas are recorded for further analysis. BS is an effective tool for opening out the thinking horizon and focusing on a problem to generate ideas that may be used for improving the process. It can also be used with the cause-and-effect tool. A large number of ideas are generated that may be used in numerous situations. Some objectives of BS and their classification is presented in Fig. 1.7.

Pareto analysis can be used to analyse the ideas generated by the BS. The causes having the greatest impact are identified through analysis The data are represented in the form of a bar chart ranked in the descending order of their frequencies. This analysis typically shows that 80% of the effect corresponds to 20% of the causes and is therefore known as the 80/20 rule. Pareto analysis is used to examine the relative contribution of different defects/causes which lead to rejection. The relative frequency of different categories of defects is ranked by means of the Pareto diagram. The quality-control team can then assess the contribution of various defects and prioritise accordingly for their removal. Figure 1.8 illustrates the use of Pareto analysis to analyse faults in a printed circuit board related to soldering. It is clear from Fig. 1.8 that the pin hole is the principal defect, followed by the open joint and so on. The corrective actions may then be prioritised accordingly.

1.4 Future trends

SPC has become one of the most commonly used tools for maintaining acceptable and stable levels of quality in modern manufacturing. The modern

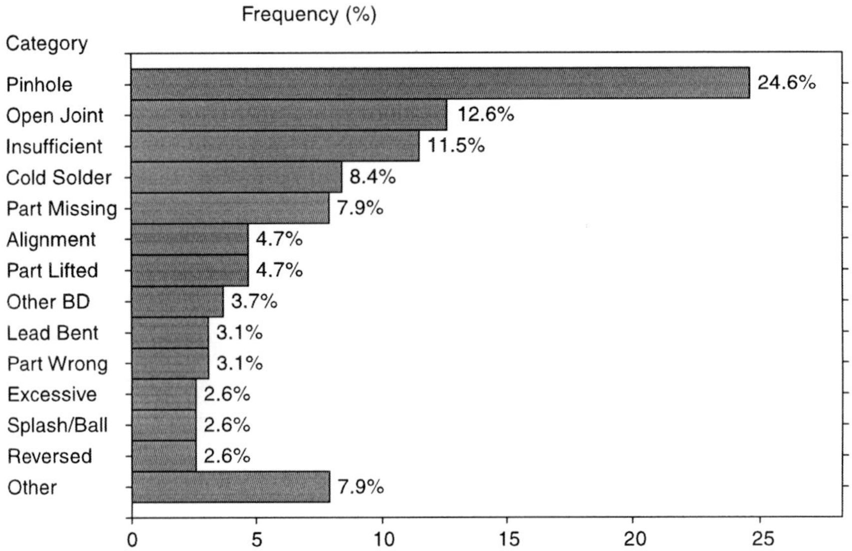

1.8 Pareto diagram showing the frequency of defects.

manufacturing environment is focused on computer integrated manufacturing and the challenges lie in developing advanced computer algorithms and process controls to implement the SPC tasks automatically.

Currently, the focus is on unit process-control methods such as run-2-run (R2R), unit process development and transfer and improvements in the methods to ensure component functionality and reliability. Considerable potential has been identified in the manufacturing of health-related systems and various health-monitoring systems have been developed or are in the development stages.

Much work is being done on the process of prediction and the improvement of product parameters and yield. New methods which help in process improvement, such as virtual metrology have been developed, incorporating control density improvement and the reduction of measurement operations. Models for data visualisation and analysis are in progress and still more effective models related to process improvement are to be developed.

The modern manufacturing world is demanding more precise and accurate methods for meeting industrial expectations. Advanced process control methods are always necessary across a variety of applications. More sophisticated methods of fault diagnosis are therefore being developed by researchers. Sensor implementation and integration with numerically controlled machines are developing rapidly. Investment in sensor technology that provides real time information for modern computer integrated manufacturing is increasing and more research is under way to meet the requirements of industries worldwide.

1.5 References

Bushe, G. R. (1988), 'Cultural contradictions of statistical process control in American manufacturing organizations', *Journal of Management*, Vol. 14, No. 1, pp. 19–33.

Guh, R. S., Tannock, J. D. T. and O Brien, C. (1999), 'IntelliSPC: a hybrid intelligent tool for on-line economical statistical process control', *Expert Systems with Applications*, Vol. 17, No. 3, pp. 195–212.

Ishikawa, K. (1985), *What is total quality control? The Japanese Way*, Prentice-Hall, Englewood Cliffs, NJ, USA.

Kuo, C. H. and Huang, H. P. (1997), 'Colored timed PETRI NET based statistical process control and fault diagnosis to flexible manufacturing systems', *Proceedings of the 1997 IEEE International Conference on Robotics and Automation Albuquerque*, April 1997, New Mexico, USA.

Montgomery, D. C. (1996), *Introduction to Statistical Quality Control*, 3rd Ed. Wiley, New York, USA.

Medsker, L. R. (1995), *Hybrid Intelligent Systems*, Kluwer Academic Publishers, Boston, MA, USA.

2
Basic principles of control systems in textile manufacturing

S. S. SAHA, Government College of Engineering & Textile Technology, Berhampore, India

DOI: 10.1533/9780857095633.1.14

Abstract: Robust control of machines and processes of textile manufacturing has become paramount during last few decades in order to cope with the stringent quality requirements. This chapter presents a brief account of basic control systems so that practising textile engineers can control processes effectively and efficiently. Details of various control strategies and their application for speed control have been considered in this chapter, along with design guidelines. Use of modern techniques of digital control and control with artificial intelligence has also been discussed in brief. A case study describing the application of two different techniques in textile manufacturing has also been presented with performance comparison.

Key words: closed-loop control, PID-controller, transfer function, intelligent controller, digital control.

2.1 Introduction

The demand for automatic control is becoming increasingly important in various industries. In manufacturing processes it ensures that certain physical parameters such as temperature, pressure, speed or voltage are maintained at optimum values in relation to other variables. In other words, the responsibility of control engineering is to bring these physical parameters to certain pre-defined values, set points (SP), and then maintain them. The design of a control system only makes sense if there is some objective or value that needs to be achieved to ensure a system performs as required, for example, maintaining a particular speed of rotation. Fundamental to any control system is the ability to measure the output of a system and to take corrective action if the output deviates from its desired value. In driving a vehicle for example, the action of steering is controlled by the feedback from driver's senses. The eyes and ears transmit signals to the brain, which interprets this signal and transmits a signal to the arms to turn the steering wheel, adjusting the actual direction of movement to bring it in line with the desired direction. Thus, steering an automobile constitutes a feedback-control system.

An early example of a control system is the flyball governor, developed by James Watt in 1788 for controlling the speed of a steam engine. Another significant date in the history of automatic feedback-control systems is 1934 when Hazen's paper 'Theory of Servomechanisms' was published in the Journal of the Franklin Institute, marking the beginning of intense interest in this new field. The word 'servomechanism' actually originated in this paper. The late 1950s saw the development of conventional classical control theory. The concept of an intelligent control system was first used in the 1990s with the development of fuzzy logic controllers, artificial neural networks, etc.

2.1.1 Principles of process control: open-loop and closed-loop control systems

The simplest form of control is open-loop control. In an open-loop control system an actuator is used to directly control the system output without utilising any information about the physical output. The actuator signal to the system is, therefore, unaffected by and unaware of the actual output of the system. The response of an open-loop system is only dependent on the characteristics of the system itself rather than the relationship between the output signal and system input.

One example of an open-loop control system is a separately excited d.c. motor, which is intended to drive the shaft of a machine tool as shown in Fig. 2.1. The spindle of the machine tool is required to be rotated at a constant speed, which is determined by the machine operator. In this system, for a fixed magnetic field, a required value of d.c. voltage (V_t) is applied to the armature of the d.c. motor to produce the desired motor speed. The output voltage (V_t) of the d.c. regulator (d.c.–d.c. converter) is determined by the d.c. supply voltage (V_S) and operation duty cycle (k) of the converter. Very often the reference input or set point (SP) is directly related to the desired value of system output and thus the desired output parameter can be obtained by adjusting the SP by the operator. In the example of Fig. 2.1,

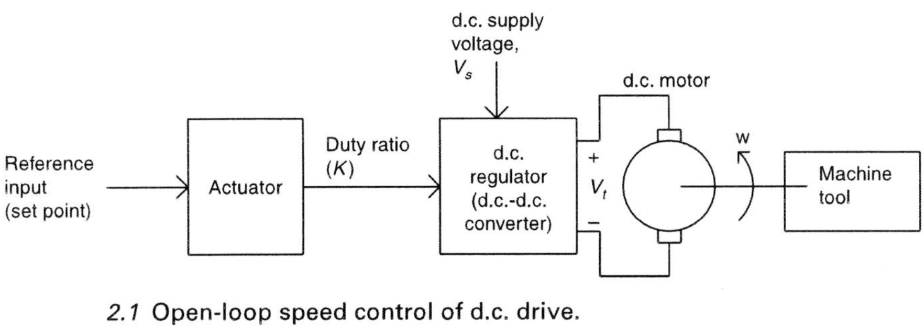

2.1 Open-loop speed control of d.c. drive.

the motor, along with the d.c. regulator, forms the system, the duty cycle is the input quantity, which is set by the actuator and the speed of the shaft is the output parameter.

The terminal voltage (V_t) applied to the d.c. motor by the d.c.–d.c. converter can be expressed as follows:

$$V_t = k \cdot V_S$$

where V_t = motor terminal voltage, k = duty ratio of the d.c.–d.c. converter and V_S = d.c. supply voltage.

In a separately excited motor with a fixed magnetic field excitation, like the one shown in Fig. 2.1, the speed of the motor (output) is a function of both the d.c. motor terminal voltage and the load torque. The machine tool will maintain a constant speed, provided there are no changes to either of these inputs. If the d.c. supply voltage (V_S) to the regulator is varied however, then the d.c. motor terminal voltage (V_t) will be disturbed, and if the load on the motor is changed the current drawn by the motor from the d.c. regulator will change too. In either circumstance the speed of the motor (system output) will change and, in an open-loop system, there is no feedback to correct the input levels and reset them to the desired settings. The machine tool will therefore, maintain the new constant speed until either input function is altered.

In contrast to open-loop control systems, closed-loop control systems utilise the additional actions of measurement and feedback of the actual output quantity and compare it with the desired reference quantity at input. Using this feedback the actuator input signal can thereby be changed to achieve the desired output. In closed-loop systems then, the output value has a direct effect upon the input quantity. In the d.c. motor example, a person could be assigned the task of checking the actual speed of the motor (output) and comparing it with the desired reference speed (input). If the output did not have the desired value, the person could alter the duty ratio, k, of the d.c. regulator to achieve the desired motor speed. Introducing a person into the system provides a means through which the output is fed back and compared with the input. Any necessary changes can then be made in order to ensure that the output matches the desired value. The feedback action therefore controls the effective input to the dynamic unit. To improve the performance of closed-loop systems, so that the output quantity is as close as possible to the desired quantity, the person in this example can be replaced by any mechanical, electrical, or other form of feedback control unit capable of performing the same task.

A closed-loop control system can be represented by a simplified block diagram containing all the essential components as shown in Fig. 2.2. In this example the sensor senses the physical parameter at the output and converts

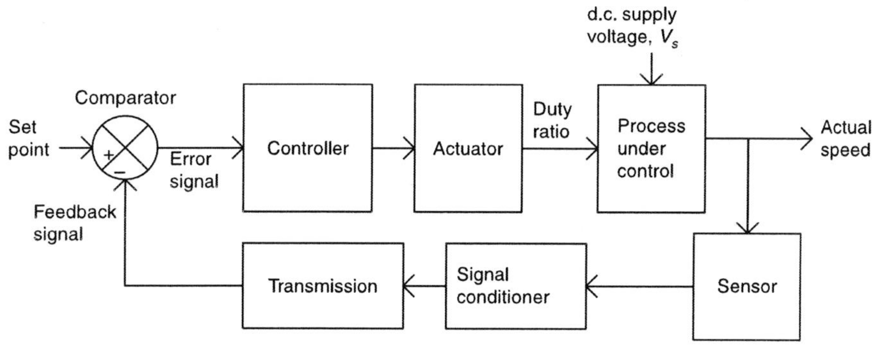

2.2 Closed-loop speed control of d.c. drive.

it into an equivalent electrical signal. This signal is then conditioned and transmitted to the comparator. The comparator compares the feedback signal to the reference signal, which is set to the desired output. The difference between these two signals is called the error signal. Based on the polarity and magnitude of the error signal, the controller sends a signal to the actuator detailing the necessary corrective measures to be taken. The actuator then adapts accordingly to attain the desired output of the system. In this way, the system can achieve the desired output regardless of disturbance from outside. The actuator in an automated process may in fact consist of a combination of several actuators, each providing an output that drives the following one in a predetermined sequence. Once feedback is applied, the system becomes a closed loop. Closed-loop systems can achieve far greater accuracy than open-loop systems, although they rely entirely on the accuracy of the comparison between the desired and the actual output values and, therefore, on the accuracy of the measured output value.

2.1.2 Feedback and feed-forward control

In a feedback-control system, the corrective action starts only if there is deviation of the output variable from its expected value. In the d.c. drive in Fig. 2.1 for example, if the d.c. supply voltage changes, then the speed of the motor deviates from the reference speed and the comparator generates an error signal for the controller. The controller now acts accordingly and sends the necessary signal to the actuator to minimise the deviation. If the system works with a long time constant, then it will be some time before any corrective action takes place.

If, however, the change in d.c. supply voltage can be measured as soon as it occurs and the information is forwarded directly to the controller, then necessary action can be taken immediately by sending a signal to the actuator before any deviation in output behaviour. Thus, a system with feed-forward control can take necessary corrective action as soon as any measurable disturbance occurs in the system. In this way feed-forward control can minimise the transient error even if large and frequent disturbances occur in the system. The limitation of feed-forward control, however is that this method can only account for measurable disturbances. It would therefore, be advantageous to include a feedback control loop in addition to feed-forward control, if there is any possibility of unmeasurable disturbance occurring in the system.

2.2 Components of control systems

The sequential tasks performed by a control system are measurement of actual output quantity by suitable sensor or transducer, transmission of the measured variable after proper conditioning, comparison of the signal with the desired reference quantity at input and then performing the task of taking corrective action by the controller, if the output deviates from its desired value. Thus, a control system is composed of various basic components.

2.2.1 Sensors/transducers

In most control systems, the appropriate physical parameter of a system (e.g. displacement, speed, temperature, pressure, etc.) intended to be controlled at the desired level is sensed and converted into an equivalent electrical quantity, by a device called a transducer. Some commonly used transducers are:

Position transducer:	Potentiometer, LVDT, optical encoder, synchros.
Motion transducer:	LVDT.
Temperature transducer:	Thermocouple, thermistor, RTD.
Force transducer:	Strain gauge, load cell.
Flow transducer:	Orifice plate, venture tube, Pitot tube.
Pressure transducer:	Diaphragm, bellows, Bourdon tube.

2.2.2 Signal conditioner and transmitter

In many applications, the transducer is used as one of the four arms of a Wheatstone bridge in order to properly sense the electrical signal obtained from the transducer. The transducer signal is then obtained as a difference in voltage across the arm containing the transducer. The electrical signal generated by transducers is often too low in voltage and power level to be transmitted and compared by the controller. To increase the strength of this signal, instrumentation amplifiers, such those shown in Fig. 2.3, are used. The instrumentation amplifier uses an operational amplifier (OPAMP) as a differential amplifier tool to sense the difference between the signal from

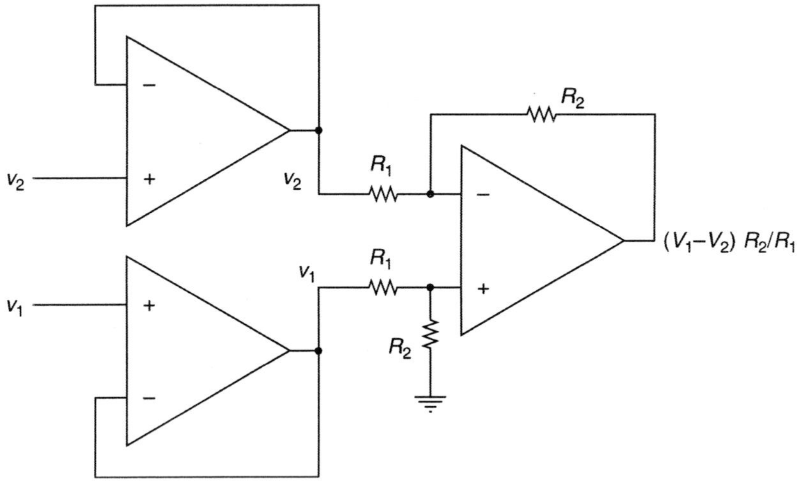

2.3 Instrumentation amplifier.

the transducer and the reference signal, while rejecting the common mode signal. The two OPAMPs used at the inputs of the instrumentation amplifier act only as unity gain buffer. They electrically isolate the part of the transducer directly connected to the actual physical system, often functioning at a high voltage, from the expensive control unit, thereby eliminating the risk of extensive damage to the control unit (e.g. computer system, its peripherals, etc.) and of fatal accidents to operating personnel. The input OPAMPs also have high input resistance and thus avoid causing any electrical loading to the low-power transducer signal.

In most process industries, the transducers are installed at the load (process) end and the controllers are installed at the central control unit, which may be far away from the load point. Transmission of the transducer signal over a long distance may result in a considerable loss in signal quantity, random electromagnetic interference (EMI) from neighbouring power sources and radio frequency interference (RFI) noise from communication systems. All these factors may alter the quality of the actual feedback signal. In such cases, therefore, it is necessary for signal transmission to be made through shielded cables, which are grounded at regular intervals so that vast majority of noise signals pass into the ground through a low resistance path without causing any interference to the transducer signal.

2.2.3 Comparator and controller

The comparator and controller form the central and intelligent section of the control system. The comparator has two inputs: the SP indicates the reference value of the parameter and the process variable (PV) indicates the

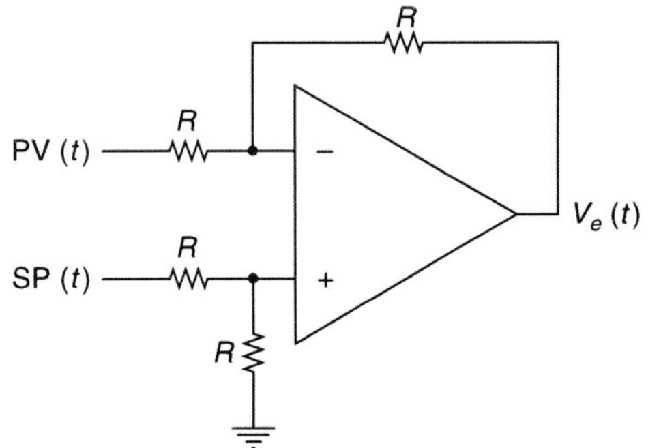

2.4 Unity gain error amplifier.

actual value of the parameter. The comparator compares these two signals to obtain the error signal, $V_e(t)$. Based on the polarity and magnitude of the error signal, the controller drives the actuator to modify the process in such a way as to make the PV equal to the SP. Thus, the comparator always consists of an error amplifier followed by the main controller, though its type might vary depending on the process requirements. An error amplifier can easily be created electronically using an OPAMP in subtraction mode, as in Fig. 2.4. The output of the OPAMP is the error signal given by $V_e(t)$ = SP – PV.

2.2.4 ON–OFF controller

The ON–OFF controller's output is either fully ON or fully OFF, based on the PV and whether it is below or above set point. For practical applications, however, an ON–OFF controller should always have a dead band or hysteresis to prevent excessive cycling or chattering. Whether the controller is ON or OFF is dictated by the error signal. When the error signal exceeds a specified limit, the controller turns ON, and if the error signal falls below another specified limit the controller turns OFF. An electronic version of an ON–OFF controller using an OPAMP is illustrated in Fig. 2.5.

In this example, the controller in its OFF condition only switches to ON when the error signal V_e exceeds the level of E_2, given by

$$E_2 = \frac{R_2}{R_1 + R_2} V_{\text{Sat}}$$

where V_{Sat} is the saturation output voltage of OPAMP at OFF state and vice versa; once the controller is turned ON, it will turn OFF again only when the error signal, V_e falls below the level of E_1 given by

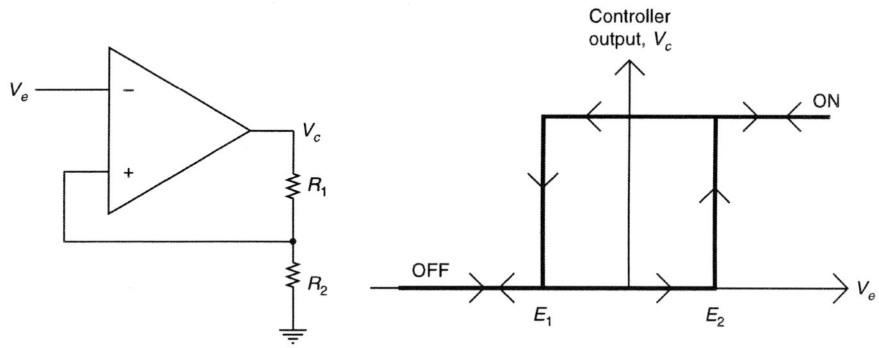

2.5 ON–OFF controller.

$$E_1 = -\frac{R_2}{R_1 + R_2} V_{\mathrm{Sat}}$$

2.2.5 Proportional (P-)controller

In a P-controller the output is proportional to the error signal, V_e. A P-controller using an OPAMP is illustrated in Fig. 2.6, along with its response for a unit step error signal. As the OPAMP used in this configuration acts inversely, a second OPAMP is used in unity gain inverting amplifier mode to re-invert the inverted signal.

The output of first stage OPAMP (P-controller) is given by

$$V'(t) = -K_p V_e'(t);$$

where $K_p = R_2/R_1$; K_p is called the proportional constant.

The output of second stage of OPAMP (unity gain inverting amplifier) is given by

$$V_C(t) = -V(t) = K_p V_e(t)$$

The P-controller is very fast and responds immediately and proportionally to the error signal, that is, according to variations in the PVs relative to the reference set point. If the error signal is zero, then a zero signal will be sent by the P-controller to the actuator. This is not desirable, since a non-zero signal is required for the actuator to provide an output. Referring back to the d.c. drive example in Fig. 2.1, when the speed of the d.c. motor

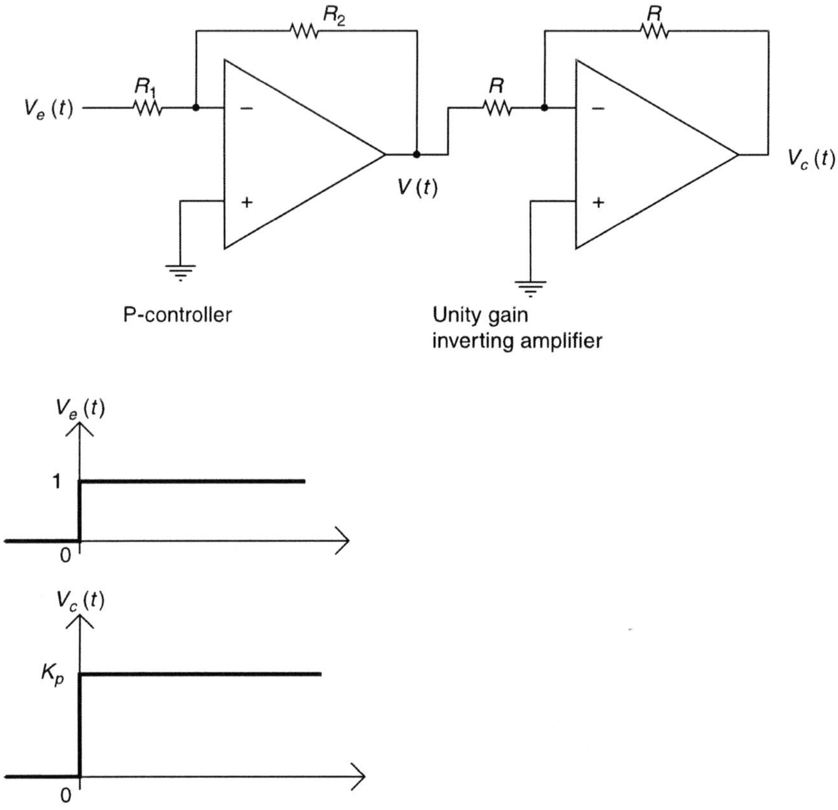

2.6 Proportional controller and its response for unit step input.

deviates from the reference speed, the error signal, representing the difference between the reference speed and the actual speed, sends a proportional signal to the actuator. The actuator then modifies the speed of the motor accordingly. If the motor attains the reference speed, however, then the error signal will be zero, which will produce a zero signal for the actuator and mean no output. To ensure that the actual speed follows the reference speed in a steady state, a specified non-zero signal for the actuator must exist to ensure constant output. P-controllers, therefore, have a specified non-zero value error signal called a 'steady-state error' or 'off-set' error of the P-controller. The steady-state error of the P-controller can be greatly reduced by selecting a large gain for the controller, but this may also cause system oscillation and reduce the stability of the system, as shown in the following example.

If we model the d.c motor of Fig. 2.1 with following parameters it can then be considered as test model in subsequent sections for computer simulation:

$$\text{Armature resistance, } r_a = 0.5\,\Omega$$
$$\text{Armature inductance, } L_a = 1.5\,\text{mH}$$
$$\text{Frictional coefficient of load, } B = 0.0001\ \text{Nm/rad/s}$$
$$\text{Inertia constant of load and motor, } J = 0.00025\ \text{Nm/rad/s}^2$$
$$\text{E.m.f. constant of motor, } K_e = 0.05\,\text{V/rad/s}$$
$$\text{Torque constant of motor, } K_t = 0.05\ \text{Nm/A}$$

In order to verify the dependence of performance on the proportional constant, the P-controller has been tested under a closed-loop control with small and large values of K_p. The response (output) for unit step input (set point) is shown in Fig. 2.7. It can be seen that, for a small value of K_p, the system output has a steady-state error, that is, the output is much less than unity at steady state. However, with large value of K_p, the steady-state error is minimised, but at the cost of system oscillations at the output.

2.2.6 Integral (I-)controller

The I-controller works on the principle that the rate of change in output is proportional to that given by the error signal. An OPAMP based I-controller and its response to a unit step error signal is shown in Fig. 2.8. The output of the integral controller is given by

$$V_c(t) = -V(t) = K_i \int V_e(t)\, dt;$$

where $K_i = 1/R_iC_i$; K_i is called the integral constant.

An I-controller will respond rapidly to a large error signal, swiftly changing its output to correct the error. As the error gets smaller, however, the controller's output will change more slowly. Once the error signal has been reduced to zero, there will not be any change in the controller's output. Thus, the integral controller is a relatively slow controller and it acts until the error signal disappears. The I-controller's performance in the closed-loop control of a d.c. drive has been verified through computer simulation. The step input response of the system is shown in Fig. 2.9.

2.2.7 Proportional-integral (PI-)controller

The I-controller has very poor transient response. Upon the occurrence of a step error signal the P-controller will immediately respond by stepping

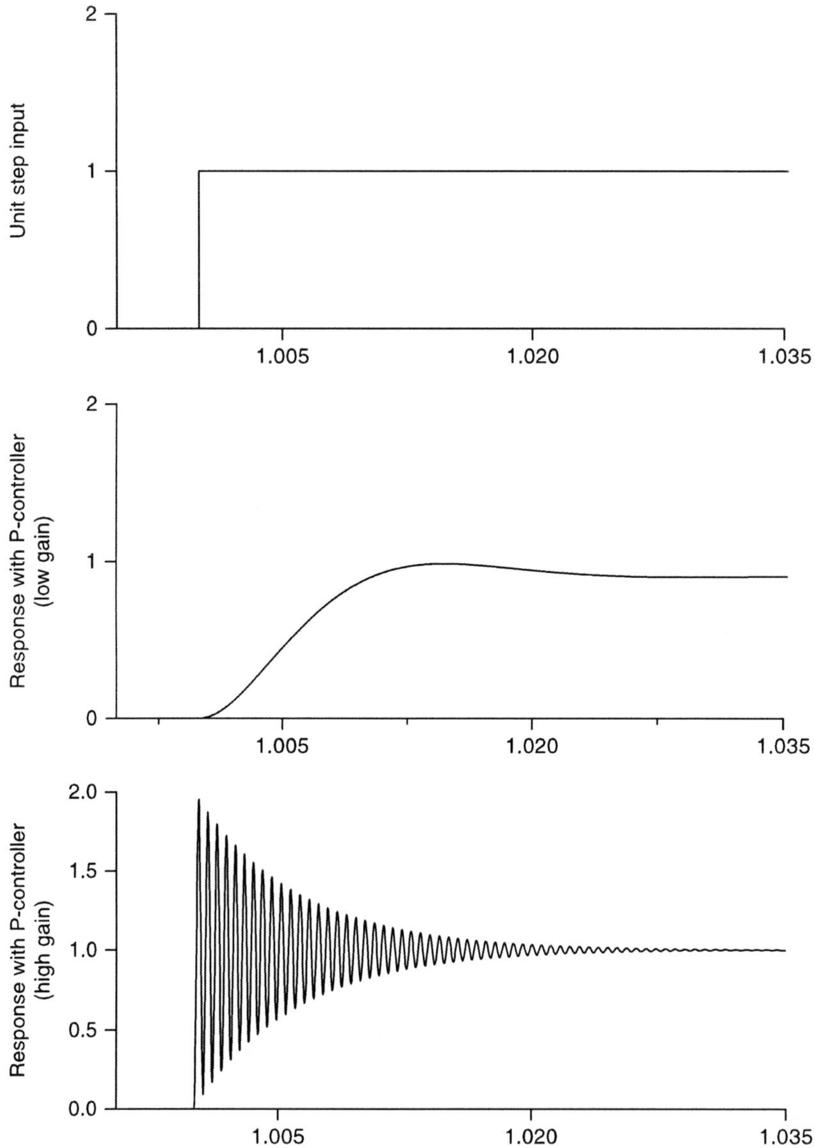

2.7 Step response of proportional controller with small and large value of K_p.

its output proportionally. The I-controller, however, will start changing its output continuously until all steady-state error has been eliminated. Thus, a P-controller is more useful for attaining a fast transient response, whilst an I-controller can minimise the steady-state error. The advantages of both

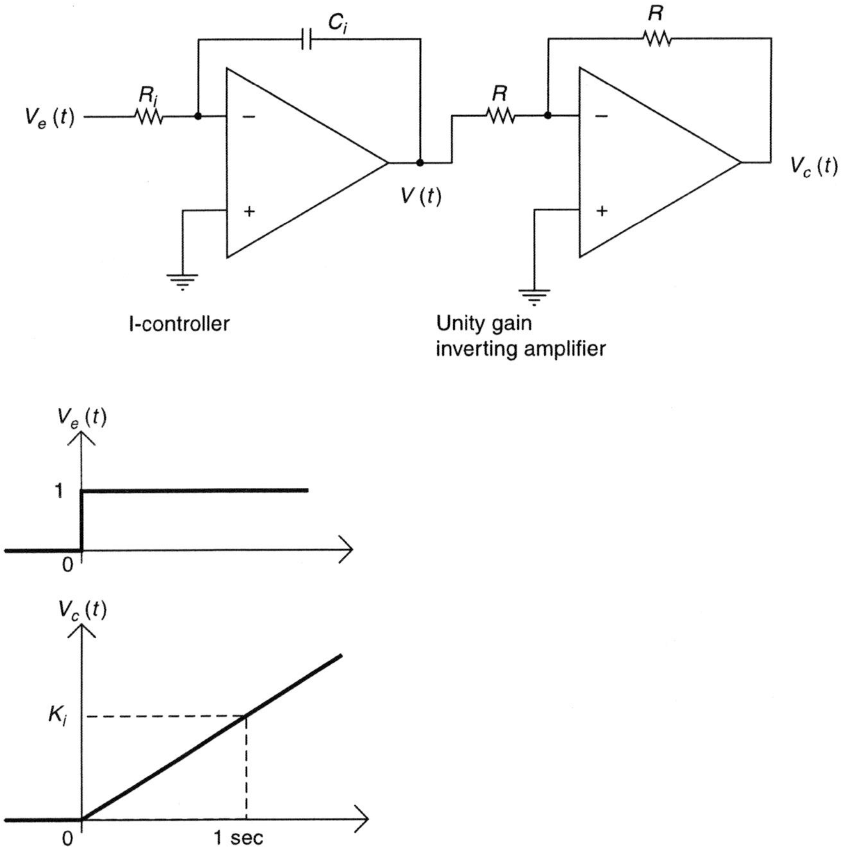

2.8 Integral controller and its response for unit step input.

these controllers can be obtained when their features are combined into a single controller to form a PI-controller, as shown in Fig. 2.10 along with its step input response. A PI-controller can also be developed by paralleling separate blocks of a P- and an I- controller. This configuration has the advantage of easy and independent adjustment of K_p and K_f.

The output of a PI-controller is given by

$$V_C(t) = -V(t) = K_p V_e(t) + K_i \int V_e(t)\,dt;$$

where $K_p = R_f/R_i$ and $K_i = 1/R_iC_i$; K_p and K_i are proportional and integral constant, respectively.

The response of the PI-controller for closed-loop control of the test d.c. drive has been simulated and is presented in Fig. 2.11.

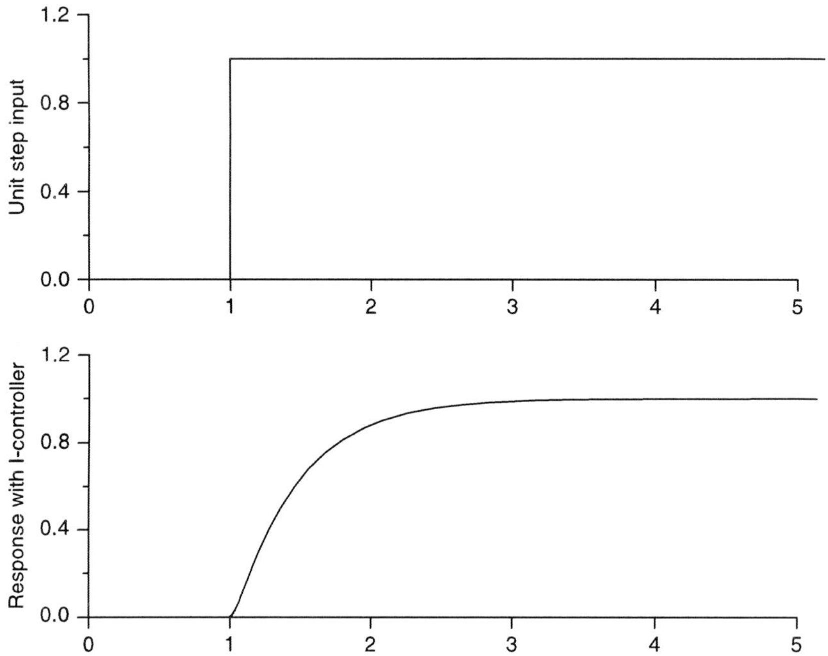

2.9 Unit step response of integral controlled d.c. drive.

2.2.8 Derivative (D-)controller

The PI-controller can provide a fast transient response as well as a good steady-state error. However, processes with large inertia (motors, etc.) require some form of additional kick, along with their response to the step error signal, in order to overcome their inertia and to respond much faster to error steps than can be achieved with a P-controller. This improved transient response can be provided by a D-controller. The output of a D-controller is proportional to the rate of change in the error signal. Figure 2.12 shows an electronic D-controller and its unit step response. The D-controller produces an output only when there are changes in the error signal. However, the output of the D-controller is limited to V_{Sat}, the saturation voltage of a practical OPAMP. Thus, even a system with a large constant error will have no output other than at the instant of the occurrence of the error signal. D-controllers are, therefore, only used in combination with a P- or a PI-controller.

The output of the D-controller is given by

$$V_c(t) = -V(t) = K_d \frac{V_e(t)}{\mathrm{d}t};$$

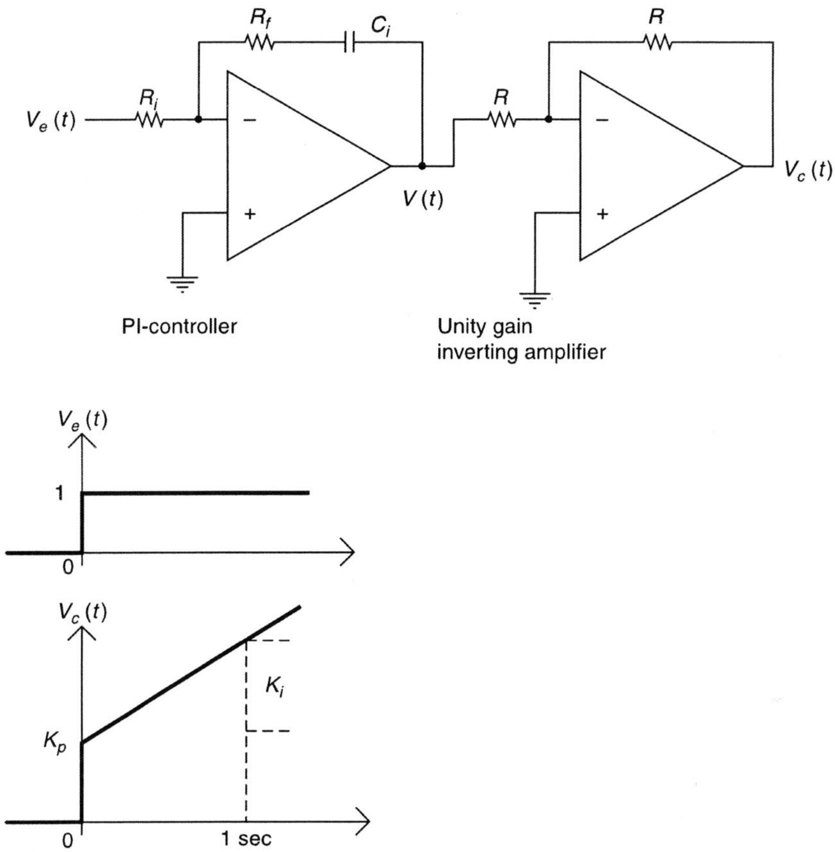

2.10 PI-controller and its response for unit step input.

where $K_d = R_d C_d$; K_d is called the derivative constant.

2.2.9 Proportional-integral-derivative (PID-) or three term controller

A three term or proportional-integral-derivative (PID)-controller can be formed by combining a P-controller, an I-controller and a D-controller to create a single controller or a parallel combination of three separate controllers (P, I and D), as shown in Fig. 2.13 along with its unit step response. PID-controllers provide a rapid transient response and eliminate the steady-state error. The PID-controller is the dominant control structure used in industrial control due to the fact that it offers a practical and inexpensive solution to the majority of processes that can be approximated by low-order dynamics.

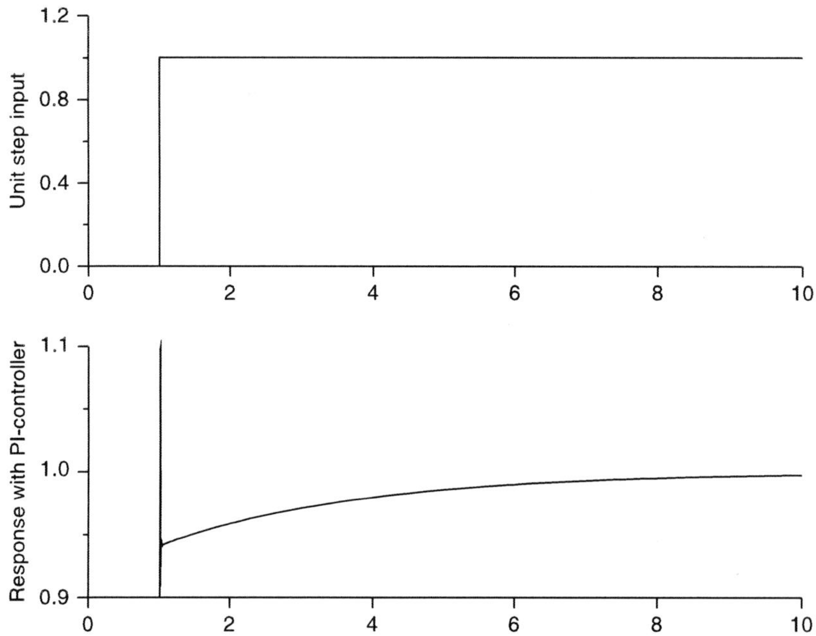

2.11 Unit step response of proportional-integral controlled d.c. drive.

The performance of a PID-controller for closed-loop control of the d.c. drive has been verified under test conditions. The simulated unit step response is shown in Fig. 2.14.

2.3 The control system design process

One disadvantage of using a feedback-control system is the oscillations at the output, which are otherwise not present as long as the system operates in open-loop mode. The oscillations are due to an attempt by the controller to change the system output to minimise the error signal as much as possible. Thus, the output swings first in one direction and then to the other, thereby gradually reducing the variation. The advantages of using closed-loop systems are obtained therefore at the expense of system stability.

Before implementation of a closed-loop control system, a suitable controller first needs to be selected and properly designed so that it can tightly regulate the output within appropriate limits. The controller is intended to send the required signal to the actuator based on all possible error signals. So, if the controller is not selected or designed properly, then, on occurrence of an error signal, it may over-compensate the system output by sending a signal to the actuator of a magnitude much higher or lower than actually

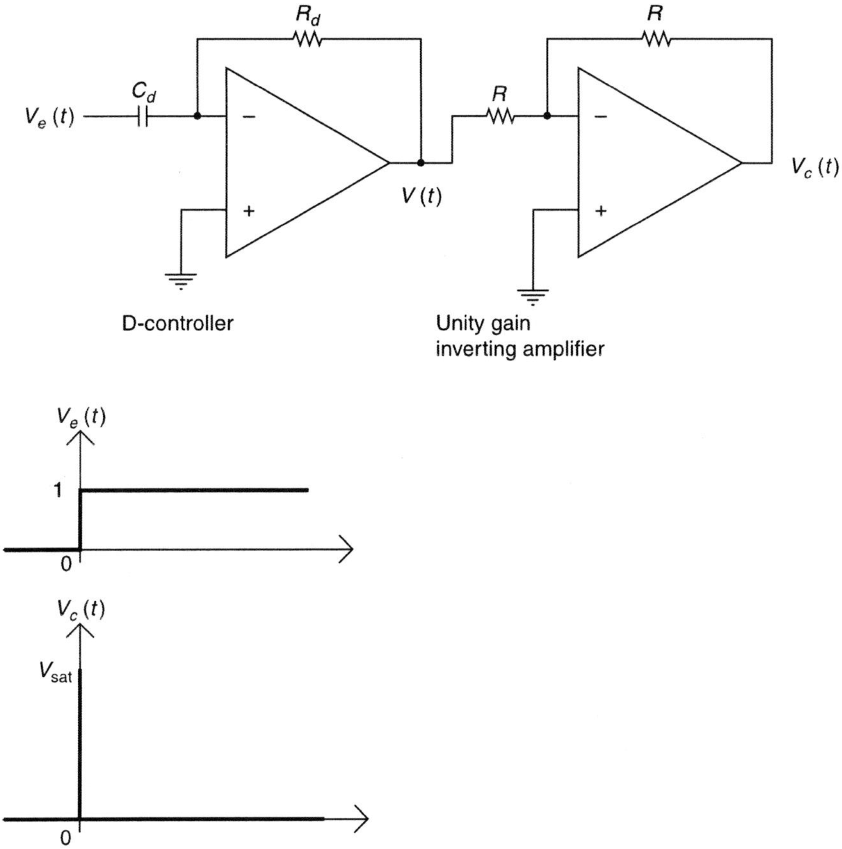

2.12 Derivative controller and its response for unit step input.

required. Any over-compensation of output will further increase the error signal and as a result, the faulty controller may further deviate the output from the expected range. This cumulative effect may ultimately bring the system out of stabilisation. Therefore, before actual implementation of the controller, the closed-loop control system needs to be simulated, tested and analysed using all the worst-case error signals.

2.3.1 The transfer function in process control systems

During the simulation and testing of a closed-loop control system all the blocks of the system need to be represented by their transfer functions, which characterise the input–output relationship of each component. The transfer function is defined as the ratio of the Laplace transform of the output (response function) to the Laplace transform of the input (driving

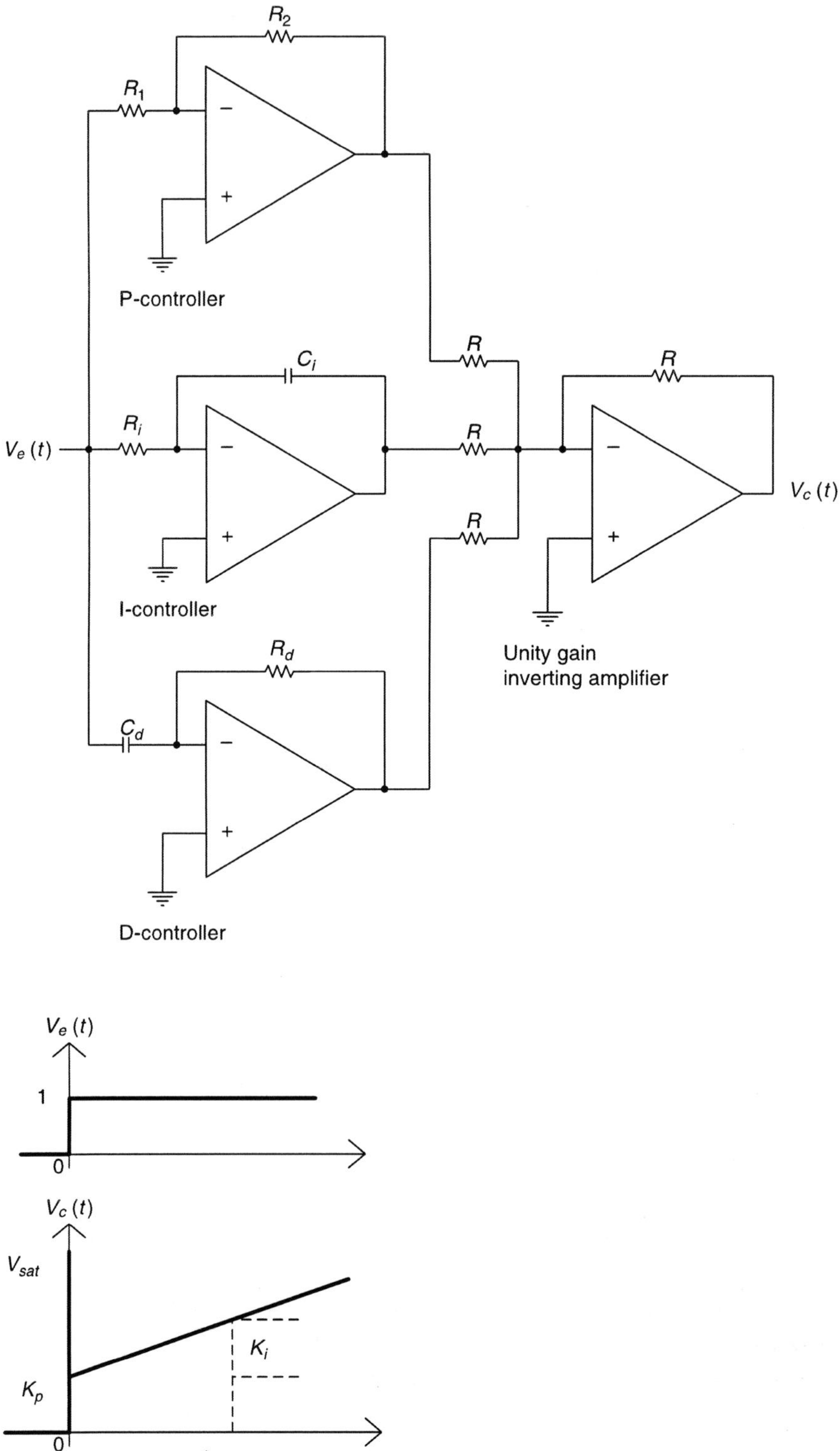

2.13 PID-controller along with unit step response.

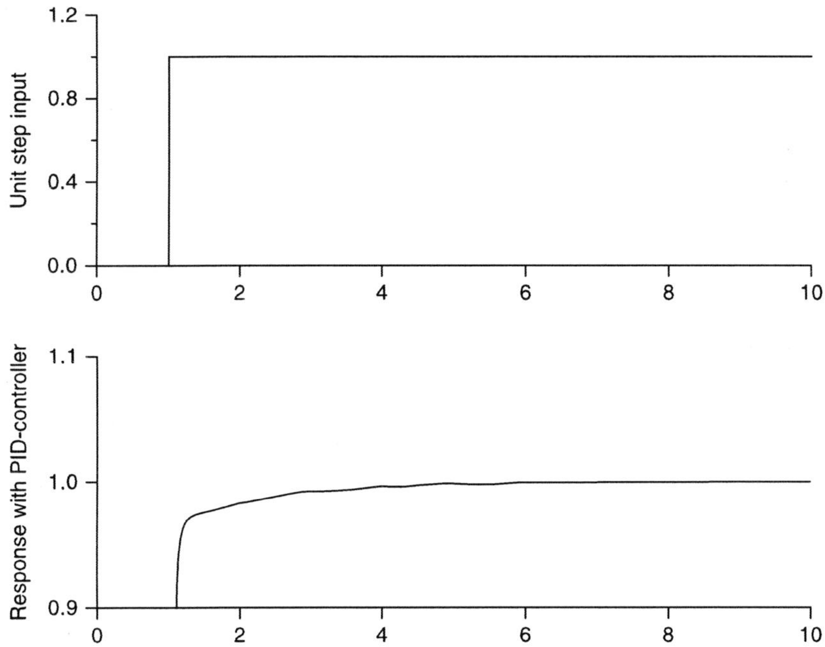

2.14 Unit step response of PID-controlled d.c. drive.

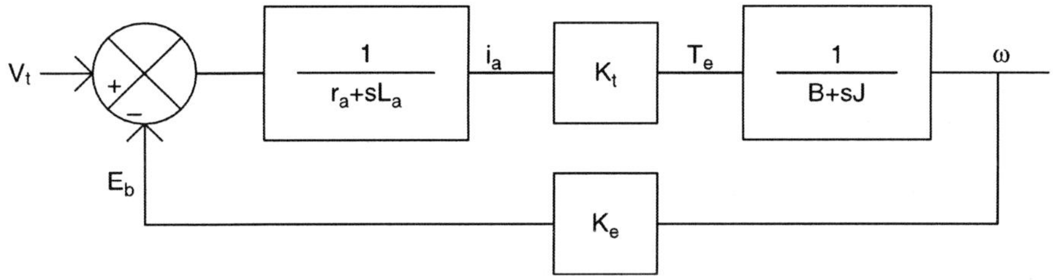

2.15 Functional block diagram of d.c. motor with transfer function of each block.

function) under the assumption that all initial conditions are zero, and it is obtained from the system mathematical model that is a set of linear, differential equations describing the system dynamics and the functions performed by each component in that system. Figure 2.15 shows the block diagram of the dc drive system along with the transfer function of each block.

As an example, the dynamic behaviour of the d.c. motor with a fixed magnetic field excitation of Fig. 2.1 can be expressed by the following differential equations:

$$V_t - E_b = r_a i_a + L_a \frac{\mathrm{d}i_a}{\mathrm{d}t}$$

where

V_t = Motor terminal voltage

$E_b = K_e\,\omega$ = Back e.m.f. developed across the motor

K_e = E.m.f. constant of motor

ω = Angular velocity of motor rotation

r_a = Armature resistance

L_a = Armature inductance

i_a = Armature current

Again, the dynamic torque expression of the motor is given by

$$T_e = T_L = B\omega + J\frac{\mathrm{d}\omega}{\mathrm{d}t}$$

where $T_e = K_t\,i_a$ = Electromagnetic torque developed by d.c. motor

T_L = Loead torque

B = Frictional coefficient of load

J = Load inertia

Thus, the open-loop control system of Fig. 2.1, can be represented by a mathematical model comprising the above differential equations and the transfer function of the complete system can be defined,

$$\text{Transfer function} = \frac{\omega(s)}{V_t(s)} = \frac{K_e}{(r_a + sL_a)(B + sJ) + K_e K_t}$$

2.3.2 Determination of controller key parameters

The design of a control system is an important and specific part of engineering design. The goal of control system design is to obtain the correct configuration, specification and determination of key parameters necessary for the controller to meet the specific system requirements.

The first task of control system design is modelling the process, which means expressing the physical system under examination in terms of a mathematical model that can be easily dealt with and understood. A model can often be made up of a complicated collection of identities and equalities even for apparently simple systems, and particularly when attempting to account for all possible eventualities. Control system implementation with such types of models can lead to an extremely costly design exercise,

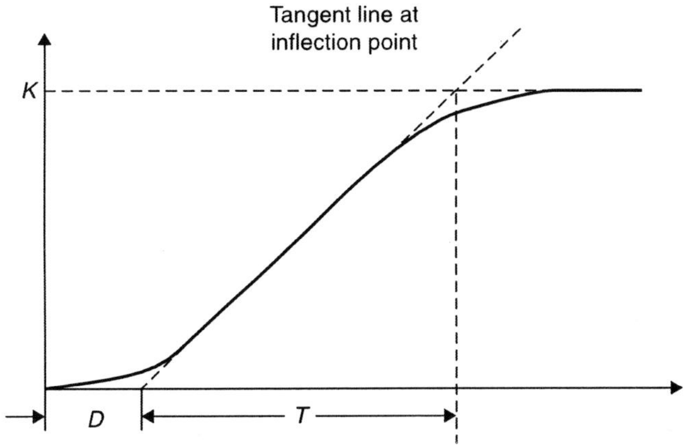

2.16 Determination of delay time (D) and time constant (T) from S-curve.

due to the time taken to attain an optimum performance in the presence of many complexities. Conversely, if the model describing a physical system is over simplified, such that important properties of the system are not included, then it may lead to an incorrect analysis and an inadequate controller design. A certain amount of common sense and practical experience is therefore necessary when forming a mathematical model of a physical system, in order to decide which characteristics of the system are important within the confines of the particular set-up in which the system has to operate.

Once a physical system has been modelled, the Laplace transform method is invoked to determine the transfer function of the system, which is used in subsequent design and analysis stages.

The next step in control system analysis is the design stage, in which a suitable control strategy is selected in order to achieve the desired system performance. The previously obtained system model is greatly used during this stage. Adjustment of the parameters of the controller is carried out in this stage through successive iteration until the performance specifications are met. If the performance meets the specifications, then the design is finalised. Once the design is complete, the controller is often implemented into the hardware. Upon arrival at the site where the system will be used, however, the parameters of the controller are often retuned in order to optimise its performance.

The design of controller parameters K_p, K_i and/or K_d can be carried out with optimum system performance using classical control system-design techniques such as Routh-Harwitch criteria, Root locus analysis, Bode plot and Nyquist criteria. Use of these techniques involves considerable mathematics. An easier, but relatively less accurate empirical method, the 'Step

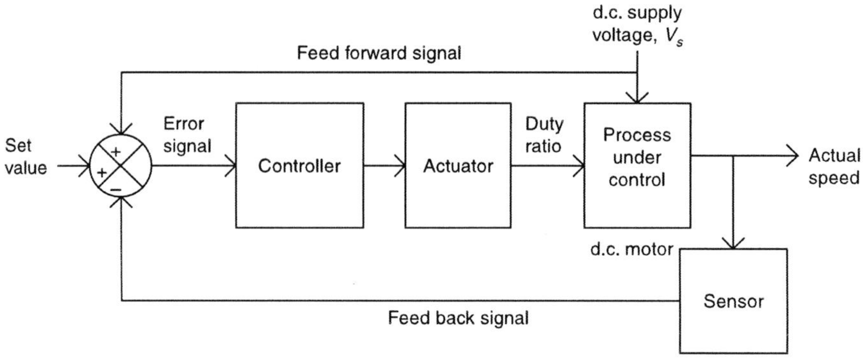

2.17 Combined feed-forward and closed-loop feed-back control system.

Response Process Reaction Method' suggested by Ziegler-Nichols[4] is still in use for many process control systems whose open-loop step response is nearly S-shaped as shown in Fig. 2.17. The S-shaped curve may be characterised by two constants, 'delay time' (D), also called 'transportation lag', and 'time constant' (T). The delay time and time constant are determined by drawing a tangent line at the inflection point of the S-shaped curve and determining the intersections of the tangent line with the time axis. Once these two parameters are determined, the parameters of the PID-controller can be determined from the empirical relations given in Table 2.1.

The process reaction method cannot, however, be used in systems in which the open-loop step response has a large overshoot. For such systems, another empirical method, called the 'continually cycling method', suggested by Ziegler-Nichols[4] can be useful. In this process, the system is first tested in a closed loop with only a P-controller. The gain of the P-controller is increased until the system controlled output oscillates continually with constant amplitude. At this stage, the controller gain is called the 'ultimate gain' (K_{Cr}) and the time period of oscillation is called the 'ultimate period' (T_{Cr}). Once K_{Cr} and T_{Cr} have been established, the parameters of the PID-controller can be calculated, as shown in Table 2.1.

2.4 Digital control systems

The extraordinary development of digital computers (microprocessors, microcontrollers) and their extensive use as controllers in a variety of fields and applications, has brought about important changes in the design of control systems. The superior performance, low cost and design flexibility now available in various types of control systems has significantly increased

Table 2.1 Ziegler-Nichols PID parameter tuning rules

Type of controller	K_p	K_i	K_d
Ziegler-Nichols PID parameter tuning rules based on step response method			
P-controller	$\frac{T}{D}$	0	0
PI-controller	$0.9\frac{T}{D}$	$0.27\frac{T}{D^2}$	0
PID-controller	$1.2\frac{T}{D}$	$0.6\frac{T}{D^2}$	$0.6\ T$
Ziegler-Nichols PID parameter tuning rules based on critical gain and critical period			
P-controller	$0.5\ K_{Cr}$	0	0
PI-controller	$0.45\ K_{Cr}$	$0.54\frac{K_{Cr}}{T_{Cr}}$	0
PID-controller	$0.6\ K_{Cr}$	$1.2\frac{K_{Cr}}{T_{Cr}}$	$0.075\ K_{Cr}T_{Cr}$

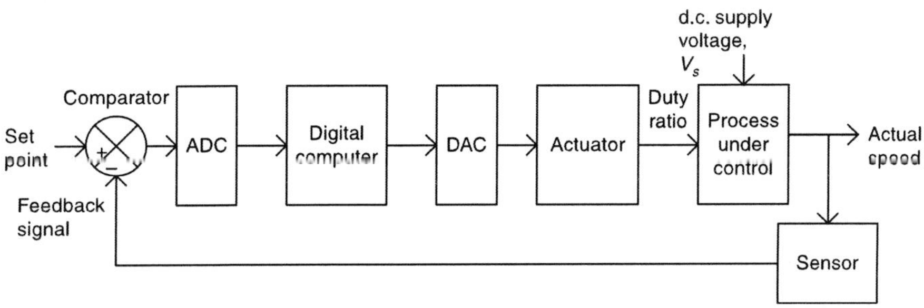

2.18 Block diagram of closed-loop digital control system.

the popularity of digital controllers over analogue controllers in many applications.

In principle, a digital control system is similar to an analogue control system. Here, the analogue controller block is replaced with a digital computer, which performs the task of generating the required signal for the actuator, as shown in Fig. 2.18. Since digital computers deal only with binary numbers, an analogue-to-digital (A/D) converter is required before the controller stage for converting the analogue error signal into digital form. It does so by sampling the analogue signal at periodic intervals and then holding over the sampling interval. Conversely, a digital-to-analogue (D/A) converter is also required after the controller stage as the actuator operates on the analogue signal.

The digital controllers have several advantages over analogue schemes, some of which are as follows:

- Flexibility in control action: modification of an analogue controller can only be made through rewiring and replacement of a component, whereas a simple change to the computer program is all that is required for the modification of a digital controller.
- Reduced cost involvement for addition of further control loops: the addition of extra loops in digital controllers merely requires any extra hardware to be connected up to the same computer, whereas additional hardware loops would otherwise be required for an analogue controller.
- Improved user interface: digital controller information can be displayed graphically on a monitor when required, as opposed to the analogue alternative of employing a large panel for displaying limited information.
- Adaptive control: digital controllers can be modified online, that is, even when the system is in operation.
- Cost effective: due to the rapid development of VLSI technology, the cost of digital controllers is decreasing in comparison to that of analogue controllers, especially for applications requiring high accuracy and optimum performance.

2.5 Intelligent control systems using soft computing

Artificial intelligence (AI) has been used in industrial applications since the early 1960s – mostly in the area of expert knowledge-based decision making for the design and monitoring of industrial products or processes. Fuzzy logic

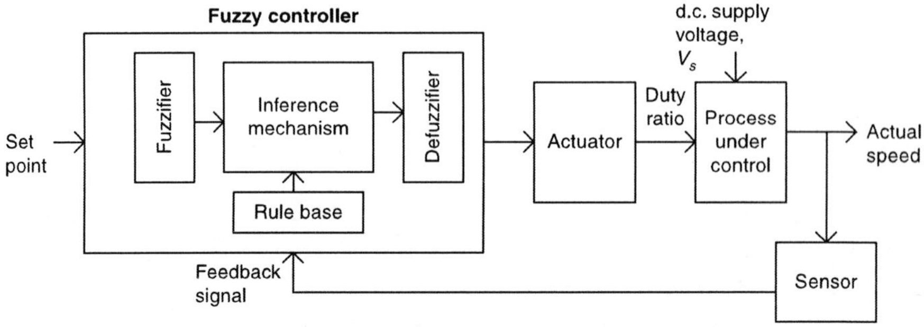

2.19 Block diagram of closed-loop fuzzy control system.

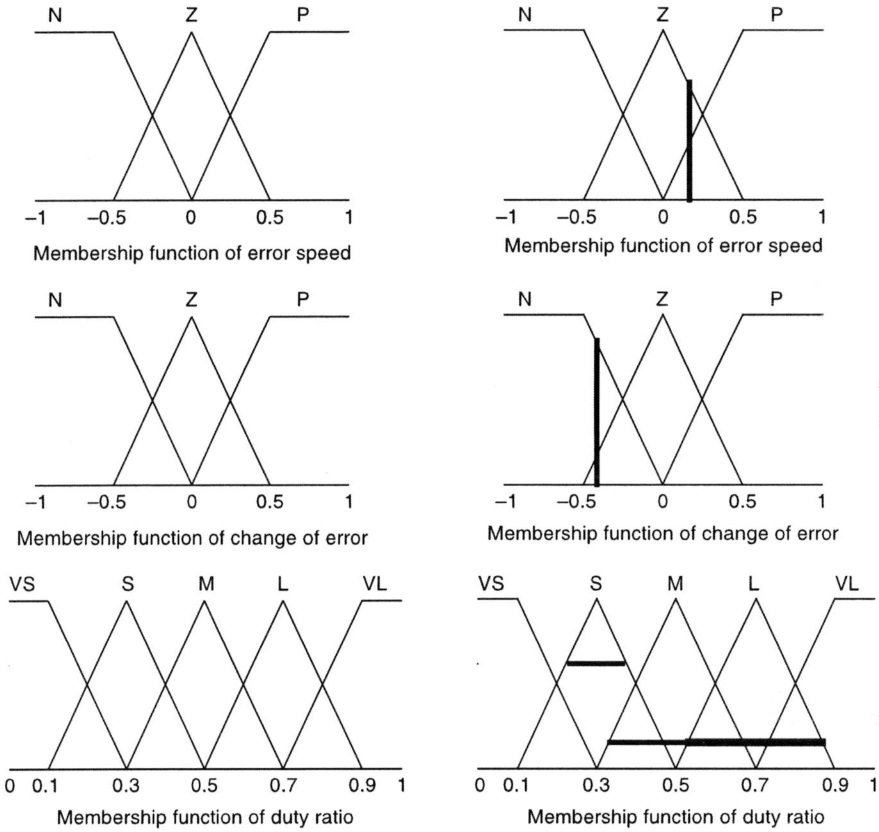

2.20 Membership functions of input and output variables.

control (FLC) has gained great popularity in many applications due to its ability to control nonlinear, uncertain systems even in processes for which no mathematical model is available. FLCs are based on fuzzy logic and constitute a method of converting linguistic expressions into a suitable control strategy through the interaction of a rule base, which controls the behaviour of the system. There is, however, no well-defined or systematic method for designing and tuning the FLC for cases where expert knowledge is not available.

The application of a FLC for speed control of a separately excited d.c. motor is shown in Fig. 2.19. The fuzzy controller block contains four components: fuzzifier, rule base, inference engine and defuzzifier.

The fuzzifier transforms the analogue (continuous) input variables into membership functions. A membership function is a typical curve that converts the crisp inputs into fuzzy membership values (from 0 to 1), indicating the level of acceptability of the inputs to a fuzzy set. Membership functions can take various forms such as triangular, trapezoidal or Gaussian of which a triangular membership function is the simplest and most commonly used. In the example of speed control of a d.c. motor by a fuzzy controller, the selected input variables are error speed (deviation of actual speed from reference speed) and change of error speed. These are transformed into membership functions of negative (N), zero (Z) and positive (P) as shown in Fig. 2.20. The output variable of the controller for the actuator is the duty ratio, which is represented by the membership functions of very small (VS), small (S), medium (M), large (L) and very large (VL).

The next block of the fuzzy controller is the fuzzy inference mechanism, which maps input linguistic variables into output linguistic variables on the basis of fuzzy rules.

A fuzzy rule consists of two parts, the first of which is an 'IF' or 'antecedent' statement and the second is a 'THEN' or 'consequent' statement. The antecedent part joins the input membership functions with logical operators such as 'AND' or 'OR'. The consequent part deals with the output membership functions. Fuzzy rules provide quantitative reasoning, which relate input fuzzy sets to output fuzzy sets. In the present example, the following base rules are formulated based on experts' knowledge for speed control of a d.c. motor.

IF error speed is negative AND change of error is negative THEN duty ratio is very small.
IF error speed is negative AND change of error is zero THEN duty ratio is small.
IF error speed is zero AND change of error is negative THEN duty ratio is small.
IF error speed is zero AND change of error is zero THEN duty ratio is medium.
IF error speed is zero AND change of error is positive THEN duty ratio is large.
IF error speed is positive AND change of error is zero THEN duty ratio is large.
IF error speed is positive AND change of error is positive THEN duty ratio is very large.

The output of each fuzzy rule is also called a 'fuzzy set'. All output fuzzy sets are aggregated into a single fuzzy set typically by a 'max' operator. This step is known as 'aggregation'. Finally, the resulting fuzzy set is resolved into a single crisp number by 'defuzzification'. There are several methods of defuzzification such as 'centre of gravity' (centroid), 'centre of sums', 'mean of maxima' and 'left-right maxima'. The calculation of crisp output in the centre of gravity method[10] is achieved using the following mathematical expression:

$$x^* = \frac{\int \mu_A(x)\,x\,\mathrm{d}x}{\int \mu_A(x)\,\mathrm{d}x}$$

where x^* is the defuzzified output and $\mu_A(x)$ is the output fuzzy set after the aggregation of individual implication results.

In the present example, an arbitrary error speed of 0.3 and a change of error of –0.4 is assumed for the d.c. drive under consideration. An error speed of 0.3 cuts the zero error speed membership function at 0.6 and the positive error speed membership function at 0.4. Again, the change in error speed of –0.4 cuts the zero change of the error membership function at 0.2 and the negative change of error membership function at 0.8. These conditions fire third, fourth and sixth rules and correspond to the duty ratio of small membership function of 0.6, medium membership function of 0.2 and large membership function of 0.2. After defuzzification the intended duty cycle was found to be 0.42.

2.6 Application of control systems in textile processing

The scope of potential applications of various control systems in currently used textile machines and processes is enormous. Control of the warp tension is paramount in weaving as the end-breakage rate and machine efficiency is largely dependent on it. Mehmet, Hakan and Cengiz[11] demonstrated the use of PID and fuzzy controller in the let-off system of weaving. In this system, the floating backrest of a loom is pressed in a downward direction as the tension increases in the warp sheet. A proximity sensor is used to monitor the movement of the backrest and an encoder is used to identify the angular position of the main shaft. The signal generated by the proximity sensor is compared with the preset reference value to generate the error signal. This signal is then transmitted to the servo motor, which rotates the weaver's beam to release the warp sheet. Both PID and fuzzy controllers have been used for this type of work and a comparative performance analysis reveals that it is fuzzy controllers that achieve the minimum warp tension as well as the lowest warp breakage rate. Some modern drawframes used in the spinning process also use such soft computing-based control systems to achieve the highest quality results.

2.7 References

1. Ronald S. Burns (2001), *Advanced Control Engineering*, Butterworth-Heinemann, Oxford, U.K.
2. Kevin Warwick (1996), *An Introduction To Control Systems*, World Scientific Publishing Co. Pte. Ltd., Singapore.

3. Manfred Schleicher and Frank Blasinger (2003), *Control Engineering: A Guide for Beginners,* JUMO GmbH & Co. KG, Fulda, Germany.
4. Katsuhiko Ogata (1996), *Modern Control Engineering*, Prentice-Hall of India Pvt. Ltd, India.
5. John J. D'Azzo, Constantine H. Houpis and Stuart N. Sheldon (2003), *Linear Control System Analysis and Design with Matlab*, Marcel Dekker, Inc., USA.
6. Ali Zilouchian and Mo Jamshidi (2001), *Intelligent Control Systems Using Soft Computing Methodologies,* CRC Press, USA.
7. J. Michael Jacob (1989), *Industrial Control Electronics: Applications and Design*, Prentice-Hall International Inc., USA.
8. Friedrich Frohr and Fritz Orttenburger (1992), *Introduction to Electronic Control Engineering*, Wiley Eastern Limited, London.
9. I. J. Nagrath and M. Gopal (1992), *Control Systems Engineering*, Wiley Eastern Limited, India.
10. Kevin M. Passino and Stephen Yurkovich (1998), *Fuzzy Control*, Addison Wesley Longman, Inc., California, USA.
11. Dayik Mehmeta, Calis Hakanb and Kacacan M. Cengiz (2009), Adaptive control of let-off system in weaving, *The Journal of The Textile Institute*, Vol. 100, No. 2, March 2009, pp. 186–194.

3
Testing and statistical quality control in textile manufacturing

A. DAS, Indian Institute of Technology Delhi, India

DOI: 10.1533/9780857095633.1.41

Abstract: This chapter discusses the use of statistical tools, such as measurement of variability, differences between means, significance of variables, control charts and hypothesis testing in textile industries. It is necessary to test the fibre, yarn and fabric for their quality while purchasing in bulk in the industry. An entire batch cannot be accepted by testing only one or two samples. Acceptance sampling technique is a reliable method of selection of samples from bulk. The quality of the product or process can be identified by control charts. Certain controllable variables such as count of yarn, strength, etc. are taken for analysis and their importance is found by statistical significance tests. So in this chapter, different statistical techniques and their applications in textile industry are clearly explained with relevant examples.

Key words: sampling error, control charts, fabric testing, hypothesis testing, significance test.

3.1 Introduction: statistical quality control

The modern textile industry is a complex, high technology industry facing numerous competitive challenges. Markets are becoming more complex. Short life cycles are common, and demands for rapid response and just-in-time delivery are growing. As competition continues to increase, textile companies must rely more on superior quality, innovative products and rapid response to customer needs, to secure markets and continue to grow. To cope with these challenges, many textile companies have implemented quality management initiatives, to reduce costs and increase both product quality and customer satisfaction.

One of the most recent trends in automation is the use of continuous monitoring systems for on-line quality measurement and for recording defects. The continuous monitoring of a process often produces more data than manufacturers are equipped to use profitably, and continuous correction of a process based on a data stream may actually decrease product quality or mask problems needing attention.

Statistical quality control (SQC) is designed to sample a large population on an infrequent basis. Quality assessment therefore only takes place on a small portion of the total product. It has been argued that the use of continuous monitoring would mean that there would be no need for SQC. This attitude assumes that the sole function of SQC is to catch the defective product before it reaches the customer (i.e. acceptance sampling), and ignores the potential for statistics as a tool for product improvement. In recent years, those SQC techniques that worked well for final product quality control have been applied to both the materials being processed and to process conditions. This procedure is now known as statistical process control (SPC). However, some SPC techniques that work with infrequent sampling may not be useful when very frequent or continuous sampling is necessary. Nevertheless, SPC is currently widely used in the textile industry.

3.2 Basic measurement concepts in statistical quality control

SQC comprises the set of statistical tools used by quality control professionals. It can be divided into three broad categories:

i. **Descriptive statistics:** These are used to describe quality characteristics and relationships. This group includes the mean, standard deviation, range and distribution of data.
ii. **SPC:** This involves inspecting a random sample of the output from a process and deciding whether the characteristics of the products in the sample fall within a predetermined range. SPC is used to determine whether the process is functioning properly or not.
iii. **Acceptance sampling:** This involves random inspection of a sample of goods. Based on the results of the sample, a decision is made as to whether a batch of goods should be accepted or rejected.

The tools in each of these categories provide different types of information for use in quality analysis. Descriptive statistics are used to describe certain quality characteristics, such as the central tendency and variability of observed data. Although descriptions of specific characteristics are helpful, they are not enough to identify whether there is a problem with quality. Acceptance sampling can help to solve this problem. However, although acceptance sampling is helpful in deciding on acceptability after the product has been produced, it does not aid in identifying a quality problem during the production process. To do this it is necessary to use tools from the SPC category.

Variation in the quality of manufactured textiles is inevitable. The manufacturing processes currently in use are not capable of producing completely identical products. However, inspection of all of the raw materials and finished goods is impossible, because:

- *The standard test is destructive in nature.* For example, a fabric manufacturer buys yarn from a spinning mill. It has been settled between the two parties that each consignment of yarn delivered to the fabric manufacturer should have an average linear density inside the tolerance range 40 ± 1 Tex. When the batch of yarn is delivered, it would be impractical to test the whole consignment for whether or not the average linear density lies within the tolerances, as the standard test for linear density is destructive. There would therefore be no product left to work with.
- *The population size is too large.* For example, a menswear manufacturer marks the size of the trousers he produces according to the waist size. In order to design the trousers, therefore, he must know the average waist size of the men in the population to whom he is hoping to sell the trousers. To determine this average exactly, the waist size of every man in the population would have to be measured, which would of course be prohibitively expensive and time-consuming.
- *The rate of production is too high to examine every product.* For example, a garment manufacturer knows from past experience that usually 2% of the garment blanks he produces are defective. He is content with this level of defective items but he does not want it to increase, so he aims to control the level, that is, to detect quickly any increase in the number of defective items being produced so that remedial action can be taken. To do this, garment blanks must be inspected, but the rate of production is too high to examine every single blank.

The only reasonable method of addressing the above problems is to examine a small fraction of the population or output, on the assumption that the results of the sample are representative of the untested population or output.

However, there are still some questions that arise, for example:

- Using sample data, what can be said about the average value of the population from which the sample was drawn?
- How many standard tests should be carried out?
- How should the test results be used to describe whether or not the population meets the required specifications?
- How should these results be used to detect a change in the preset levels?

When there is variation within a population and only a sample has been examined, our knowledge of the population is incomplete and uncertain. Statistical methods deal with this by also measuring the degree of uncertainty.

The statistics can be summarised according to the data:

- Measures of central tendency
- Measures of variability

3.2.1 Measures of central tendency

Central tendency represents the average of set of values or data. The average is the general term which can be specifically defined by three different terminologies namely, mean, median and mode.

Arithmetic mean

This is a commonly used term in the industry that has the advantages of (1) being simple to understand, (2) being easy to calculate, and (3) using all the measurements. It is therefore the most popular method used to locate a distribution.

- *Sample mean*

$$\bar{X} = \frac{\sum_{i=1}^{n} X_i}{n} = \frac{X_1 + X_2 + \cdots + X_n}{n}$$

- *Population mean*

$$\mu = \frac{\sum_{i=1}^{N} X_i}{N} = \frac{X_1 + X_2 + \cdots + X_N}{N}$$

where X_i = observed value, n = sample size, N = population size.

Median

This is the middle value when the data are arranged in increasing or decreasing order. The median divides the area under the frequency curve into two equal parts.

$$M = L + \left(\frac{h}{f}\right)\left(\frac{N}{2} - C\right)$$

where L is the lower limit of the median class, f is the frequency of the median class, h is the class interval, C is the cumulative frequency of class preceding the median class.

Example:
Suppose 15 threads have been tested for single thread strength in grams and the values have been noted down in order of increasing strength:

174 178 180 181 184 186 186 187 189 191 191 193 195 196 196

The median is the eighth value, that is, 187 g.

Should there be an even number of values, then the mean of the two middle values is taken to determine the median:
147 149 151 151 152 153 153 154 155 156

The median is the sum of the fifth and sixth values divided by 2, that is (152 + 153) ÷ 2 = 152.5 g.

Mode

The mode is the value occurring most frequently in the data.

$$\text{Mode} = L + h\frac{(f_1 - f_0)}{(2f_1 - f_0 - f_2)}$$

where L is the lower limit of the modal class, f_1 is the frequency of the modal class, f_0 is the frequency of class preceding the modal class, f_2 is the frequency of class preceding the modal class, h is the class interval.

The median and the mode are not usually used in the textile industry.

If the frequency curve is symmetrical the mean, median and mode will coincide. Where the curve is moderately asymmetrical, there is an interesting approximate relationship between the three values (Fig. 3.1):

$$\text{Mode} = \text{Mean} - 3(\text{Mean} - \text{Median})$$

3.2.2 Measures of variation

a. *Range* is the difference between the largest and the smallest observations in the sample:

$$\text{Range} = X_{\text{largest}} - X_{\text{smallest}}$$

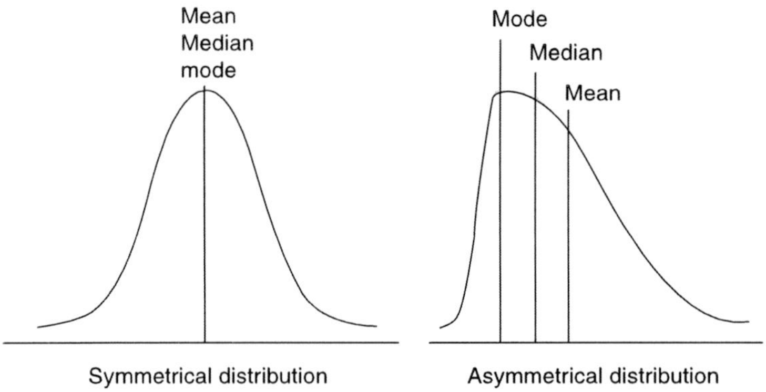

3.1 Symmetric and asymmetric frequency curve.

Table 3.1 Deviations in the garment blank lengths

x	$x - \bar{x}$	$d = \lvert x - \bar{x} \rvert$	d^2
54.5	0.9	0.9	0.81
53	−0.6	0.6	0.36
55.7	2.1	2.1	4.41
51.8	−1.8	1.8	2.64
54.2	0.6	0.6	0.36
52.4	−1.2	1.2	1.44
		$\Sigma d = 7.2$	$\Sigma d^2 = 10.62$

b. *Variance* is measured as follows:

$$\text{Sample variance, } S^2 = \frac{\sum_{i=1}^{n}\left(X_i - \bar{X}\right)^2}{n-1}$$

$$\text{Population variance, } \sigma^2 = \frac{\sum_{i=1}^{N}\left(x_i - \mu\right)^2}{N}$$

As an example, the deviations for the sample of garment blank lengths are given in Table 3.1. From Table 3.1, sample variance for the garment blank length is found as $S^2 = 10.62/(6 - 1) = 2.124\ \text{cm}^2$

c. *Standard deviation (SD)*:

- Sample standard deviation is found using the following formula:

$$S = \sqrt{\frac{\sum_{i=1}^{n}\left(X_i - \bar{X}\right)^2}{n-1}}$$

- Population standard deviation is found using:

$$\sigma = \sqrt{\frac{\sum_{i=1}^{N} (X_i - \mu)^2}{N}}$$

Since $S^2 = 2.124$ cm^2, we find that the standard deviation of the garment blank length data is $S = (2.124)^{1/2} = 1.46$.

d. *Coefficient of variation* uses the following formula:

$$\text{CV} = \left(\frac{S}{\bar{X}}\right) 100\%$$

For the garment length data, $\bar{x} = 53.6$ cm and $s = 1.46$ cm, so the coefficient of variation is $C = 100 \times 1.46/53.6 = 2.72\%$

Example:
Of a large batch of hand towels, 5% are under 60 g in weight and 40% are between 60 and 65 g. Assuming the normal distribution, find the mean weight and standard deviation.

Solution: If we consider the normal distribution of variant X, that is, mass of hand towel:

The area that will lie to the left of $X = 60$ is 0.05
The area between $z = 0$ and $z = (60 - \bar{x})/\sigma$ is $0.50 - 0.05 = 0.45$

From the table, the value of z_1 corresponding to this area is 1.645. Therefore,

$$\frac{(60 - \bar{x})}{\sigma} = -1.645 \quad \text{(i)}$$

Also, the mass of 40% of the hand towels is between 60 and 65 g, the area between the ordinates at $X = 60$ and $X = 65$ is 0.4 or the area to the left of the ordinate at $X - 65$ is 0.45. The area between $z = 0$ and $z_2 = (65 - \bar{x})/\sigma$ is $0.50 - 0.45 = 0.05$. The value of z_2 corresponding to this area is 0.13.

$$\frac{(65 - \bar{x})}{\sigma} = -0.13 \quad \text{(ii)}$$

Solving equations (i) and (ii):

$60 - \bar{x} = -1.645\sigma$
$65 - \bar{x} = -0.130\sigma$

Result: $\sigma = 3.3$ and $\bar{x} = 65.429$.

Statistical techniques are important tools for effective process control and innovative solutions to problems. The main focus of statistical techniques is to avoid defects that are produced in the manufacturing process. Experiments designed to assess the advantages of novel types of processing or to determine optimal conditions also fall into the category of SQC. Statistical techniques are also very useful in determining sample size, deciding rate of recurrence of inspection, deciding natural limits of variation of the process, testing conformity of sample to specification provided and so on.

In any product line, no two articles are perfectly identical. For example, it is impossible to find two bundles of yarn with exactly the same count, strength, evenness, length, etc. This is due to raw material variation. The product quality depends on the raw materials used in the process and the level of technical enhancement attained during manufacturing. Machines and tools will sustain wear and tear through use and it is neither practical nor cost-effective to remove or repair the machines after every small occurrence. Therefore, a certain margin for error must be built into the manufacturing process.

3.2.3 Sampling error

It is worth noting that there is room for variation in the accuracy of samples. For example, although the average count spun by a mill can be determined as 40^sNe, the count in individual leas could fall anywhere between 35^sNe and 45^sNe in 95% of the leas produced. The higher the number of samples tested, the closer the average sample count comes to the actual count spun by the mills. The disparity between the actual count (40) and the count estimated from the sample (35–45 assuming the sample size is one lea), is termed the 'standard error' due to sampling.

The standard error in the estimated count from a sample of 'n' is given by $\pm 2\left(\text{CV}\%/\sqrt{n}\right)$ at 95% confidence limits.

It is important to consider the sampling error while interpreting any computed data. For example, for a CV of 4.5%, if the average count of a sample of 36 leas is 60 then the count for the department as a whole would be, at 95% confidence, within

$$60 \pm \frac{(2 \times 4.5)}{\sqrt{36}} = 60 \pm 1.5$$

3.3 Interpretations: critical difference

With the advent of a large number of testing instruments, it has become possible to test various aspects of the quality of textile materials and therefore present a large volume of data to the manager for decision-making.

Table 3.2 No. of tests and CD % for various fibre properties

S. No	Fibre property	No. of tests	CD (% mean)
1	2.5% span length	4 combs/sample	4
2	Uniformity ration	4 combs/sample	5
3	Micronaire value	4 plugs/sample	6
4	Fibre strength at 3 mm gauge length	10 breaks/sample	5
5	Maturity coefficient	600 fibres/sample	7
6	Trash content	8 tests/sample	7

Table 3.3 Sample size and CD for yarn properties

S. No	Yarn property	No. of tests	CD (% mean)
1	Lea count	40	2.0
2	Strength	40	4.0
3	Single yarn strength – Uster	100	2.8
4	Evenness – U%	5	7.0
5	Twist – Single yarn	50	3.4
6	Twist – Double yarn	50	2.0
7	Yarn appearance	5 boards 10 readings	Half a grade or 5 grade index

However, interpretation of the test data is made more complex due to the instrumental and sampling errors associated with the data, as discussed above. Experimental work has been carried out by research associations to assess the variability of each property of fibre and yarn and to fix the levels of variability in average values, which manufacturers could then accept as a base while dealing with test data. This variability has been expressed in terms of critical difference. With help of such values the test data can be interpreted meaningfully.

Critical difference (CD) is a measure of the difference between two values that arises solely due to natural or unavoidable causes. When the difference between two values exceeds the CD, then the two values are said to be statistically different. The CD depends upon the CV% and the number of tests carried out to determine the quality characteristics.

The values of CD for various fibre and yarn properties are given in Tables 3.2 and 3.3. These values are based on the recommended number of tests for each fibre or yarn property as given in the same labels. If, however, the number of tests carried out to determine a specific property differs from the recommended number, then the CD would vary from the values reported. Under such circumstances, a new CD must be computed using the formula:

$$\text{New CD\%} = \text{CD\% (Tables 3.2 and 3.3)} \times \left(\frac{\sqrt{N_1}}{\sqrt{N_2}} \right) \qquad [3.1]$$

where N_1 = number of tests recommended in Tables 3.2 and 3.3, N_2 = number of tests actually conducted.

It may be noted that

i) an increase in CV% would increase CD and
ii) an increase in the number of tests (N_2) would decrease CD.

3.3.1 Fibre test data

In the following section some of the major problems encountered commonly in the mills are discussed and solutions are suggested.

Consistency between basic and delivery samples

A mill has received a basic sample of Shankar-6 cotton, whose micronaire value on testing was found to be 4.0. The delivery sample had a micronaire value of 3.8. The mill manager is interested to know whether they could accept the delivery sample. Four tests were carried out on both the basic and delivery samples to determine the micronaire value.

The extent of variation present between the two samples is computed as follows:

$$\text{Difference in micronaire values between two samples} = 4.0 - 3.8 = 0.2$$

$$\text{Average micronaire values of two samples} = \frac{(4.0 + 3.8)}{2} = 3.9$$

$$\text{Difference expressed as \% of the average} = \left(\frac{0.2}{3.9}\right) \times 100 = 5.13\%$$

$$\text{The CD for micronaire values as per Table 3.2} = 6\%$$

Since the actual difference of 5.13% is less than the CD of 6%, the delivery sample could be considered as not significantly different to the basic sample, as far as the micronaire value is concerned. The mill manager could therefore accept the delivery sample.

In above example, the CD is computed based on the mean of the micronaire values of the basic and delivery samples. This is because when two samples are compared to determine their difference, the allowable limits for the CD should be fixed based on the average of the two samples.

However, if data from a test is to be compared with a specific value, a different procedure should be adopted, as shown in the example below.

A mill wants to purchase cotton of 3.7 micronaire value to spin 60^s count. The cotton sample received from a supplier was tested for micronaire and it was found to be 4.0 (on the basis of four tests). The mill technician now needs to know whether the cotton sample conforms to his requirements or not.

Difference in micronaire values between the specific value (3.7) and the actual value (4.0) = 4.0 – 3.7 = 0.3

$$\text{Difference expressed as a \% of the specific value} = \left(\frac{0.3}{3.7}\right) \times 100 = 8.1\%$$

Since the mean value is being compared with a specific value, the CD should be calculated on the basis of the specific value.

The CD for micronaire value (as in Table 3.2) = 6%

Since the actual difference of 8.1% is higher than the CD of 6%, the mill could not purchase cotton from that supplier. In the same way, data on the fibre length, strength, fineness, maturity and trash could be analysed with the proper use of the CD values reported in Table 3.2.

3.3.2 Yarn test data

The following sections cover some of the common problems and their respective solutions in the spinning division of most of the mills.

Pinion changes in spinning frame

Mill 'D' spins 60^s Ne using a specific variety of cotton. A random sample of 40 cops is taken from a spinning frame and the count is checked by taking one lea from each cop. The average count is found to be 62^s Ne. The problem is to ascertain whether changing the draft wheel in the spinning frame is necessary to correct the count. (The change wheel in the frame has 50 teeth.)

Difference between actual and nominal count = 62 – 60 = 2

For reasons explained earlier, the percentage difference is calculated on the basis of the nominal count.

Thus the difference expressed as a percentage of the nominal count = (2/60) × 100 = 3.3%

Since the difference of 3.3% is greater than the CD value of 2% (Table 3.3), the count of yarn from the particular spinning frame (62^s Ne) is significantly different from the nominal count (60^s Ne).

But while making wheel changes to correct the count, the limitation imposed by mechanical factors should also be taken into account. A change of pinion is advantageous only when the true deviation from the desired count exceeds $C/2A$, where C is the nominal count and A is the number of teeth in the change pinion.

In the present problem,

$$\frac{C}{2A} = \frac{60}{(2 \times 50)} = 0.6, \text{ i.e., } 1\%$$

Since the actual difference of 3.3% is greater than the sum of the CD and mechanical errors (2% + 1%), wheel changes could be recommended in the spinning frame to correct the count.

Yarn evenness between samples

Mill 'E' produces 40^s yarn. While testing two samples from different spinning frames for evenness, the U% values are found to be 13.8 and 15.0 respectively on the basis of ten observations in each case. It is then necessary to assess whether the yarn produced in the frames is even or not.

Because ten tests were done to assess $U\%$, the CD of 7% given in Table 3.3 needs correction.

Using formula [3.1]

$$\text{New CD\%} = \frac{(7 \times \sqrt{5})}{\sqrt{10}} = 5\%$$

The difference between the two samples expressed as a percentage of the average = (1.2/14.4) × 100 = 8.3%

Since the actual difference of 8.3% is higher than the CD of 5%, it can be concluded that the two yarn samples are not even.

Average and minimum CSP

Mill '*F*' produces 50^s P/V yarn for the export market. The mill intends to produce yarn with a minimum CSP of 2500 and CV strength of 6%. The mill would like to decide the average CSP it has to achieve so that the minimum CSP is 2500.

The relationship between the average and minimum CSP is given by the formula:

$$\text{Average CSP} = \frac{(\text{Minimum CSP} \times 100)}{(100 - 3\text{CV})} \qquad [3.2]$$

$$\text{Average CSP} = \frac{(2500 \times 100)}{(100 - 18)} = 3049$$

Thus, if the mill aims at an average CSP of 3049 with a CV of strength not greater than 6.0%, then it can expect to produce a yarn with a minimum CSP of 2500.

3.4 Interpretations: '*t*' tests, '*F*' tests and the chi-square method

The following section covers the tests for analyzing the statistical significance of the data set, namely, the '*t*' test, '*F*' test and chi-square method.

3.4.1 '*t*' and '*F*' tests

To analyse properties for which CD values are not stipulated, for example, hairiness or CV of twist, special tests need to be applied.

Two such tests are:

i) '*t*' test: to compare two mean values,
ii) '*F*' test: to compare two variances (square of standard deviation).

Hairiness between samples

A mill produces 80^s P/C yarn. While testing two yarn samples for hairiness, one each from the RF1 and RF2 ring frames, the number of hairs per 1000 m are found to be 8000 and 10 000. The CVs of hairiness are therefore 30% and 40% respectively (on the basis of 20 tests for each sample). The mill wants to know whether the RF2 ring frame is producing a more hairy yarn.

In order to decide whether the two samples differ in terms of hairiness, the '*t*' value of significance is to be applied as no CD value for hairiness is reported.

$$t = \frac{(X_1 - X_2)\sqrt{n}}{\sqrt{S1^2 + S2^2}} \qquad [3.3]$$

where X_1 = average hairiness of sample 1, X_2 = average hairiness of sample 2, n = number of tests carries out for sample 1 and sample 2, S_1 and S_2 = standard deviations for sample 1 and sample 2, respectively.

From the value of CV and mean, the SD values can be deduced.
Thus, $S_1 = 2400$ and $S_2 = 4000$

$$t = \frac{(10000 - 8000)\sqrt{20}}{\sqrt{2400^2 + 4000^2}} = 1.9$$

The calculated value of 't' should then be compared against the standard value of 't', which is available in any standard statistics book. For this, it is necessary to know the degrees of freedom in the problem concerned. The degree of freedom is $2(n - 1)$, where 'n' is the number of tests carried out for samples 1 and 2.

In the present case:

$$\text{Degrees of freedom} = 2\ (20 - 1) = 38$$

The value of 't' for 38 degrees of freedom is 2.0.

Since the calculated value, 1.9, is less than 2.0, it can be concluded that the yarns from both frames are similar with regard to hairiness.

Twist variability among samples

Mill 'H' produces 40^s yarn. On testing two samples drawn from two spindles for twist, the standard deviations of twist were found to be 1.31 and 2.85 based on 50 and 60 tests respectively. The mill therefore wants to find out whether the two samples differ in terms of their twist variation.

As two standard deviations are to be compared, the 'F' test needs to be conducted, as follows:

$$F = \frac{S_1^2}{S_2^2} \qquad [3.4]$$

S_1 = standard deviation of twist for one sample (sample 1) for which the number of tests conducted is n_1;

S_2 = standard deviation of twist for the other sample (sample 2) for which the number of tests conducted is n_2.

S_1 and S_2 are to be chosen in such a manner that 'F' is always greater than 1.

In the present problem,

$$S_1 = 2.85 \qquad S_2 = 1.31$$

$$n_1 = 60 \quad \text{and} \quad n_2 = 50$$

$$F = \frac{2.85^2}{1.31^2} = 4.7$$

The value of F (refer to the 'F' table in any standard statistics book, for $df_1 = 59$ and $df_2 = 49$) is 1.6 where df_1 = degrees of freedom for sample 1 and df_2 = degrees of freedom for sample 2.

Since the calculated value of F (4.7) is greater than the 1.6, it is confirmed that the two yarns differ significantly in their twist variability.

Application of 'F' test: auto levellers in cards

Due to the incorporation of auto levellers in the high production cards in a mill, the CV% of card sliver decreased from 4 to 3.5. It is now necessary to determine whether this difference is statistically significant. The mean value of a hank of card sliver is 0.2.40 readings were taken to measure CV%.

As two CV values are to be compared, the 'F' test must be employed:

$$\mathrm{SD}_1 = \frac{\mathrm{CV1} \times \mathrm{mean}}{100} = \frac{4 \times 0.2}{100} = 0.008$$

$$\mathrm{SD}_2 = \frac{\mathrm{CV2} \times \mathrm{mean}}{100} = \frac{3.5 \times 0.2}{100} = 0.007$$

$$\frac{\mathrm{SD}_1^2}{\mathrm{SD}_2^2} = \frac{0.008^2}{0.007^2} = 1.31$$

This value is to be compared with the values of 'F' in statistical tables. Since 40 readings were taken to assess CV%, the degrees of freedom are:

$$n_1 = 40 - 1 = 39 \quad \text{and} \quad n_2 = 40 - 1 = 39$$

The value of F (for $n_1 = 39$ and $n_2 = 39$) = 1.53

Since the calculated value of 'F' (1.31) is lower than that given in tables (1.53), it can be inferred that auto levellers do not improve the CV% of card sliver.

3.4.2 The χ^2 (chi-square) method

This method is used when there is no prior knowledge of the distribution of the test values.

$$\chi^2 \text{ is given by the formula } \frac{(O-E)^2}{E}$$

where O and E are observed and expected values, respectively.

Comparison of end breaks

A mill maintains an average end breakage rate of 150 breaks per 1000 spindle hours in 60^s Ne. When the mixing was changed, the breakage rate increased to 220 breaks per 100 spindle hours. In order to determine whether the change of mixing increased the breakage rate, the following can be used (breaks were observed for 1000 spindle hours):

In the present problem $O = 220$ and $E = 150$

$$\chi^2 = \frac{(220-150)^2}{150} = \frac{4900}{150} = 32.7$$

The χ^2 (chi-square) value (refer to the 'χ^2' table in any standard statistics book) for 1 degree of freedom is 3.84.

Since the actual value of $\chi^2(32.7)$ is greater than 3.84, it could be concluded that the end breaks did increase due to the change in mixing.

Comparison of neps during carding

A mill processes cotton fibre through a card-A and card-B. Neps in the card web were assessed by taking ten readings from each of the cards. The neps were found to be 120 and 80 per 10 g respectively. The mill now needs to find out whether the card-A really does generate more neps.

We have here two observed values for neps (120 and 80) but the expected number of neps from each card is unknown. Under the circumstances, an assumption is made that the expected value is the mean of the two observed values, and the following formula is used:

$$\chi^2 = \frac{(O-E)^2}{E} \text{ will reduce to } \frac{(A-B)^2}{A+B}$$

where A and B are the two observed values.

$$\chi^2 = \frac{(A-B)^2}{A+B} = \frac{(120-80)^2}{120+80} = 8$$

χ^2 (chi-square) value for 1 degree of freedom = 3.84.

Since the actual value of $\chi^2(8)$ is greater than 3.84, it is confirmed that the card-A generates more neps.

Note

i) The data collected should be used in their original form and not be reduced to any standard form such as percentages, etc.
ii) Either the observed or expected values should be numerically greater than 200 if a 20% difference between them is to be statistically significant.

3.5 Decision-making using control charts

Process control forms a key part of the overall effort to maintain quality in yarn manufacture. The properties of the material as determined at different stages of processing act as indicators to ascertain whether a correction at any specific stage of processing is necessary or not. The application of appropriate statistical methods would facilitate a proper understanding of the data. SPC methods extend the use of descriptive statistics to monitor the quality of the product and process. They can be used to determine the amount of variation that is common or normal and to monitor the production process to make sure production stays within this normal range; that is, that the process is in a *state of control*. The most commonly used tool for monitoring the production process is a control chart. Different types of control charts are used to monitor different aspects of the production process.

A control chart (also called a process chart or quality control chart) is a graph that shows whether a sample of data falls within the common or normal range of variation. A control chart has upper and lower control limits that separate common from assignable causes of variation. A process is defined as out of control when a plot of data reveals that one or more samples fall outside the control limits.

Control charts are one of the most commonly used tools in SPC. They can be used to measure any characteristic of a product, such as the count or hank, strength, U%, etc. The different characteristics that can be measured by control charts can be divided into two groups: variables and attributes.

Variable data are data that can be measured on a continuous scale such as a thermometer, a weighing scale or a tape rule.

Attribute data are data that are counted as having a certain attribute or value; for example, as good or defective, or as possessing or not possessing a particular characteristic.

3.5.1 $\bar{X}$ and R charts

$\bar{X}$ and R charts are control charts that are commonly used to control average ($\bar{X}$) and variation – range (R) in terms of data. They are used to control

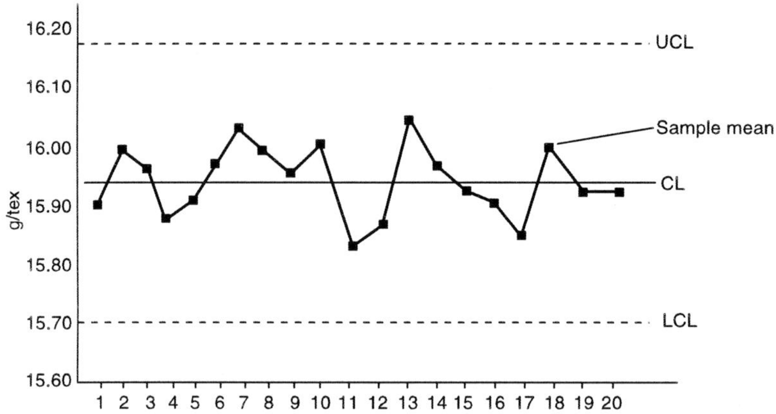

3.2 $\bar{X}$ chart.

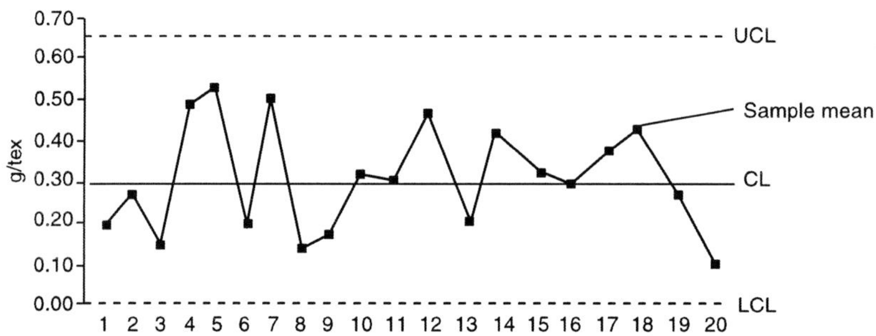

3.3 $\bar{R}$ chart.

the hank or count of textile material, strength, U%, etc. The method for creating these charts is given below:

a. Collect data over a fixed period of time, testing 20 to 25 samples per day in sub-groups of 4 to 5.
b. Calculate the average ($\bar{X}$) and range ($\bar{R}$) the range for each sub-group, followed by the overall average.
c. Calculate the upper and lower limits of the group ranges: $D_4\bar{R}$ and $D_3\bar{R}$, where D_4 and D_3 are factors that can be read from the statistical tables.
d. If any group range is higher than $D_4\bar{R}$ or lower than $D_3\bar{R}$ omit these values and recalculate the $\bar{R}$ and the upper and lower limits. Repeat this procedure until all the group ranges fall within the limits.
e. The control limits to be fixed for the averages are $X \pm A_2\bar{R}$ The A_2 values can be found in any standard statistical book.
f. Plot the centre line and control limits, then add the sample number and data to the X-axis as well as the average and range to the Y-axis.

After setting up the charts for $\bar{X}\bar{X}$ and $\bar{R}$, the daily test values should be plotted in the chart as and when the tests are carried out. Figures 3.2 and 3.3 provide examples of $\bar{X}\bar{X}$ and $\bar{R}$ charts.

3.5.2 Using p and C charts

The p chart is used to record the fraction of defectives; that is, the proportion of excessively high or low weight laps or machine stoppage. The C chart is used to record the average number of defects. The control limits are as follow:

$$p \pm \frac{\sqrt{\bar{p}\left(1-\bar{p}\right)}}{n}$$

where p is the average fraction of defectives; and

$$c \pm 3\sqrt{c}$$

where c is the average number of defects per unit.

The method of calculating average values and fixing control limits is the same as described above for the $\bar{X}$X-bar and $\bar{R}$R-bar charts.

Example for p chart

In a blow room situation, for the average lap weight to be under control, the proportion of heavy laps should be 0.5. If 100 laps are produced in one shift, then the control limits for the proportion of heavy laps would be

$$0.5 \pm 3\frac{\sqrt{0.5 \times 0.5}}{100} = 0.5 \pm 0.15$$

If the proportion of heavy laps lies between 35% and 65%, then the lap weight is under control.

Example for C chart

If the average rate of end breaks in a ring frame is 16, then the control limits are $16 \pm 3\sqrt{16} = 16 \pm 12$.

If the rate of end breaks in any test exceeds 28 or falls below 4, it should be concluded that the breakage rate has significantly increased or decreased.

3.6 Decision-making: hypothesis testing

A statistical hypothesis is an assumption about a population parameter that may or may not be true. In order to determine whether or not it is true, the most accurate method would be to examine the whole population. However, this is usually impractical, so a random sample needs to be generated and used as a representation of the whole population. The sample is then examined to determine whether it is consistent with the statistical hypothesis. If it is not, then the hypothesis is considered to be untrue.

There are two types of statistical hypothesis; the null hypothesis (H_0), which states that the observations made from two samples are due to chance; and the alternative hypothesis (H_1 or H_a), which suggests that the observations from the samples are influenced by a non-random cause.

3.6.1 Hypothesis tests

The decision to accept or reject a null hypothesis is described as hypothesis testing. It consists of four steps:

1. State the hypotheses (the null and alternative hypotheses are mutually exclusive).
2. Create an analysis plan (how the sample data will be used to evaluate the null hypothesis).
3. Analyse the sample data.
4. Interpret the results (accept or reject the null hypothesis based on the sample data analysis).

3.6.2 Decision errors

A hypothesis test can result in two different types of error. A Type 1 error occurs when the researcher rejects a null hypothesis that then turns out to be true. The probability of this occurring is called the 'significance level' and is denoted by α. A Type 2 error occurs when the researcher accepts a null hypothesis that then turns out to be false. The probability of this occurring is called the 'power' and is denoted by β.

3.6.3 Decision rules

Decision rules are included in the analysis plan, and are used to determine whether to reject the null hypothesis. Statisticians describe these rules with reference to either a P-value or a region of acceptance. A P-value measures the strength of the evidence in support of a null hypothesis (e.g. the probability of observing the test statistic assuming the null hypothesis is true); whereas the region of acceptance is a range of values into which the test statistic must fall if the null hypothesis is to be accepted. The region outside the region of acceptance is called the region of rejection. If the test statistic falls into this region, it is said that the hypothesis has been rejected at α level of significance. Where there is only one region of rejection (e.g. on one side of the sampling distribution), the test of the hypothesis is referred to as a 'one-tailed' test. Where the region of rejection occurs on both sides of the sampling distribution, the test is referred to as 'two-tailed'.

3.7 Decision-making: significance testing

Significance testing is used to determine whether there is enough evidence to reject the null hypothesis; that is, if the difference between the assumed value in the null hypothesis and the observed value in the experiment is large enough to reject the possibility of the result being due to chance. The 'significance level' is the value used to specify how large the difference has to be in order to reject the null hypothesis. This is usually 1% or 5%. The other 99% or 95% is described as the 'confidence level'.

3.7.1 Test for a single mean (large sample available)

This is performed in situations in which a significance test for a single mean is appropriate. In these cases, the researcher conducting the experiment has to decide between two possibilities: either the mean = μ_0, or it has not changed. See Example 1.

3.7.2 Test for a single mean (small sample available)

The significance test described in the previous section is suitable for use in two different cases: when the value of standard deviation (σ) is already known, and when the sample is sufficiently large to enable a good estimate 's' of σ to be drawn. The latter condition is generally encountered when the sample size is greater than approximately 30.

When the sample size is smaller than this, the sample estimate 's' may not be precise enough, meaning that some corrections will have to be made to take into account potential errors in the calculation of α that may occur due to the use of 's' instead of σ.

The modification to the test is simple and straightforward. It is necessary to consult the t-table rather than the normal distribution tables in this case.

The following calculation should be made:

$$t_0 = \frac{(\bar{X}_0 - \mu_0)}{s/\sqrt{n}}$$

and the observed value compared with the tabulated values for $k = n - 1$ degrees of freedom.

See Example 2.

3.7.3 Test for the difference between two means (independent samples)

Suppose there are two populations, both with unknown mean μ_1 and μ_2, and it is necessary to test the null hypothesis that their difference is equal to a specified value μ_s. Many practical situations involve the testing of equality of means, that is, $\mu_s = 0$, but any other value can also be specified. The alternative hypothesis could be either single sided or double sided.

Independent samples of size n_1 and n_2 are taken, randomly chosen from the two populations, the means of which are $\bar{x}_{10}$ and $\bar{x}_{20}$. Therefore $(\bar{x}_{10} - \bar{x}_{20})$ is a point estimate of the difference $\mu_1 - \mu_2$.

The value of α can be found by computing:

$$U = \frac{([\![\bar{x}_{10}]\!] - \bar{x}_{20}) - \mu_S}{\sqrt{(\sigma_1^2 / n_1) + (\sigma_2^2 / n_2)}}$$

and comparing the resulting value with normal distribution tables. But this calculation requires knowledge of the unknown population variances $\sigma_1^2 - \sigma_2^2$. If the samples are reasonably large, that is, n_1 and n_2 are greater than approximately 30, then σ_1^2 and σ_2^2 may be replaced by their sample estimates s_1^2 and s_2^2. The equation then becomes:

$$U = \frac{([\![\bar{x}_{10}]\!] - \bar{x}_{20}) - \mu_s}{\sqrt{(s_1^2 / n_1) + (s_2^2 / n_2)}}$$

When this equation has been used to obtain U, the value of α can be found from the standard tables.

However, sample sizes are usually small, that is, less than 30. In this situation, the above methodology can lead to errors in the significance level. It is therefore necessary to assume two things in such cases: one, that the two populations are normal; and two, that their variances are equal.

Here, the common variance is given by the equation:

$$s^2 = \frac{(n_1 - 1)s_1^2 + (n_2 - 1)s_2^2}{n_1 + n_2 - 2}$$

which is used to calculate:

$$t_0 = \frac{k(\bar{x}_{10} - \bar{x}_{20}) - \mu_s}{\sqrt{\frac{1}{n_1} + \frac{1}{n_2}}}$$

Therefore in this case, $k = n_1 + n_2 - 2$ degrees of freedom.
See Example 3.

3.7.4 Test for the difference between two means (matched samples)

If the samples available for the tests are not chosen independently from the populations of which the means are being compared, that is, the samples are matched in pairs, then a different method is used.

See Example 4.

$\bar{X} = 336 / 12 = 336/12 = 28 = 450/15 = 30 = 71.6$

3.7.4 Test for single variance

If the results of inspection of a sample set reveal a high level of variability among the sample units, the manufacturer or user may call for some change in the manufacturing process of the item on which the tests were carried out. Following the alterations, the manufacturer or user will obviously wish to check whether or not the changes introduced in the process parameters have resulted in some reduction in the variability of the sample items or not. In such a case, the variance of the new sample set is compared with a preset or standard variance value that falls within the acceptable range.

See Example 5.

3.7.6 Test for difference between two variances

Following a modification to the process as described in the previous example, it may sometimes be necessary to compare the variability of two sample sets and not the variance of the new sample set from any preset standard. In such a case, the variances of two sample sets, one taken before the modifications in process and the other taken after the modifications, are compared. The results will show whether there has been a significant reduction in variability; that is, whether the process modifications have had a positive result or not.

See Example 6.

$\bar{X} = 2.256 = 2.256, \mu = 2.5$

3.7.7 Test for a single proportion

Variation is inherent to all textile products. It is common practice to set norms or specifications for the end product in a manufacturing process; however it is never possible to ensure that all products have exactly the same dimensions or characteristics. Some proportion of the bulk production is bound to be defective in some way or other. Therefore, an acceptable level of defective items in a specific amount of bulk is always preset, on the basis of which it can be assumed prior to any manufacturing process that $x\%$ of the products will probably be defective. But in some cases, the proportion of defective items from a manufacturing cycle is substantially more than $x\%$. It is then necessary to carry out a further test to conclude whether the process has in fact developed a fault.

See Example 7.

$= 22.078$, $\mu = 217.2$

Examples

Example 1: *Single mean with a large sample (n greater than 30)*

One hundred ring bobbins are tested for count and the mean count is found to be 34.2s. The frame should be spinning 34s. If the standard deviation of the sample is 0.62, can we conclude that the frame is really spinning off count?

Answer

Step 1: Calculate the standard error of the mean by using the standard deviation of the sample as an approximation of the SD of the population:

$$\text{SE} = \frac{\text{SD of population}}{\sqrt{n}} = \frac{0.62}{\sqrt{100}} = 0.062$$

Step 2: Calculate the difference between the nominal mean and the sample mean and divide by SE.

$$= \frac{\text{Nominal mean} - \text{Sample mean}}{\text{Standard error of the mean}} = \frac{34 - 34.2}{0.062} = 3.2$$

Step 3: Compare the value obtained from step 2 with the values 1.96 and 2.58:

$3.2 > 2.58$

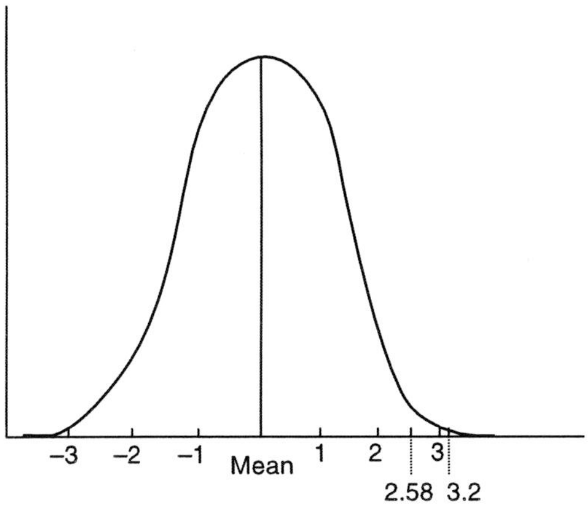

3.4 Significance testing of means (Horizontal scale in units of SE).

Conclusion: The difference in count is statistically significant at the 1% level. The frame is spinning 'off count'.

To reach a conclusion we have noted that the value 3.2 is greater than 2.58. This value (3.2) represents the difference between the nominal and sample means expressed in terms of the standard error.

A glance at Fig. 3.4 will show that the chance of the sample mean lying outside the limit (nominal mean ± 2.58 SE) is only one in 100 if the population from which the sample was drawn had a mean equal to the nominal mean. Once it has been concluded that there is a real difference, the ring frame should be adjusted. The chance of this conclusion being incorrect is only 1 in 100.

Example 2: *Single mean with a small sample*

Twelve ring cops were tested for count and the mean was found to be 94.2^s. The standard deviation of the twelve results is 2.2 counts. If the nominal count is 92^s, then the ring spinning process is too fine.

Answer

Step 1: Estimate the standard error of the mean by dividing the standard deviation of the sample by the square root of the number in the sample.

$$SE = \frac{2.2}{\sqrt{12}} = 0.635$$

Step 2: Calculate the value of t:

$$t = \frac{\text{Nominal mean} - \text{Sample mean}}{\text{Standard error}} = \frac{92 - 94.2}{0.635} = 3.47$$

Step 3: Consult the t-table for $v = n - 1$, then compare the value of t obtained in step 2 with the 5% and 1% values for t.

$$V = 12 - 1 = 11$$

$$t = 2.201 \text{ at the 5\% level}$$

$$= 3.106 \text{ at the 1\% level}$$

$$3.47 > 3.106$$

Conclusion: Since 3.47 is greater than 3.106, we can conclude that the ring is spinning too fine because the difference between the sample mean and the nominal mean is significant at the 1% level.

Example 3: For 10 randomly chosen samples of yarn, manufactured by two different card settings, the effect on neps per 1000 km is as follows:

Setting A: 10, 6, 16, 17, 13, 12, 8, 14, 15, 9
Setting B: 7, 13, 22, 15, 12, 14, 18, 8, 21, 23, 10, 17

It is necessary to test whether card settings A and B differ significantly in their effect on neps per km.

Solution: Let us take the null hypothesis that A and B do not differ significantly with regards to their effect on neps per km.

Applying the t-test:

$$t = \left(\frac{X_1' - X_2'}{S\sqrt{\{(1/n_1) + (1/n_2)\}}} \right) \qquad \text{(i)}$$

$$\text{where } S = \frac{\sqrt{\left[\sum (X_1 - X_1')^2 + \sum (X_2 - X_2')^2\right]}}{n_1 + n_2 - 2}$$

Table 3.4 Nep count and variability with card settings A and B

Yarn manufactured by setting A			Yarn manufactured by setting B		
Neps per km, X_1	$(X_1 - X_1')$	$(X_1 - X_1')^2$	Neps per km, X_2	$(X_2 - X_2')$	$(X_2 - X_2')^2$
10	−2	4	7	−8	64
6	−6	36	13	−2	4
16	4	16	22	7	49
17	5	25	15	0	0
13	1	1	12	−3	9
12	0	0	14	−1	1
8	−4	16	18	3	9
14	2	4	8	−7	49
15	3	9	21	6	36
9	−3	9	23	8	64
			10	−5	25
			17	2	4
$\Sigma X_1 = 120$	$\Sigma(X_1 - X_1') = 0$	$\Sigma(X_1 - X_1')^2 = 120$	$\Sigma X_2 = 180$	$\Sigma(X_2 - X_2') = 0$	$\Sigma(X_2 - X_2')^2 = 314$

Calculating the required values:
Mean neps per km value with setting *A*

$$X_1' = \frac{\Sigma X_1}{n_1} = \frac{120}{10} = 12 \text{ neps per km}$$

Mean neps per km value with setting B

$$X_2' = \frac{\Sigma X_2}{n_2} = \frac{180}{12} = 15 \text{neps per km}$$

Putting values in equation (i),

$$S = \sqrt{\frac{(120 + 314)}{(10 + 12 - 2)}} = 4.66$$

$$t = \left\{\frac{(12 - 15)}{4.66}\right\} \sqrt{\frac{1}{10} + \frac{1}{12}} = 1.51$$

Degree of freedom, $v = n_1 + n_2 - 2 = 10 + 12 - 2 = 20$.

For $v = 20$, the table value of t at 5% level is 2.09. The calculated value is less than the table value and hence the experiment provides no evidence against the hypothesis. It can therefore be concluded that there is no

Table 3.5 Weight of cones and variability after conditioning under two different process conditions: A and B

Condition A			Condition B		
x	$x - \bar{x}'$	$(x - \bar{x}')^2$	Y	$y - \bar{y}'$	$(y - \bar{y}')^2$
25	−3	9	44	14	196
32	4	16	34	4	16
30	2	4	22	−8	64
34	6	36	10	−20	400
24	−4	16	47	17	289
14	−14	196	31	1	1
32	4	16	40	10	100
24	−4	16	30	0	0
30	2	4	32	2	4
31	3	9	35	5	15
35	7	49	18	−12	144
25	−3	9	21	−9	81
			35	5	25
			29	−1	1
			22	−8	64
$\Sigma x = 336$	$\Sigma(x - \bar{x}') = 0$	$\Sigma(x - \bar{x}')^2 = 380$	$\Sigma y = 450$	$\Sigma(y - \bar{y}') = 0$	$\Sigma(y - \bar{y}')^2 = 1410$

significant difference between the two card settings as regards their effect on the value of neps per km.

Example 4: Below are listed the weight gains (in g) of cones after conditioning, when processed under two different process conditions: A and B:

Condition A: 25, 32, 30, 34, 24, 14, 32, 24, 30, 31, 35, 25
Condition B: 44, 34, 22, 10, 47, 31, 40, 30, 32, 35, 18, 21, 35, 29, 22

It is necessary to test whether the two conditions differ significantly as regards their effect on increase in cone weight.

Solution: Null hypothesis, H_0: $\mu_x = \mu_y$; that is, there is no significant difference between the mean increases in weight under conditions A and B.

The alternative hypothesis, H_1: $M_X \neq M_Y$ (two-tailed)

$$\bar{x} = \frac{336}{12} = 28$$

$$\bar{y} = \frac{450}{15} = 30$$

$$S^2 = \sqrt{\frac{\Sigma(x_i - \bar{x})^2 + \Sigma(y_i - \bar{y})^2}{n_1 + n_2 - 2}} = 71.6$$

$$n_1 = 12$$

$$n_2 = 15$$

Under the null hypothesis,

$$t = \frac{\bar{x}X - \bar{y}}{\sqrt{S^2\left[(1/n_1) + (1/n_2)\right]}}$$

$$= \frac{28 - 30}{\sqrt{71.6\left[(1/12) + (1/15)\right]}}$$

$$= -0.609$$

Tabulated $t_{0.05}$ for $(12 + 15 - 2) = 25$ degrees of freedom is 2.06.

Analysis: since the calculated value of t is less than the tabulated value, H_0 may be accepted at a 5% level of significance and we may conclude that the two conditioning parameters do not differ significantly as regards their effect on weight gain.

Example 5: A polyester fabric is stitched using two different types of sewing threads: a 100% polyester sewing thread and a polyester-cotton blend sewing thread. Two random samples of 11 and 9 stitched fabrics show the sample standard deviation of the seam strengths as 0.8 and 0.5 respectively. Assuming that the seam strength distribution is normal, it is necessary to test the hypothesis that the true variances are equal at the 10% level. [Assume that $P(F_{10,8} \geq 3.35) = 0.05$ and $P(F_{8,10} \geq 3.07) = 0.05$.]

Solution: We want to test the null hypothesis, H_0: $\sigma_x^2 = \sigma_y^2$ against the alternative hypothesis, H_1: $\sigma_x^2 \neq \sigma_y^2$ (two-tailed).

The following values are given: $n_1 = 11, n_2 = 9, S_x = 0.8, S_y = 0.5$.

Under the null hypothesis:

$F = S_x^2/S_y^2$ follows F distribution with $(n - 1, n - 2)$df

$$n_1 s_x^2 = (n_1 - 1) S_x^2$$

or

$$S_x^2 = \left(\frac{n_1}{n_1 - 1}\right) s_x^2 = \left(\frac{11}{10}\right) \times (0.8)^2 = 0.704$$

Similarly,

$$S_y^2 = \left(\frac{n_2}{n_2 - 1}\right) s_y^2 = \left(\frac{9}{8}\right) \times (0.5)^2 = 0.28125$$

Therefore,

$$F > \frac{0.704}{0.28125} = 2.5$$

The significant values of F for the two-tailed test at the level of significance α = 0.10 are:

$$F < F_{10,8}\left(\frac{\alpha}{2}\right) = F_{10,8}(0.05) \qquad \text{(a)}$$

and

$$F = F_{10,8}\left(1 - \frac{\alpha}{2}\right) = F_{10,8}(0.95)$$

We are given the tabulated significant values:

$$P\left(F_{10,8} \geq 3.35\right) = 0.05 \quad \text{i.e.} \quad F_{10,8}(0.05) = 3.35 \qquad \text{(b)}$$

Also $P\ (F_{8,10} \geq 3.07) = 0.05$
that is, $P\ (1/F_{8,10} \leq 1/3.07) = 0.05$
that is, $P\ (F_{10,8} \leq 0.326) = 0.05$ (c)
that is, $P\ (F_{10,8} \geq 0.326) = 0.95$

Therefore, from a, b and c, the critical values for testing the null hypothesis against the alternative hypothesis at the level of significance $\alpha = 0.10$ is given by:

$$F > 3.35 \quad \text{and} \quad F < 0.33$$

Since the calculated value of F (=2.5) lies between 0.33 and 3.35, it is not significant and so the null hypothesis of the equality of the population variances may be accepted at the level of significance of $\alpha = 0.10$.

Example 6: The breakages in a ring frame (while spinning 30^s) count across the doff are given successively as doff completes. According to mill practice, the breakage percentage for this count is 2%. What should the variance of population be at the acceptable (1%) level of significance?

Table 3.6 End breakages rate in a ring frame at various positions of doff

Doff %	Breakage per 100 spindles per hour (%)
3–15	2.72
15–25	3.63
25–35	2.72
35–45	2
45–70	0.92
70–90	0.8
90–100	3

Table 3.7 Waste obtained at various speed frames

M/c no.	701	702	703	704	705	706	707	708	709	710
Waste (g/batch)	20.08	22.86	20.56	24.08	22.5	24.04	24.00	20.20	20.00	22.46

Solution: Assuming that there is no significant difference in the breakage percentage at various positions of doffs (the null hypothesis):

Sample size, $n = 7$
Because the sample is small, the t-test must be used here.
Degree of freedom, $v = n - 1 = 7 - 1 = 6$

$$\bar{x} = 2.256, \ \mu = 2.5$$

$$s = 1.068$$

$$t = \bar{x} - \frac{\mu}{\left(s / \sqrt{n}\right)}$$

The tabulated value of t is $t_{0.005,6} = 3.71$
The calculated value of t is $(2.256 - 2)/(1.068/\sqrt{7}) = 0.634$
Since the calculated value of t is less than the tabulated value, the null hypothesis is acceptable at a 1% level of significance. There is no significant difference in the breakage percentage at various positions of doff.

Example 7: The waste (in g per batch) found at various speed frames in a spinning unit are as follows:

Do the values vary significantly at a 5% level of significance? Mill practice in this case is that waste produced at a speed frame is generally 21 kg per batch.

Solution: Assume the null hypothesis that there is no significant difference in the waste obtained at various speed frames.

Sample size, $n = 10$
Because the sample is small, we will use the t-test here.
Degree of freedom, $v = n - 1 = 10 - 1 = 9$

$$\bar{x} = 22.078, \quad \mu = 21$$

$$s = 1.72$$

$$t = \frac{\bar{x} - \mu}{\left(s / \sqrt{n}\right)}$$

The tabulated value of t is $t_{0.025,9} = 2.26$
The calculated value of t is $(2.256 - 2) / (1.068 / \sqrt{7}) = 0.634$

Since the calculated value of t is less than the tabulated value, therefore null hypothesis is acceptable at 5% level of significance. There is no significant difference in the waste obtained at various speed frames.

3.8 Testing fibre and yarn properties

3.8.1 Fibre properties

Fibre properties are very important and should be considered in each stage of the textile manufacturing process. There are many different fibre properties, of which the following are most important to textile technicians:

fibre length,
fibre strength and elongation,
fibre fineness,
fibre maturity.

Fibre length is one of the most important fibre characteristics, and has an influence on factors such as spinning limit, yarn strength and evenness, and product appearance. It can also influence productivity in textile manufacturing. For example, shorter fibres (e.g. < 4–5 mm) will usually be lost as waste in the manufacturing process. Fibres of 5–15 mm in length tend to contribute to the fullness of the yarn rather than its strength, whereas fibres above 12–15 mm are long enough to contribute to yarn strength and survive carding without significant shortening.

Fibre strength is another important characteristic, with the minimum strength for a useful textile fibre being approximately 6 cN/tex. Fibre bundles are usually tested for strength using HVI instrumentation, which uses the following scale of values:

- 32 and above = very strong,
- 30–32 = strong,
- 26–29 = base,
- 21–25 = weak,
- 20 and below = very weak.

In most fibres (with the exception of polyester and polypropylene), strength depends on the moisture content, which will change depending on the ambient conditions.

Fibre elongation can be divided into three different categories: permanent elongation, in which the fibre is stretched and does not return to its original shape on relaxation; elastic elongation, in which the fibre does return to its original shape; and breaking elongation, the point at which the fibre breaks. Elongation is measured as a percentage of the original length of the fibre. Elasticity is very important in textile fibres, as textile products must have the ability to return to shape after deformation and stretching, for example, in the elbow of a garment. The fibre elongation should therefore be at least 1–2%. Synthetic fibres often have much higher elongation values, for example, 15–30%, although a very high elongation value can make spinning and drafting more difficult. The relationship between strength and elongation is expressed in a 'stress–strain' curve. Each fibre has a typical curve, and it is therefore important when blending fibres that those with similar curves are chosen.

Another of the main fibre characteristics is *fineness*. The fineness of the fibres in a yarn determines how many fibres appear in a cross-section of the yarn, which in turn determines the strength and evenness of the yarn. There are usually over 100 fibres in yarns used for new spinning technologies, although there can be as few as 30 in yarns for other applications. As well as the yarn strength and evenness, fibre fineness has an influence on the spinning limit and such factors as the drape and lustre of the fabric. The fibre fineness can also influence productivity, in terms of the end breakage rate. Fineness is determined by the relation of mass to length, as follows:

$$\text{Tex} = \frac{\text{Mass (g)}}{\text{Length (km)}} \quad \text{or} \quad \text{dtex} = \frac{\text{Mass (dg)}}{\text{Length (km)}}$$

The final fibre characteristic to be discussed in this section is *fibre maturity*. A natural fibre (e.g. cotton) consists of a cell wall and lumen, and the fibre is said to be more mature as the cell wall thickens. A moisture-swollen cotton fibre is considered 'mature' when the cell wall comprises 50–80% of the fibre cross-section; 'immature' when it comprises 30–45%; and 'dead' when it is less than 25%. It is important that cotton stock does not include too many immature or dead fibres, as these do not have adequate strength or stiffness and so can lead to problems such as loss of yarn strength, variable dye uptake and processing difficulties.

3.8.2 Yarn properties

As well as fibre properties, it is also important to routinely test yarn properties such as fineness/count, twist, strength, evenness and surface integrity.

Yarn fineness can also be described as yarn count, number or size, and as with fibre fineness, can be variable as yarn does not always have a circular cross-section. To address this problem, 'linear density' or mass per unit length is used to measure the fineness of yarn. This can be done using direct or indirect systems of measurement. Direct count systems such as the 'tex' (Ne) determine yarn count as the mass of a unit in length of yarn, for example, the mass in grams of 1 km of yarn. Other systems such as the 'kilotex', which measures mass in kg per kilometre, and the denier, which measures mass in grams of 9000 m, are also used. Indirect count systems determine yarn count as 'units of length' per 'unit of weight'; for example, hanks (840 yards) per pound (lb) or kilometres (km) per kilogram (km).

Yarn twist is the number of spiral turns in a yarn in order to bind the constituent fibres together. Twist is described using three main parameters: direction, level (turns per unit of length) and factor. Twist direction is especially important when twisting two single yarns together to form a ply yarn. The aesthetic and strength requirements of the ply yarn determine whether the twist is Z on Z or S on Z. Forming a ply yarn in this way will result in contraction or extension of the length of the ply yarn compared to the two single yarns, depending on the direction of the twist. This later influences the appearance of the fabric made from the yarn, and can be used for example to create different surface effects. It can also influence the fabric stability or torque.

While linear density is the most practical way to express yarn fineness, as discussed above, it is sometimes necessary to know the yarn diameter, for example, when determining the structural features of a fabric (e.g. width or thickness). The following equations can be used to express the yarn radius as a function of the yarn density:

Direct count (tex):

$$d = k_1 \sqrt{\frac{\text{tex}}{\rho}}$$

Indirect count (cotton count):

$$d = \frac{1}{k_2\sqrt{\rho \mathrm{Ne}}}$$

The value of the yarn diameter therefore depends on both the linear and volumetric densities of the yarn.

One of the main characteristics of yarn quality is its strength, and there have been many studies devoted to this subject. It is an important parameter throughout textile manufacture; for example, a weak yarn can cause defects and even breakages in spinning and weaving, and processes such as dyeing and sizing can alter the strength and mechanical behaviour of yarn. It is therefore very important to determine and test yarn strength at regular intervals. The stress–strain behaviour of yarn has a direct effect on the properties of the end fabric; for example, a strong yarn will produce a strong fabric.

Finally, testing yarn density is an important factor in textile manufacture. Yarn consists of fibres and air pockets and the packing fraction (φ), which determines the yarn bulk density, is calculated as follows:

$$\varphi = \frac{V_f}{V_y}$$

V_f = volume of fibres in yarn and V_y = volume of yarn (fibres + air).

The packing fraction indicates the air to fibre ratio; for example, a packing fraction of 0.5 means that there are equal amounts of air and fibre in the yarn (most spun yarns have a packing fraction of well above 0.5). The yarn density again has a direct effect on the properties of the end fabric, as a high packing fraction can create a stiff, weak yarn, whereas a low packing fraction can create a yarn with minimal surface integrity (i.e. it is likely to disintegrate during processing). Yarn density also has an effect on fabric performance, in terms of properties such as flexibility, dimensional stability, strength, insulation and absorption.

The type of fibre used in the yarn and the spinning technique (e.g. ring-spun vs. rotor-spun) can also have an effect on the yarn density.

3.9 Testing fabric properties

It is important to test fabric properties throughout the production process, as part of quality control. Testing the fabric properties at regular intervals ensures that they are suitable for their intended purpose. The following are a few of the important fabric properties that need to be checked regularly:

fabric weight,
fabric thickness,
fabric strength,
fabric abrasion resistance.

3.9.1 Fabric weight

Fabric weight is expressed as the weight of the fabric in grams per m^2. It has no limits but does affect the many of the fabric properties.

Fabric weight is a fundamental property that needs to be controlled during the manufacturing process in order to avoid economic loss, for example, by buying heavier fabric than is necessary for the product being manufactured.

Fabric weight, that is, GSM, influences other fabric properties such as thickness, flexural rigidity, bending rigidity, drape, air permeability and thermal properties. For example, the lighter the fabric, the lower its bending rigidity.

3.9.2 Fabric thickness

In order to determine the thickness of a compressible material such as textile fabric, the precise measurement of the distance between two parallel plates should be measured when they are separated by the cloth. A known arbitrary pressure between the plates should be applied and maintained.

It is useful to measure fabric thickness, in order to check the material against the specification.

Fabric thickness is also useful in studying fabric properties such as thermal insulation, resilience, dimensional stability, fabric stiffness, abrasion and total handle value.

It is also useful when studying fabric geometry.

3.9.3 Fabric strength

Fabric strength can be divided into three components: its resistance to tensile force, its resistance to tearing force and its resistance to bursting force. Whether the strength of the fabric is measured in all these components depends on the type of fabric and its end use.

Tensile strength and elongation

Tensile strength and elongation depend upon the following factors:

Raw material.

Yarn strength.

Fabric construction (Weave: plain weave is stronger than other types, for example floats-twill or satin weave. Density: Low density causes weave slippage which results in seam slippage).

Finish applied to fabric. (For example, a resin finish improves seam slippage.)

Adverse of finishing process.

Where the fabric is to be subjected to tension (e.g. for industrial purposes), tensile strength is necessary. The tensile strength of a fabric should always be several times greater than the maximum stress likely to be encountered in use, because the strength of most fabrics will diminish during their lifetime due to rubbing, flexing and chemical attack. Fabric tensile strength is also useful in monitoring various relevant process parameters during production. For example, if a shortfall in the fabric tensile properties is observed during production, this observation can help to identify potential causes in the process.

Fabric tear strength

The tear strength of the fabric refers to its resistance to tearing force. Usually fabric tears when it is snagged by a sharp object and the immediate small puncture is converted into a long rip. This is probably the most common type of strength failure, so testing fabric tear strength is very important.

Resistance to tearing force is of great importance in clothing fabrics and in technical fabrics such as those used for parachutes. Tear tests are not suitable for knit fabrics and non-woven fabrics.

It is important to measure tear strength for flat sheet-like materials such as fabric, plastic films, paper and leather. Outdoor clothing and uniforms are examples of clothing where tearing strength is of importance.

Fabric bursting strength

Bursting strength is a method of measuring strength in which the material is stressed in all directions at the same time and is therefore more suitable for materials such as knitted fabrics, lace or non-woven.

Fabrics used in parachutes, filters, sacks and nets are simultaneously stressed in all directions during service. In service, a fabric is more likely to fail by bursting than by a straight tensile fracture; one example is the stress present at the elbows and knees of clothing. During the test, fabric fails across the direction that has the lowest breaking extension.

3.9.4 Fabric abrasion resistance

Abrasion or wear is the wearing away of any part of a material by rubbing against another surface. Textile material becomes unserviceable for several reasons. The first is abrasive wear. Carpets are often discarded because of extensive wear due to abrasive action over a period of time. Measuring the abrasion resistance of fabric is therefore important for customer acceptance and satisfaction, as well as manufacturing practicalities.

The following are factors that can cause abrasion of fabric during usage:

Friction between two fabrics, such as the rubbing of a jacket or coat lining on a shirt, pant pockets against pants fabric, and so on.

Friction between the cloth and external objects such as furniture or seat covers.

Friction between fibres and dust or grit in a fabric, resulting in the fibres being broken or cut. This is an extremely slow process and may take several years before it becomes apparent. It might be observed in home textiles.

The abrasive resistance of a fabric should therefore be tested and controlled, in order to achieve good quality fabrics.

3.10 References

1. G. A. V. Leaf (1984), *Practical statistics for the textile industry*, The Textile Institute, Manchester, U.K.
2. Douglas C. Montgomery, George C. Runger (2010), *Applied Statistics and Probability for Engineers*, John Wiley & Sons.
3. E. L. Grant and R. S. Leavenworth (1998), *Statistical Quality Control*. 6th Ed. McGraw-Hill, New York, USA.
4. A. Abouelela Radwan, H. Abbas, H. El deeb, S. Nassar, *A statistical approach for textile fault detection*, IEEE International conference on Systems, Man, and Cybernetics 2000, vol. 4, 2857–2862.
5. Vincent A. Voelz, *HypothesisTesting*, August 30, 2006, Math Bio Boot Camp 2006, University of California, San Fransisco, pp. 2–11.
6. T. V. Ratnam and K. P. Chellamani (1999), *Quality Control in Spinning*, T. V. Ratnam, K. P. Chellamani 3rd Ed. SITRA, Coimbatore.
7. J. E. Booth (1996), *Principles of textile Testing*, 3rd Ed. CBS Publishers and Distributors.
8. V. K. Kothari (1999), *Testing and Quality Management*, IAFL Publications, New Delhi.
9. http://www.scribd.com/doc/15627537/Yarn-Characteristics.
10. http://www.rieter.com/en/rikipedia/articles/technology-ofshort-staple-spinning/raw-material-as-a-factor-influencing-spinning/

Part II

Process control in fibre production and yarn manufacture

4
Process and quality control in cultivating natural textile fibres

M. ZIMNIEWSKA, Institute of Natural Fibres and Medicinal Plants, Poland, I. FRYDRYCH, Technical University of Lodz, Poland, J. MANKOWSKI, Institute of Natural Fibres and Medicinal Plants, Poland, and W. TRYWIANSKA, Gdynia Cotton Association, Poland

DOI: 10.1533/9780857095633.2.81

Abstract: This chapter discusses process control in vegetable fibre production from the cultivation of fibrous plants to fibres ready for application in further technologies. The chapter is focused on the processing of different types of natural fibres occurring in fruits, leaves and stems. The key parameters of agronomic as well as technological factors of early processes determining the quality of vegetable fibres are also discussed in this chapter.

Key words: process control, cotton, bast fibres, fibrous plants.

4.1 Introduction

There are a wide variety of different natural fibres, which, in a general classification system, are divided into vegetable and animal fibres. Vegetable fibres are produced as a result of photosynthesis in fibrous plants and the general term used is lignocellulosic fibres, although some of the fibres, including cotton, have little or no lignin. The growing conditions of fibrous plants, fibre extraction and the preliminary fibre processes all have a significant influence on fibre quality. The use of natural fibres has a very long history (Czerniak, 1998), dating back to about 8000 BC when people developed methods of obtaining, processing and utilising fibre according to end-users' needs. Achievements of civilisation and dynamic development of agricultural techniques, engineering and chemicals have led to huge improvements in natural fibre quality and performance. Introducing elements of process control supports maintaining the correctness of technological operations in natural fibre production. This chapter focuses on describing process control in the production of natural fibres such as cotton, and strict lignocellulosic fibres: flax, hemp, sisal and jute, presented in Table 4.1.

4.2 Control of cotton fibre quality

Cotton is the main raw material used in the textile industry and accounts for about 50% of the processed fibres produced worldwide. It belongs to the

Table 4.1 Tabular presentation of selected natural fibre

Fibre	Botanical name	Plant photo	Longitudinal view of fibres	Image of fibres cross section
Cotton	*Gossypium hirstum*			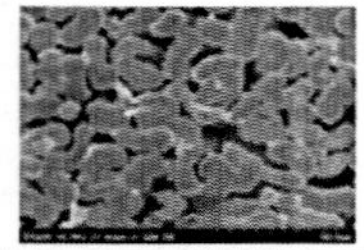
Flax	*Linum usitatissimum*			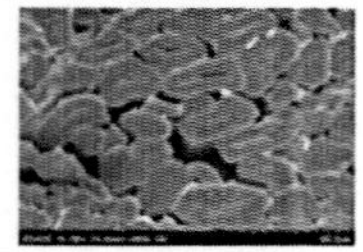
Hemp	*Cannabis sativa*			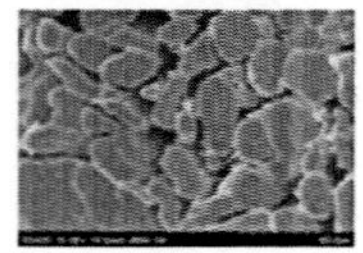
Sisal	*Agave sisalana*		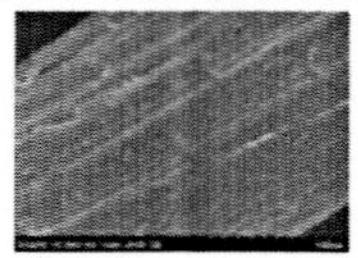	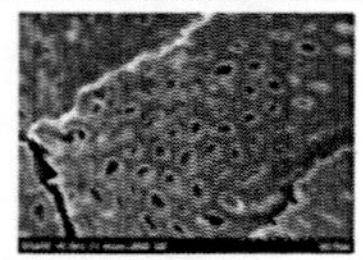
Jute	*Corchorus capsularis*	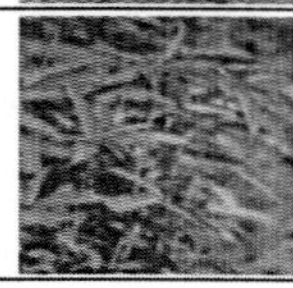		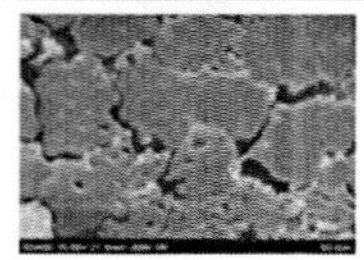

Source: Research work INF & MP Poznań.

category of natural, vegetable, mono-cellular fibre. Cotton is characterised by a high variability of properties within a single origin and, additionally, a high differentiation between the fibre properties cotton of different origin is observable. Basic fibre parameters such as length, maturity, strength and elongation, depend on many factors including the botanical type of cotton used, the cultivation region, the soil and weather conditions in which the plant is growing and maturing, the irrigation system used, and methods of picking and ginning.

It has been established as a result of analytical research, confirmed by experiment (Balls, 1928; Zurek, 1948; Zurek and Nawrocka, 1976; Zurek and Zakrzewski, 1984), that quality is dependent on such fibre parameters as fibre length, uniformity, short fibre content (SFC), fineness and maturity, as well as immature and dead fibre content, tenacity and strain at break, colour, trash content, neps and seed-coat neps (SCN), visible fibre matters (VFM), stickiness, etc. Fibre quality evaluation can be carried out for the purposes of classification and trading, and for technological and scientific applications. It can be organoleptical or instrumental. Organoleptical evaluation has been used for classification and trading purposes for many years and is still, to a lesser extent, used today. Instrumental evaluation is mainly used for technological and scientific purposes.

4.2.1 Factors influencing cotton fibre growth and quality

Basic fibre qualities such as length, maturity, strength and elongation, depend on the botanic type of cotton (*Gossypium barbadense, G. hirsutum, G. herbaceum and G. arboretum*), the cultivation region (Australia, Argentina, USA, Central Asia, China, etc.), soil type (cotton requires clay-sand soil), the manner of growing (as a herb, shrub or tree), the weather conditions in which the plant is maturing, the method of irrigation, the method of harvesting (by hand or mechanical) and the method of ginning (saw or roller gins).

For example, for *G. hirsutum*, only half the flowers form the bolls and, of those, only half the bolls produce a standard value fibre. *G. barbadense*, on the other hand, produces twice the number of ripening bolls as *G. hirsutum* although its smaller fibre content contribute to a smaller total crop. Cotton cultivation is possible only in areas where certain conditions of both temperature and humidity are fulfilled. In good temperature and moisture conditions the seeds germinate 5–7 days after sowing. In less favourable conditions seeds germinate after 10–15 days. Between 8 and 12 days after germination the first leaf appears. The first flowers will appear 2–3 months after germination. The period of florescence of one flower is 1–2 days. After blossoming the seed capsule, or boll, forms and ripens. Figure 4.1a shows an example of a boll. The ripening period of bolls from early flowers is about 40–70 days, and for later flowers about 80 days. Very late flowers never ripen – when harvested their fibres and seeds are dead.

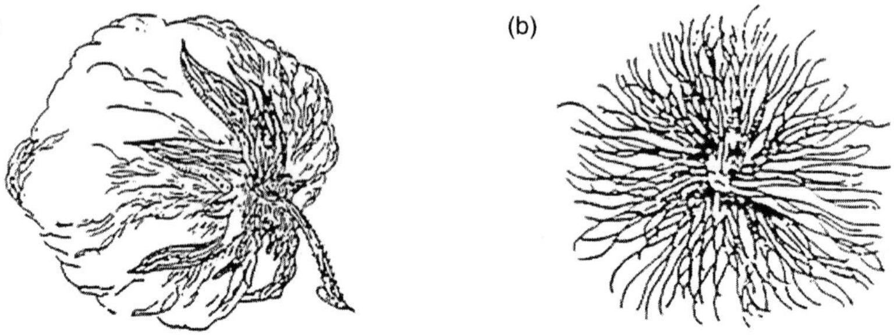

4.1 (a) Completely opened boll, (b) hairy seed.

The full cycle of cotton development is:

- for *G. hirsutum* 125–150 days,
- for *G. barbadense* 145–160 days.

The weight of hairy seeds (Fig. 4.1b) in one boll also varies from variety to variety:

- for *G. hirsutum* 3–12 g,
- for *G. barbadense* –2.5–4 g.

On the day of florescence the cells of the outer layer of fibre, called the epidermis, starts to lengthen. The time of longitudinal growth is about 18–20 days. With these hairs appear other hairs that are shorter and thicker, the undercoat. The undercoat does not appear on every seed, and in some cases it appears only in the narrow end of the seed.

The properties of the fibres in the boll depend on the placement of the boll on the shrub (the most valuable fruit is generally found at the lower part of the plant) as well as on the fibre placement on the seed. Cotton seed has a shape like a drop. The basic characteristics of cotton fibres differ depending on where they are positioned on the seed: the longest fibres are at the broad end of the seed, the chalaza, and the shortest are found at the narrow end of the seed, the micropyle. The thickest fibres are on the micropyle and the finest on the chalaza. The sizes of seeds and the properties of fibres also depend on the seed location in the boll. The biggest seeds with the best built hairs are in the middle of the boll. Cotton fibre maturity and strength can be affected if immature bolls are opened.

After appropriate time for growing and reaching the full length, the process of building the fibre secondary wall starts. This takes 25–30 days. The cellulose for the secondary wall is deposited in 24 h cycles. If cotton is cultivated in a steady temperature, the secondary wall has a continuous structure. The thickness of the secondary fibre wall depends very much on the surrounding conditions. In good weather conditions the secondary wall will be thicker than the fibre wall produced if the growing conditions are too dry. During the development of the secondary wall the quantity of protoplasm and sugar in the fibre diminishes.

Even with cottons of an identical genetic quality, the fibre quality may vary due differences in the method of harvesting and ginning. Some aspects of the cotton fibre quality will be superior for cotton that has been processed through a manual harvesting and ginning system versus a completely mechanical system. Mechanised cotton harvesting and processing systems tend to do more damage to fibres than manual systems. The resulting differences will be in the length and nep content of the fibres.

4.3 Indexes for cotton fibre quality

The term quality expresses the degree of excellence. Sometimes this degree of excellence is expressed by the amount of money someone wants to pay for it. For example, the price received by cotton producers is determined largely by the cotton breeders' secondary market, the textile processors. The price paid by processors is usually determined by the quality of the cotton lint as estimated by various parameters such as fibre length, strength, fineness, colour and lack of foreign matter.

Sometimes the term quality means an ability to fulfil certain needs. In the case of cotton fibres this last definition seems to be the most appropriate. The quality of cotton fibres is determined by measuring the properties of the fibres in relation to several parameters expressing the ability of the yarn to fulfil the functions it was designed for. For cotton producers, the quality of the cotton will be determined by values such as micronaire, upper half mean, uniformity index, strength, elongation, SFC and trash content. For spinners, quality is determined by ability of the cotton fibre to fulfil the requirements for successful processing and production and a good quality end product of yarn that is of high strength and of a good evenness. The emphasis on strength has largely been prompted by changes in textile technology that allow for more rapid processing but place greater physical stress on the cotton fibre.

Foulk et al. (2007) found that different cotton fibre properties are important for different spinning systems. For example, for ring spinning the key properties are length, strength, fineness and friction. For open-end spinning the key properties are strength, fineness, length, cleanliness and friction. Finally, for the vortex system the key properties are length, fineness, strength, friction and cleanliness. Textechno (2007) noted that the important properties for air-jet spinning are fineness, neps, trash and dust content, strength, length and uniformity and friction. The selection of the optimum cotton fibres for a textile mill end product is challenging, and more an art than a science.

Some of the quality parameters of cotton, such length, micronaire, strength and grade, directly affect the market price (Moore, 1996). Others, such as length uniformity, elongation, SFC, nep content and seed-coat fragment content, can have long term effects on cotton's attractiveness relative to synthetic fibres. Breeders need also to be aware of these parameters.

At the beginning of the twentieth century, when cotton was assessed for quality organoleptically, the United States Department of Agriculture (USDA) established cotton grade standards. By the end of the twentieth century, and continuing into the beginning of the twenty first, increased laboratory assessment of quality led to the establishment of the indices detailed below to express the intrinsic value of the tested cotton.

Korickij (1983) introduced the so-called complex quality index (CQI) expressing the cotton spinnability, its geometric property index, cotton yield during the spinning process and its price.

$$\text{CQI} = \frac{A_i B I^4}{C} \qquad [4.1]$$

where

A_i – coefficient equal to 0.0108 for long staple cottons or 0.0141 for medium staple cottons (derived for the Russian cottons),
C – cotton fibre price,
B – cotton yield during the spinning,
I – geometric property index calculated according to the formula:

$$I = 0.1 l_m \text{UI}\left(\frac{1 - \text{SFC}}{100}\right)\text{Mat}\,(\text{Tt})^{-0,5} \qquad [4.2]$$

where

l_m - weighted mean length of cotton fibres,
Tt - cotton fibre fineness,
Mat - cotton maturity,
SFC - short fibre content.

The most simplified form of cotton fibre quality is the so-called fibre quality index (FQI) expressed by the following equation:

$$\text{FQI} = \frac{(\text{Str*len})}{\text{Tt}} \qquad [4.3]$$

where

Str – cotton fibre strength,
Tt – cotton fibre fineness,
len – cotton fibre length.

If HVI mode of fibre testing is used then the above expression is changed as follows:

$$FQI_{HVI} = \frac{UHML.UI.FS}{FF} \quad [4.4]$$

where FQI_{HVI} is HVI fibre quality index, UHML is upper half mean length, UI is length uniformity index, FS is fibre bundle tenacity and FF is micronaire.

Majumdar (2005) proposed the modified version of FQI, using multi-criteria decision making techniques, in the following form:

$$MI_{AHP} = \frac{FS^{0.27}.FE^{0.039}.UHML^{0.291}.UI^{0.145}}{FF^{0.11}.SFC^{0.145}} \quad [4.5]$$

where FE is fibre elongation and SFC is short fibre content.

Some other approaches are based on the regression models connecting fibre properties with the parameters characterising the spinning ability or spun yarn quality. The spinning consistency index (SCI) proposed by Uster is expressed as a function of the HVI properties by the following regression equation:

$$SCI = -414.67 + 2.9FS + 49.17UHML + 4.74UI - 9.32FF + 0.65Rd + 0.36(+b) \quad [4.6]$$

where Rd is reflectance degree and b is cotton yellowness.

Generally, the higher the SCI, the higher the yarn strength and the better cotton spinnability. El Mogahzy (El Mogahzy, 1990) suggested that the actual value of cotton should be determined based on those inherent characteristics that contribute to the quality of the yarn and the textile products made from it. This approach suggests that the desirable properties of cotton are those that allow the production of the best quality yarn in a particular manufacturing system.

Based on the values of percentage relative contribution of the fibre property C_i% and the difference factor D_i a premium/discount index PDI can be determined as follows:

$$PDI = \sum_{1}^{k} (C_i \times D_i) \quad [4.7]$$

where:
k – the number of the HVI fibre properties,
C_i % – percent relative contribution expressed by the following formula:

$$C_i = 100\left(\frac{B_i^*}{\sum_1^k B_i^*}\right)R^2 \quad [4.8]$$

where: B^*_i is the standardised coefficient of ith variable ($i = 1, ..., k$) and R^2 is the coefficient of determination,

$$D_i = \frac{(X_i - \mu_i)}{\delta_i} \quad [4.9]$$

or

$$D_i = \frac{(X_i - \mu_i)}{r_i} \quad [4.10]$$

where

X_i – the value of fibre property of the bale,
μ_i – the overall mean value of the ith fibre property,
σ_i – the reference standard deviation of the ith fibre property,
r_i - the overall range of the ith fibre property.

Substituting the values of Ci in Equation [2.7] the following formulas are obtained:

For the ring spinning:

$$(\text{PDI})_{\text{RS}} = 25.95D_{\text{FL}} + 11.95D_{\text{LU}} - 14.85D_{\text{FF}} + 17.95D_{\text{FS}} \quad [4.11]$$

For open-end spinning:

$$(\text{PDI})_{\text{OE}} = 14.25D_{\text{FL}} + 12.55D_{\text{LU}} - 22.10D_{\text{FF}} + 20.10D_{\text{FS}} \quad [4.12]$$

where D_{FL}, D_{LU}, D_{FF}, D_{FS} are the difference weighting factors of HVI fibre properties.

Trash content and SFC of the input raw material are also of major importance, when the technological value of cotton is considered. These two factors (el Mogahzy, 1990) could be combined into a processing performance index I by the following formula:

$$I = \frac{1}{(\text{TC})(\text{SFC})} \quad [4.13]$$

where

TC – trash content (by weight),
SFC – short fibre content.

It is worth mentioning here how the fibre quality parameters can be measured. At the beginning there existed only simple separate instruments for fibre quality testing, so-called Low Volume Instruments such as: micronaire, fibrograph, colorimeter, stelometer, etc. Efforts have been made to improve the traditional measurement methods or to develop new ones for a better, more comprehensive and objective fibre assessment (Zurek and Frydrych, 1997). The era of automatic systems for fibre quality evaluation, so-called High Volume Instruments, began in the 1970s.

There are some differences between the measurement principles of systems, as well as between the measured parameters, because the measurement methods and instruments were invented by the different institutions from different countries. Many high volume measurement systems enable not only the determination of particular parameters, but also a complex statistical assessment (e.g. AFIS) and assured knowledge about the raw cotton necessary for an economical success of cultivation, trading and processing.

4.4 Process control in harvesting

In the twenty first century production of cotton should be more efficient than it used to be, it should characterised by the higher yields and better fiber quality to keep cotton importance on the market of textile products. To achieve this goal the harvest preparation as well as mechanical way of harvesting should be evolved. Harvest preparation process becomes more important with starting the mechanisation era in harvesting; and therefore factors of harvest control should be checked very carefully and in the responsible way.

4.4.1 Mechanical versus manual harvesting

Cotton is harvested after mature bolls have appeared on the plants. This harvesting can be carried out both by hand and mechanically. About 70% of the cotton produced globally is harvested by hand. For example, about 90% of the cotton crop in China is harvested by hand picking. Hand picking can result in cleaner cotton (i.e. less non-lint material is present) if only the seed cotton is removed from the boll. Hand harvesting also avoids other contaminants including synthetic fibres and plastics, natural organic matter from defoliation, and inorganic matter from soil. There can, however, be seed-coat fragments and stickiness (insect and plant sugars) present in the harvested cotton. Further contamination of the harvested cotton can occur during transportation of the cotton bales.

Manual picking is divided into stages. The first stage is carried out when two or three fully ripened bolls have appeared on the majority of plants. According to the scheme of plant development they will generally appear towards the bottom part of the plant. At this stage 20–35% of the raw cotton is collected. The second stage begins when the next 35–40% of bolls have matured, and the third stage is 12–15 days later. The fourth and final stage is often carried out after the ground-frosts; this final stage of picking is often completed mechanically.

Manual picking is often expensive and time consuming; this has led to the development and use of mechanical pickers. Manual picking is considered to be better because the fibres picked are usually cleaner. Mechanisation of picking is a faster process, but the fibres picked can be dirty and require additional cleaning. In order to calculate the full cost of faster, mechanical picking one must take into account the expense of the more intensive fibre cleaning which is required as part of the process. For this reason, the growers must consider which method of picking is best for them. Australia, Israel and the USA are the only countries where all cotton is harvested by machine. The driving factors behind the mechanisation of cotton harvesting in the USA include the availability and cost of labour, a desire to harvest cotton as quickly as possible in order to prevent deterioration of fibre and seed quality, and the preservation of fibre and seed quality during storage prior to ginning and throughout the ginning process. In Greece, mechanical picking accounts for more than 95% of total cotton production. The efficiency of mechanical cotton harvesting can be increased in two ways: defoliation and the growing of compact shrubs. These two methods will be discussed in the next section.

4.4.2 Defoliation

To protect the quality of cotton fibre that has been harvested mechanically it is advantageous to prepare the cotton for harvesting using chemicals. The goal of harvest-aid use is to protect the quality of the fibre and enable early harvest in order to reduce field-weathering losses, minimise trash content and staining of the lint, and allow for the safe storage of seed cotton. Defoliation should be performed prior to mechanical harvesting, because green leaves reduce the grade of the cotton, slow the harvest and increase gin costs. It is always advisable to harvest the cotton as soon as possible after a sufficient number of bolls have opened. Proper water management can also help to enhance leaf shedding. Ceasing irrigation can assist leaf drop and boll maturation in some growing areas.

A basic knowledge of crop development and maturity, as well as of harvest aids, is necessary to make a decision on the effective application of harvest-aid materials. Crop condition and air temperature will determine which defoliation materials and rates are appropriate.

For a successful harvest, defoliation must be carefully timed and carried out, as poor defoliation can lower the fibre quality. Defoliation that is done too early lowers a yield and micronaire, while defoliation that is carried out too late increases the likelihood of boll rot and lint damage or the loss of some of the crop due to weathering. It is safe to defoliate cotton when about 60% of bolls are open. Defoliation at the wrong time would stop the development of the remaining immature bolls and can lower yield and micronaire.

Boll-opening materials are often used in combination with defoliant materials to increase the percentage of the crop harvested during the first picking or to eliminate the need for a second picking. Boll maturity is very important when using a boll-opening material.

4.4.3 Compact shrubs

Defoliation is one way to increase the efficiency of mechanical harvesting. The second is plant selection, with a view to growing of more compact shrubs. Compact shrubs form bolls almost directly on the trunk (Fig. 4.2). Additionally, these selected varieties are characterised by a less dispersed blossoming time, which is advantageous in regards to unification of fibre-development range and fibre properties.

Timely harvest of the most valuable fruit – generally from the lower part of the plant – allows for a more efficient capture of yield and a higher quality crop. Cotton producers make the decision to apply harvest aids based on factors including the maturity of the plant, the condition of the crop, the prevailing water conditions and weather forecasts, and the desired harvest schedule. In countries where the cotton is trade is based on harvested seed cotton, it is important to measure the accurate trash level of machine-picked seed cotton.

4.4.4 Factors in harvest control

Fertility, water management and weed control play an important role in harvest control. The aim is to supply the crop with adequate fertility and

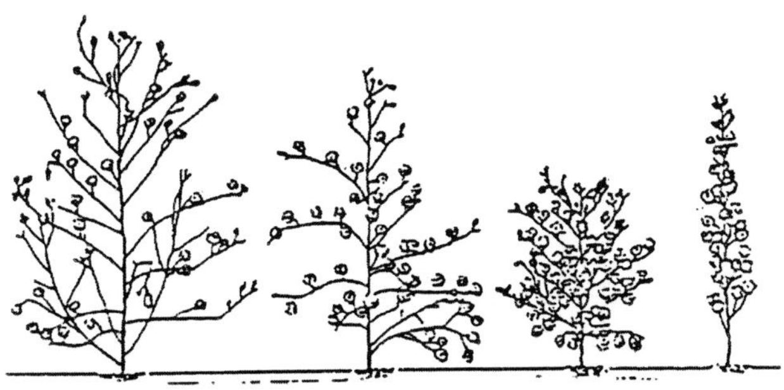

4.2 The shrub shapes.

moisture levels to obtain the optimum yield. Failure to meet the nutritional or moisture needs of the plant will negatively impact the yield potential and plant activity. Inadequate weed control will possibly impact the fibre quality. This is in addition to the loss of yield potential. Guidelines for harvest-aid products, rates and timing can often be obtained locally. However, the basic concepts apply to improving the efficacy of pre-harvest preparations. Harvest-aid products generally work better on mature cotton under warm, humid and sunny conditions.

Field evaluations involving boll opening can sometimes underestimate the boll maturity. The most reliable method is to slice open bolls with a knife. Mature bolls will be too hard and cannot be easily cut with a sharp knife. Lint will also string out when the boll is sliced, and seed coats will be dark or black. Cotton bolls are much more tolerant to wind, rain and other factors when the leaves are on the plant.

It is also recommended that the producer contact the gin prior to harvest to ensure that timely processing of the seed cotton can be achieved to help preserve the fibre quality. Producers must be able to critically evaluate every input. An understanding of the development of the plant is crucial for making strategic management decisions and maintaining profitable production.

4.4.5 Ginning

Regardless of whether it is hand or machine harvested, all seed cotton must undergo ginning. This operation minimally involves the separation of the cotton fibre from the cottonseed and baling of the fibre into transportable bundles. All commercial cotton ginning is done by machines. Depending on the machine used we can distinguish between saw and roller ginning. Saw ginning is more destructive to cotton fibres than roller ginning. Historically, saw gins were invented earlier than roller gins. Figure 4.3 shows a typical saw gin and Fig. 4.4 shows a common roller gin.

The ginning process is complex and may be divided into certain phases: opening, cleaning, proper ginning, cleaning of fibre received and pressing into bales. The opening of the cotton is carried out by means of barrel type machines. Cotton is opened by means of pneumatic conduits and passed to cleaning and ginning machines. Pneumatic conduits are equipped with devices that extract heavy particles, such as nails, pebbles, metals, which can become accidentally mixed with the cotton and can cause damage to the machines. Various types of machines are used for preliminary cleaning of the cotton prior to ginning. Cleaning is done step by step. First, trash particles are removed, then the bolls, in situations where the whole boll has been harvested, are torn apart and removed.

After the main impurities have been removed from the cotton, the proper ginning takes place. The principle of all ginning is that certain parts of the

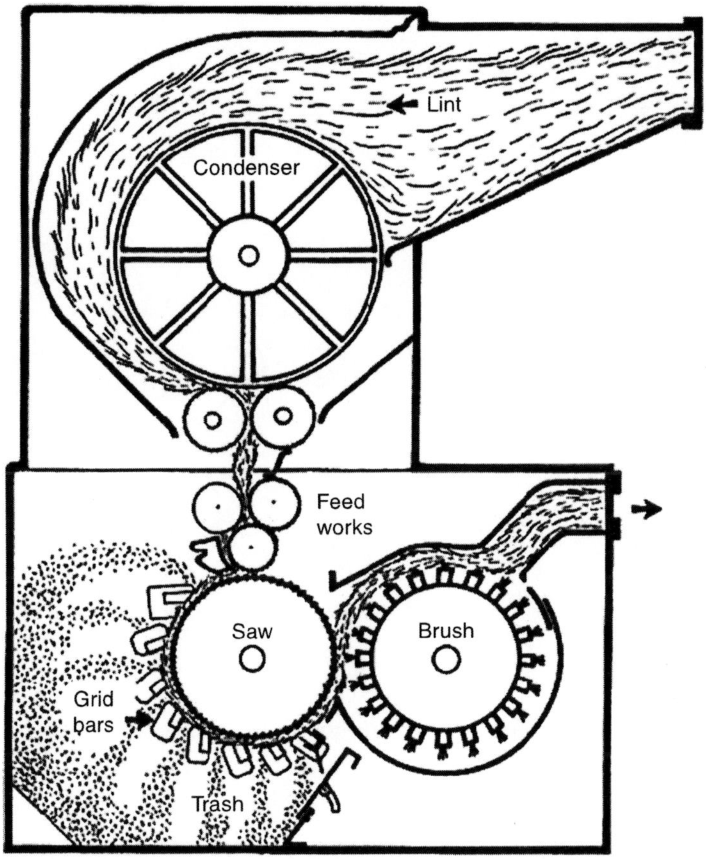

4.3 A typical saw-type lint cleaner which removes trash and blends and combs the fibre.

machine grab the fibre while others hold the seed. Under the appropriate tension the fibre is torn off the seed. This is possible because the fibre strength in the base of the seed is only 15–50% of the strength above the base. Therefore, when drawn, the fibre has been broken at its weakest point.

One of the basic requirements of successful ginning is that it damages as few seeds as possible, so that seed-coat fragments are not introduced into the lint. Seed-coat fragments interfere with the subsequent spinning process and with the strength of the yarn made from the ginned lint. Until the development of the mechanical harvester, most cotton cleaning technology had concentrated on cleaning seed cotton. Saw lint cleaners were able to efficiently remove the extra trash that was brought to the gin by mechanical harvesting.

Saw gins are used for ginning short and medium staple cotton. There are many types of saw gins but they have in common a saw-disc causing that tears the fibre off the seed. They have many shortcomings, the most important of

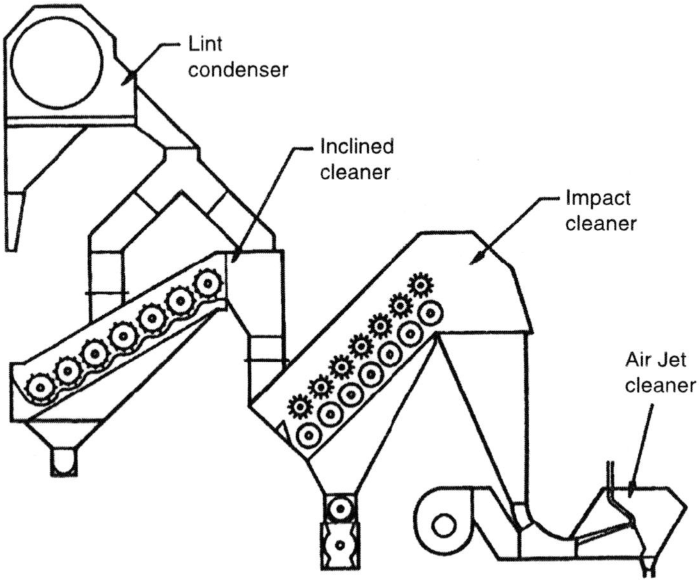

4.4 The most common lint cleaning sequence for roller ginned cotton.

which is a frequent mechanical fibre breaking as well as forming neps and curls. If the fibre is longer, there is a higher probability that it can be broken in more than one place during the ginning process. Roller gins (single, double or rotary knife) cause less shortening of the fibre during the ginning process and greatly reduce the number of breakages. Only the rotary-knife roller gin is currently used in the USA and it is effective at maintaining fibre quality (Wakelyn and Chaudry, 2010). The shortcoming of the roller gin is a lower yield than that obtained from the saw gin. Gin costs rise with increased trash levels, because a greater amount of total harvested material, including the seed cotton, is required to make a bale of cotton lint.

Uster IntelliGin for process control in ginning

Uster IntelliGin (Fig. 4.5) is a computerised, on-line cotton fibre quality measuring system that monitors, controls and optimises the cotton-ginning process (Ghorashi, 1998, 2000). It provides a cotton ginner with critical data, enabling the gin to process for weight and grade optimisation. In 2002 in the USA there were approximately 70 gins equipped with this system (Ghorashi, 2002).

Uster IntelliGin's on-line sampling produces immediate measurements of fibre colour, trash content and moisture. Information about the moisture is used to control the drying temperature to obtain the optimum fibre moisture content (7.5% for the bale). The appropriate moisture content of cotton

4.5 Uster IntelliGin.

bale means: an increase of bale mass, reduction of electrostatic charge, what facilitates a fibre transportation, reduction of pressure used at the bale pressing, what reflects in savings of energy costs, and machinery maintenance. Maintaining the moisture content on the appropriate level is very important because of the technological process (optimum dryer temperature and cleaning practices) and assures the appropriate fibre quality. Cotton is not over- or underdried or over- or under-cleaned – preserving grade and bale weight.

Data about fibre colour and trash content are used for determining the number of ginning passages. The system is easy to use and allows the ginner complete control of the process. The immediate feedback of these properties allows the ginner to make conscious decisions and get the best results from the cotton-ginning process. The system includes an automatic control mechanism that can guarantee that the ginning gives the same results throughout the week.

All the tasks are designed to optimise the moisture level of the cotton and the ginning process in order to obtain as high as possible a price for the cotton. Cotton growers want to have a high profit from their plantations, whereas the spinners are interested in the highest fibre quality. The ginners have to satisfy both of them. Using IntelliGin, cotton ginners can now customise their ginning process, so that growers can earn more income per bale and spinners are satisfied by the better fibre quality.

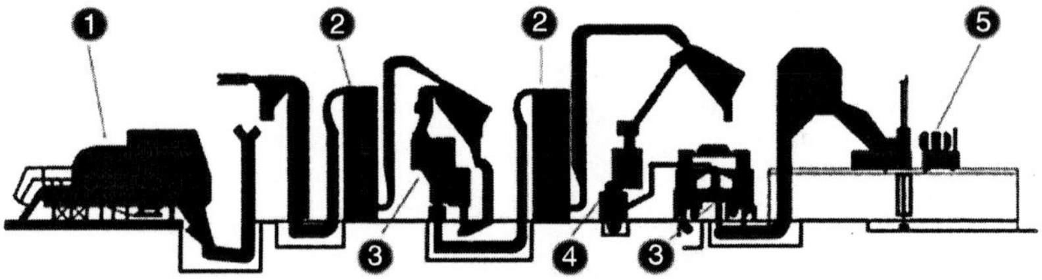

4.6 The scheme of check points of IntelliGin (booklet): 1 module of mass and moisture content, 2 drying temperature, 3 fibre cleaning valves, 4 humidity on the ginning stand, 5 final mass and moisture content in bale.

Uster IntelliGin assures an appropriate fibre quality by the:

- Increase of fibre length, because the controlled ginning reduces a fibre breakage,
- Reduction of the nep number,
- Controlled drying system assuring the optimum moisture content in fibres,
- Reduction of SFC.

Uster IntelliGin assures the continuous monitoring basing on the input data as a crop quality, fibre characteristics, methods of picking up, and climatic condition of plantation. Schematic presentation of arrangement of checking points of IntelliGin is shown in Fig. 4.6.

4.5 Control of natural lignocellulosic/bast fibre quality: climatic conditions

Natural lignocellulosic fibres are delivered from elements of dicotyledonous plants such as stems and leaves (Zychlinski, 1958). The raw material source for the textile industry is only tissues composed of fibrous cells. Their technological usability is determined by the slenderness – the ratio of the cell length to its thickness – and by strength parameters. The quality of bast fibres is determined by many different conditions classified as agronomic factors, as well as technological parameters in processing, summarised in Table 4.2.

The fibrous plants presented earlier, which are sources of bast fibre, are grown in various climatic zones. Individual requirements differ but, in general, they can be summarised as moderate temperature and high water availability. Bast fibrous plants are highly dependent on atmospheric conditions, crucial for the quantity and quality of fibre yield.

The optimal increment of fibre in stems takes place in moderate temperatures and in areas with high moisture availability. In extreme hot and dry conditions the fibre is highly lignified, thick with low divisibility. In such conditions the yields of stem fibre are reduced. For fibrous flax, for instance,

Table 4.2 Processes control parameters – factors determining the quality of vegetable fibres

Agronomic factors determining the quality of vegetable fibres	Technological factors determining the quality of bast fibres
Climatic conditions Soil conditions Cultivars of fibrous plants Quality of seed sown Tillage and fertilisation conditions, Time of sowing Sowing density Weeds, diseases, pests Time of harvest	Conditions of retting raw material: parameters of controlling the process: retting duration, water temperature, quality and pH, hydro-module Raw material moisture content Mechanical processing Chemical processing of bast fibre

the 24-hour sum of temperatures in the vegetation period is 1300–1800°C, and necessary amount of rainfall – 700 mm (Живетин, 2002). For industrial hemp these parameters are 2000°C and 500–600 mm (Живетин, 2002) respectively. Jute in growing season requires optimal temperature of 17–41°C, and 1500–2000 mm of rainfall (Franch, 2005). *Agave sisalana* can be cultivated in dry conditions, but the plant produces more new leaves during the wet season.

4.5.1 Soil conditions

Soil conditions are of significant importance, especially for helping to ensure the quality of bast fibres. The best soil conditions are secured by structural, well aerated and fertile soils with good water-storing capacity. Light (sandy) soils produce very low quality raw material (short fibre). Plants grown on soils rich in calcium produce very brittle fibre characterised by low spinnability. Generally, bast fibrous plants give better yields when nutrient content in the soil is high. It should be stressed that organic soils will yield high amounts of biomass but fibre efficiency will be low. As regards leaf fibres, *A. sisalana* can grow in poor soil, but fibre crops can be doubled if soil is richer and climate wetter (Yu, 2000).

4.5.2 Cultivars of bast crops

Main activities in bast fibrous plant breeding aim at obtaining new cultivars securing the highest yield of best quality fibre. Breeding must consider new cultivars with better resistance to diseases and pests, drought, and factors such as fertilisation level and harvest mechanisation. New cultivars must secure maximising the yield of very delicate fibre on one side and progress in agronomy and processing on the other. The cultivar has to be adapted to local agro-climatic conditions.

4.5.3 Quality of sowing seeds

The value of bast fibre crops is determined to a high degree by the quality of the sowing material, characterised by such parameters as purity, germination energy and capacity, 1000 seed weight, health condition and moisture content (Живетин, 2002). Sowing of non-qualified seeds gives no guarantee for obtaining the optimum amount of bast fibre with the required high quality, crucial for a competitive position on textile market.

4.5.4 Tillage and fertilisation conditions

Tillage is the agricultural preparation of the soil by mechanical agitation of various types, such as digging, stirring and overturning. Correct tillage in the case of bast fibrous plants should provide:

- Optimum soil structure for plants,
- Use of nutrients accumulated in the soil,
- Optimum water storage, and
- Weed control.

Only accurate tillage can secure strong and even growth, high biological and morphological uniformity and good quality and high amounts of fibre (Живетин, 2002).

When setting up a fertilisation programme for bast fibrous plants, the following factors should be taken into account:

- Soil fertility (nutrient content),
- Previous crop cultivated,
- Expected yield of fibre,
- Desired amount of fibre.

The following groups of elements are used in the fertilisation of bast fibrous plants:

a) macroelements (N, P_2O_5, K_2O, CaO, MgO),
b) microelements (B, Zn, Mo, Cu, Mn).

Nitrogen – has a yield-forming effect on above ground biomass and can prolong the growing season. An excess of nitrogen delays the flowering, can increase plant lodging and vulnerability to diseases. Excess of N has a direct negative impact on fibre content in the stem and results in quality reduction (Живетин, 2002).

Phosphorus – positive effect on plant maturity and good formation of fibrous tissue; increases yields of seed. Excessive application of P_2O_5 is limited.

Potassium – very positively influences the yield of fibrous plants but mainly the yield of fibre. K is a fibre-forming factor and, as such, is responsible for the good formation of fibre. It also improves the resistance of bast fibrous plants to diseases, pests and drought. The most significant effect of K, however, can be seen as mentioned before, in the proper formation of fibre cells and bundles, which become stronger and more divisible (Живетин, 2002).

Calcium – important nutrient and a factor regulating the soil reaction and structure. Excess of calcium in the soil reduces the quality of bast fibres, which become brittle and non-divisible.

4.5.5 Sowing time

The effect of the sowing time on the yield quantity and quality, especially on that of fibre, is very significant. Early sowing guarantees, in agronomic, technological and economical context, the optimum biomass yield and also the highest efficiency and quality of produced fibre, especially in terms of fibre strength (Живетин, 2002). Raw materials obtained from late sowings can also secure a high yield but the obtained fibre is short, weak and of low quality. Additionally, when plants are grown in warm conditions, early sowing allows for obtaining a secondary yield for example of kenaf, which can be used as a fodder (Franch, 2005).

4.5.6 Sowing density

Sowing densities of bast fibrous plants, if too high or too low, are unfavourable in terms of agronomic and technological performance. A very low sowing density causes an increase in the growth of very thick stems and a reduction in fibre content and quality (thick and coarse fibre). A sowing density that is too high will result in thin stems, susceptible to lodging and disease. When densely sown, plants become more susceptible to drought; numerous plants would stop growing too early and cause low homogeneity of raw material (Bankin *et al.* 2007). This means that it is necessity to modify the sowing density to obtain the optimum plant density for the best fibre yield and quality. Optimum planting should guarantee sufficiently thin stems that ensure obtaining high amounts of thin and divisible fibre for different industrial applications.

4.5.7 Impact of weed, diseases and pest control

Despite a wide choice of weed control for growing bast fibrous crops, either mechanical or chemical, weed control is an important problem. During the mechanical process of fibre extraction, weed biomass contaminates the fibre and thus reduces its quality. Weeds contaminate different types of fibre, especially short fibre, and make it non-spinnable (Живетин, 2002). When controlling weeds chemically, special attention should be paid to the chemicals used, as more and more consumers look for bast fibres from ecologically grown plantations.

Diseases and pests also have a significant effect on the quality of the raw material obtained and consequently on the amount and quality of fibre. Plants infected with diseases or pests produce fibre that is not formed properly, which can be coarse and of low spinnability.

4.5.8 Time of harvest

Harvesting plants at the correct stage of maturity, and the harvesting technology used, are crucial for fibre quality. Harvesting fibrous plants is performed, depending on the technological requirements, at the optimum time when fibre is properly formed, sufficiently strong, and able to form a strand of technical fibre.

Harvesting fibrous plants too early results in poorly formed fibre bundles. Fibre in such bundles is very thin and divisible but is also very weak. The fibre content is low (Venturi *et al.*, 2007). Harvesting too late increases the amount of fibre but the quality of the fibre diminishes considerably as, by this stage, the fibre is thick, of low divisibility, highly lignified and rough (Venturi *et al.*, 2007). Therefore, the optimum harvest time for bast fibrous plants is very important, and is closely connected with the direction of application.

In the case of flax, the proper degree of maturity is assessed by colour of the stalks and seed pods, which should be yellow-brown. In the case of *A. sisalana*, the time of harvesting is not critical regarding the fibre production. Leaf cutting can be conducted at a time convenient for farmers, but the interval period between cutting should be approximately one year. Sisal is usually harvested once a year, but if soil and climatic conditions are favourable it can be harvested twice a year, or three times in two years.

4.6 Process control in production

The preliminary processes of fibre production can vary because they are strongly related to the type of fibrous plants to be processed. However, in all cases the processes consist of fibre extraction and the determination of the quality of the raw materials. Bast fibres, like flax, hemp and jute, are usually retted to eliminate substances that cement the fibre to the bundles and to the rest of the tissues in the stem. Retting is not applied to sisal leaves; fibre extraction is instead conducted through the decortication process.

Table 4.3 Changes of chemical composition of raw material in fibre production processes

Raw material	Components content [%]			
	Cellulose	Lignin	Pectin	Substances dissolve in 1% solution of NaOH
Raw stem	51,42	20,86	6,18	31,7
Too short retting stem	52,80	20,78	4,92	25,3
Proper retting stem	53,26	20,20	3,26	20,6
Too long retting stem	54,32	19,82	2,73	19,3
Raw fibres	62,1	4,6	6,8	27,1
Too short retting fibres	73,2	3,9	5,2	24,1
Proper retting fibres	79,1	3,7	3,7	21,1
Too long retting fibres	81,5	3,4	2,8	20,3

4.6.1 Retting of raw material

Preliminary preparation has a considerable effect on the extraction of the optimum amount and quality of bast fibre. Retting can be conducted by two methods. The first method, common in countries of moderate temperature where rainfall and dew occur regularly, is dew retting, which is conducted in the field. After harvest, straw is left in the field where, under natural atmospheric conditions, microorganisms, mainly fungi, develop, which enzymatically decompose the pectin bonds between the woody part and the fibre. In countries with a hot climate, water retting is the most common type of retting. In these conditions mainly bacteria develop, which decompose pectin and thus enable the extraction of the fibre from the stems. Retting must be stopped at the right moment when the raw material is optimally retted. Retting duration, water temperature, water quality as well as hydro-module – the ratio of raw stem mass (kg) to water (Zychlinski, 1958) – are key parameters that have to be controlled during the retting process.

The pH of water used for retting must be controlled. In most cases the pH drops to 4.6–4.9 during retting and a rise in pH indicates over-retting and damage to the fibres. The temperature of the water used for retting has an influence on the speed of the biochemical reaction and can be within the range 20–38°C. During water retting, the moisture content of straw fibrous plants increases up to 400% and the mass of the plants decreases by 20–30%. During this time the chemical composition of the straw and fibres changes. These changes are presented in Table 4.3.

Together with changes linked to chemical composition, hydrolysis of pectin and low-polymerised hemicelluloses, changes in the structure and properties of the fibre are observed. The influence that retting time has on the properties of fibres is presented in Fig. 4.7.

For jute retting the optimum temperature is about 34°C and the optimum pH value is in the range of 6.0–8.0. Under retting (retting for too short a

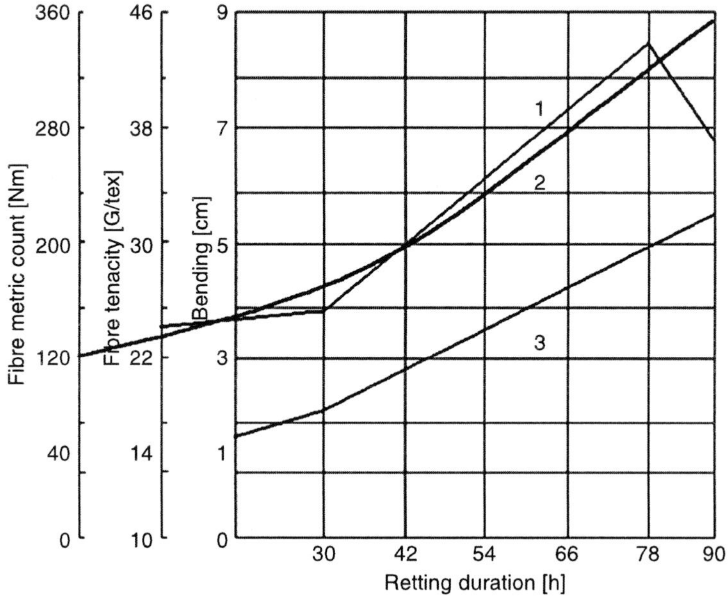

4.7 Changes of physical-mechanical properties of flax fibre in relation to straw retting time; 1 – tenacity, 2 – fibre metric count, 3 – bending.

time) can lead to difficulties with the extraction and cleaning of fibre which is characterised by low divisibility, stiffness, coarseness, and the fibre strands are very strong (Bankin *et al.*, 2007; Szalkowski, 1965). Over-retting also leads to a worsening of fibre quality, as it can result in the very advanced decomposition of pectin and hemicellulose in the fibre strands and bundles. This can increase the delicacy of the fibre but can also reduce the fibre efficiency and strength (Bankin *et al.*, 2007; Szalkowski, 1965). Controlling the degree of stem retting is achieved by evaluation of stem samples taken during the retting process. In the case of properly retted straw, the fibres can be degummed very easily from the stem, well divided with good resistance to hand-break. To ensure the high quality and high efficiency of the fibre it is important to remove (squeeze out) excess water from the retted stems. This also weakens the bond between the fibre and the woody tissue and makes mechanical fibre extraction easier (Szalkowski, 1965).

4.6.2 Moisture content in raw material

The processing straw and the extraction of fibre are also highly influenced by the moisture content in the straw. The final moisture content of dried straw after retting should not be lower than 8%. Straw with a moisture content below 7% can have significantly shorter bast fibres, although this does allow for the more effective removal of impurities. On the other hand, straw with a moisture content that is too high, exceeding 15%, which is then rolled into bales for storage, can develop mildew (Salmon-Minotte, 2000).

Dried retted straw has to be moisturised before undergoing further mechanical processes. The optimum moisture content in this case should be between 10% and 14%. Artificial moisturising of dry straw allows raw material to be obtained with the optimal moisture content in a short time. This enables continuous fibre extraction and increases both fibre quality and productive capacity (Szalkowski, 1978). Moisture content of over 20% allows for longer fibres but the impurity content (remnants of other plant tissues) is high.

4.6.3 Mechanical processing

The main objective of the mechanical processing of bast fibrous plants is fibre extraction with the maximum amount of fibre extracted of the highest possible quality to allow for further processing. Fibre extraction technology produces fibres bonded to one another in the form of strands. Obtaining technical fibre is associated with the destruction of the stem or other part of the plant containing fibre bearing tissues (e.g. sisal is obtained from leaves).

While discussing fibre extraction methods, it is essential to mention the decortication process – the extraction of green fibre without retting. The quality of the bast fibre obtained by this method is lower in comparison to the fibre obtained after retting. Decorticated fibres are thick, strong, non-divisible, and heavily contaminated by remnants of other plant tissues (wood, epidermis). This makes decorticated fibre a raw material suitable mainly for non-textile applications. The time of decortication of the jute plant is precisely determined. The best stage in the growth of the plant for decortication is 115 ± 10 days after germination (Krishnan, Doraiswamy and Chellamani, 2000). Decortication should be done immediately after harvesting. The optimum diameter for jute decortication is around 12 mm. A design of a decortication machine is shown in Fig. 4.8.

Decortication of sisal fibres is conducted for fibre extraction from fresh leaves. After decortication, the next processing stages of sisal preparation include water washing, binding, cleaning, drying and final cleaning. The product of early sisal processing is long fibres with good lustre and waste. These

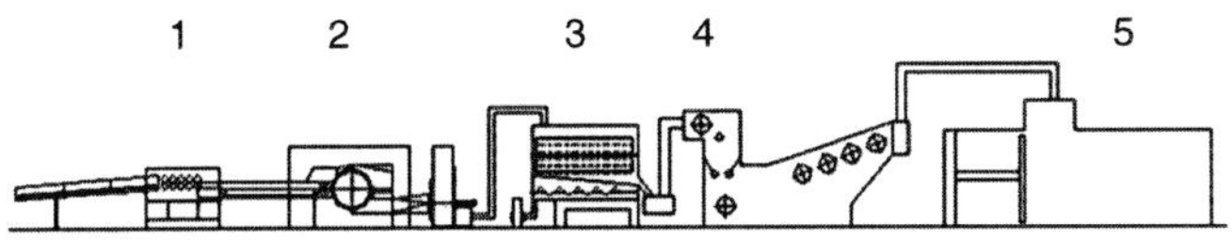

4.8 Scheme of a unit for producing bast fibres for pulp industry. 1 Breaking machine, 2 Decorticator, 3 Preliminary cleaning machine, 4 Scutcher, 5 Press.

are called kinked fibres. The fibrous stem after retting has to be mechanically processed to extract of fibre of good quality.

In the existing methods of mechanical processing, separation of the fibre from the woody tissues is done by squeezing and breaking. The stress to which raw material is exposed (stretching) during processing, present to a higher or lower extent, may cause damage or breaking of fibre, which can have a direct negative impact on its quality. The intensity of mechanical processing may lead, on one hand, to the intensive purification of the fibre and, on the other hand, to an excessive shortening of the fibre. A reversed situation may allow obtaining a higher ratio of long bast fibres but with a high content of impurities.

The technological chain of the mechanical processes of fibre extraction comprises several stages, which are suitable for long fibres or tow, among others:

- A breaking machine or a softener
- A scutching machine
- Conditioning
- Sorting fibres according to fibre type and quality.

Smoothness of the technology is related to keeping the key process parameters on the proper level. The factors within the mechanical process that have to be controlled are:

- Distance between working elements suitable for types of fibres
- Construction and diameter of feeler rolls
- Pressure of working elements on the material
- Rotary speed of rollers
- Operating speed
- Others.

The stem parameter that must be controlled during fibre extraction is its moisture content, because of its high influence of productive capacity. The effect of moisture content of raw material and scutching intensity on the fibre capacity and quality is presented in Table 4.4.

During mechanical processes, the moisture content of stem should be maintained within the range of 10–11% for the highest capacity of both scutched and hackled fibres and for good divisibility of scutched fibres. The lower level of roll rotary speed within allowed range gives higher capacity of the fibres.

Therefore, it is necessary to control the intensity of work of extraction devices so as to obtain optimum quality of raw material, depending on its final use.

Table 4.4 The effect of moisture content of raw material and scutching intensity on fibre capacity and quality exemplified on flax fibres

Moisture content [%]	Number of turns of scutching rolls [1/ min]		Capacity of scutched fibre [%]	Capacity of hackled fibre [%]	Average of metric count hackled fibre [Nm]	Average of metric count of scutching fibre [Nm]
	I section	II section				
8–9	290	270	9,3	67,5	24,6	21,0
	270	200	10,2	61,8	25,8	20,8
10–11	300	280	10,9	70,0	25,0	21,3
	280	235	12,1	72,1	24,9	21,8
11–12	310	290	10,9	67,9	26,7	21,0
	295	255	11,7	71,9	24,2	20,9
14–15	360	295	9,8	63,3	26,0	21,1
	345	260	10,5	67,4	25,7	21,0

4.6.4 Chemical processing

For some species of fibrous plants chemical processing is used to improve fibre properties, especially those important for spinning. It should be stressed, however, that each chemical process on bast fibres is linked with improvement of thinness, delicacy and divisibility but also with the risk of diminished fibre strength (Szalkowski, 1967; Franch, 2005).

4.7 Evaluating fibre quality

The first stage in the evaluation of fibre quality is conducted by sampling from the raw materials in a loose state. The key issue of fibre evaluation is ensuring proper determination of truly representative samples, the size of the sample and the manner in which it is taken. The properties of natural fibres are generally characterised by great variability, so that samples must be relatively large if useful reliability is to be assured (Morton and Hearle, 2008). Sampling from raw materials in a loose state requires special rules. Ideally, the required number of fibres should be taken one at a time from a corresponding number of places distributed at random all over the bale, bag, or whatever form the population assumes. This would be impracticable because of the amount of labour involved and it would be impossible to obtain samples free of the extent bias. A good example of a sampling procedure suitable for manufactured staple fibres is zoning shown in Fig. 4.9.

The initial sample obtained by zoning is formed into 16 tufts, and by a process of doubling, drawing, halving and discarding, these are reduced to the representative sample for measurement. The tufts are taken in pairs and repeatedly drafted by hand and recombined before being divided into

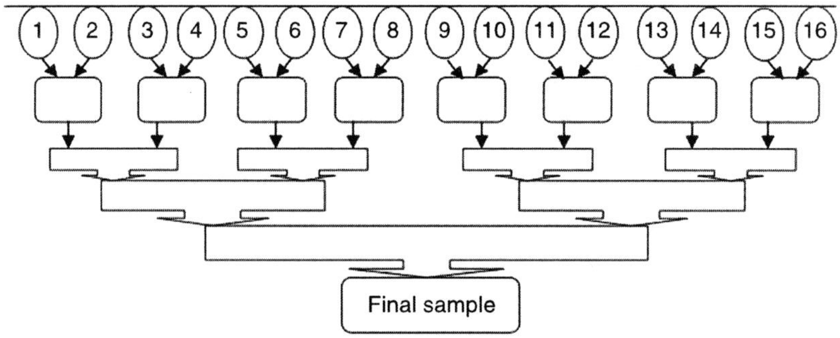

4.9 Method of sampling from raw materials in loose state.

two parts. Since the fibres are thereby not only mixed together but also effectively parallelised, lengthwise splitting is essential.

The evaluation of bast or leaf fibre is usually conducted by gravimetric methods using the knowledge and experience of fibre experts, according to classification systems proper for each type of fibre and valid for the country of origin. Long and short flax and hemp fibres are sorted into lots according to properties such as weight, fineness, softness, strength, colour, uniformity, silkiness or oiliness (greasiness), length, cleanliness, smell and handle. Fibre is classified to the best class when it is very fine, properly divided, delicate, homogenous, with oil-hand (fibre with slippery touch), with content of impurities below 3%, neutral smell, natural colour, difficult to hand-break as well as showing uniformity of all parameters (Polish Standards PN-P-80104:1997; Polish Standards PN-P-80098:1999; Batra, 1995). Sorting jute fibre is conducted after retting and decortication. Both types of jute fibres, Tossa and white, are sorted in categories according to a classification system applicable to the country of origin. The following properties of fibre are evaluated for its quality assessment: colour, fineness, strength, density, root proportion and tendering (Krishnan, Doraiswamy and Chellamani, 2000).

The main parameters determining the quality of sisal fibres according to quality standards are:

1. Length
2. Bundle strength
3. Trash content
4. Colour
5. Spot

Excellent sisal fibres are white with lustre, length greater than 95 cm, bundle strength higher than 880 N/g, trash content lower than 2.5%, and no spots.

4.8 Future trends

Despite the dynamic development of chemical fibres, vegetable fibres are still valuable raw materials for different segments of the economy, not only in the textile industry but also in the composite application. Their high importance is the consequence of such factors as wide availability, renewability, relatively low price, good mechanical properties and physiological performance. A very important aspect affecting wide use of natural fibres is their biodegradability, a feature that allows for the complete utilisation of natural fibres with no pollution of the natural environment.

Processes control of natural fibres production ensures high quality of raw materials and of the final product. The control method of bast and leaves fibre production must be developed by introduction of instrumental tests, which are more objective and repeatable, independent of the opinions of the evaluator. Wider implementation of automatic technologies into the production of fibre will improve the overall process and the efficiency of process control methods. This will in turn lead to the manufacture of better quality fibre.

4.9 References

Balls W. L. (1928), *Studies of quality in cotton*, London, Mc Millan and Co, Ltd.

Banik S., Basak M. K. and Sil S. C. (2007), Effect of Inoculation of Pectinolytic Mixed Bacterial Culture on Improvement of Ribbon Retting of Jute and Kenaf. *Journal of Natural Fibres* Vol. 4(2), pp. 33–50.

Batra S. K. (1995), '*Fibre Chemistry in Handbook of Fibre Science and Technology*', Vol. iv (edited by M. Lewin and E. M. Pearce), Marcel Dekker Inc., New York and Basel, pp. 727–807.

Bogacz E. and Kicinska-Jakubowska A. (2010), Research work INF&MP Poznan.

Czerniak L., Kirkowski R., Kozlowski R. and Zimniewska M. (1998) *The Earliest Traces of Flax Textiles in Central Europe*, Kujawy, Poland, Natural Fibres, Special Edition, 1998/1, pp. 18–19.

El Mogahzy Y. E. (1990), A Statistical Approach for Determining the Technological value of Cotton Using HVI Fiber Properties, *Text. Res. J.* No 9, pp. 495–500.

Foulk J., Gamble G., Senter H. and Meredith W. (2007), Cotton quality indices of spun yarn', Lubbock, 4th World Cotton Research Conference, ITC.

Franch R., (2005), *Bast and other plant fiber*. Woodhead Publishing Limited – Cambridge.

Frydrych I. (2005), 'Bawełna - systemy i metody oceny (Cotton – assessment systems and methods)', Scientific Editor of the Technical University of Lodz.

Ghorashi H. M. (1998), 'Uster IntelliGin – a new on-line gin process monitor and control system', *Bremen, International Cotton Conference Bremen*, Faserinstitut Bremen e.V. and Bremer Baumwollberse, 245–256.

Ghorashi H. M. (2000), 'Developments in cotton fiber quality measurements', *Bremen, International Cotton Conference Bremen*, Faserinstitut Bremen e.V. and Bremer Baumwollberse, pp. 237–244.

Korickij K. J. (1983), 'Technological economic estimation and design of textile materials quality', Moskow, Gizlegprom, Legkaja Industria.

Krishnan K. B., Doraiswamy I. and Chellamani K. P. (2000), *Jute, book: Bast fibres and other plant fibres*, Edited by R.R. Franck, Woodhead Publishing in Textiles.

Majumdar A. (2005), Determination of the Technological Value of Cotton Fibre: A comparative Study of the Traditional and Multiple-Criteria Decision Making Approaches, *Autex Research Journal*, 5(6), pp. 71–80.

Moore J. F. (1996), 'Cotton classification and quality'. USDA-ERS. The cotton industry in the United States. USDA-ERS Agriculture Economy Report No 739 Washington DC, pp. 51–57.

Morton W.E. and Hearle J.W.S. (2008), *Physical properties of textile fibres*, Four edition, Woodhead Publishing in Textiles, pp. 82–93.

Polish Standards PN-P-80098 (1999), Textile raw materials – Short hemp fibre – Requirements.

Polish Standards PN-P-80104 (1997), Textile raw materials – long dew retted flax fibre scutched and hackled – Requirements.

Salmon-Minotte J. (2000), chapter Flax, book: Bast fibres and other plant fibres, Edited by R.R. Franck, Woodhead Publishing in Textiles, pp. 95–158.

Schleth A. and Peters G (2005), Application handbook of Uster HVI, Uster Technologies AG.

Szalkowski Z. (1967), Podstawy chemicznej technologii surowców włókien łykowych. Wydawnictwo Przemysłu Lekkiego i Spozywczego – Warszawa.

Szalkowski Z. i in. (1965), Poradnik roszarnika. Wydawnictwo Przemysłu Lekkiego i Spozywczego – Warszawa.

Szalkowski Z. (1978), Poradnik Inzyniera, Wlokiennictwo, pp. 190–212.

Textechno (2007), Cotton Classifying System catalog.

Trywianska W. (1990), General information on cotton and laboratory methods of cotton testing, the handbook of GCA.

Venturi P., Amaducci, S., Amaducci, M. T. and Venturi, G. (2007), Interaction betsween Agronomic and Mechanical Factors for Fiber Crops, *Journal of Natural Fibres*, Vol. 4(1), pp. 83–97.

Wakelyn J. and Chaudry M. R. (2010), 'Cotton: Technology for the 21st Century', ICAC.

Yu C. (2000), Sisal, *Bast and other plant fibres*, edited by R.R. Franck, Woodhead publishing in Textiles.

Zurek W. (1948), Metody mierzenia i oznaczania dlugości wlokien bawelnianych (Methods of measurement and designation of cotton fibers), Lodz, Glowny Instytut Wlokiennictwa.

Zurek W. and Frydrych I. (1997), Evolution of cotton assessment – from organoleptics to electronics, Gdansk, International Cotton Conference, GCA.

Zurek W. and Nawrocka M. (1976), Dojrzałosc wlokien - dodatkowe kryterium jakosci przedzy bawełnianej (Fiber Maturity – an additional criterion of the cotton yarn quality), *Przeglad Wlokienniczy*, 30, 334–337.

Zurek W. and Zakrzewski S. (1984), The strength of cotton fibers and the degree of development of the secondary wall, *J. Text. Inst*, (5), 323.

Zylinski T. (1958), Nauka o włoknie. Wydawnictwo Przemysłu Lekkiego i Spozywczego - Warszawa.

Живетин В. В. и другие (2002), 'ён и его комплексное использование' – Ф. Г. У.П. ЦНИЛК- Москва.

5
Process control in the manufacturing of synthetic textile fibres

B. L. DEOPURA, Indian Institute of Technology Delhi, India, A. CHATTERJEE, Dr B R Ambedkar National Institute of Technology, Jalandhar India, and N. V. PADAKI, CSTRI Central Silk Board, India

DOI: 10.1533/9780857095633.2.109

Abstract: Process control and quality assurance during fibre manufacture comprise a set of policies, procedures and processes used to ensure the quality of the fibre produced at affordable cost. Process control over the input raw material properties and process parameters is exercised during fibre formation, mainly in the polymerisation, spinning, drawing, and heat-setting stages. Process control during fibre formation involves exercising control over various parameters of the material and processes; in this chapter these parameters and their significance in the quality of the fibre manufactured are explained.

Key words: fibre manufacture, melt spinning, spinning parameters, fibre quality, drawing, heat setting, polyester, nylon, PAN.

5.1 Introduction

Process control in synthetic fibre manufacture involves the active changing of process parameters in response to process monitoring to achieve the desired fibre quality (see Fig. 5.1). Process control is one element of a quality management system, which is a set of policies, procedures, and processes used to ensure the quality of the fibre produced. A quality management system not only improves the quality of the fibre produced, but also improves the sales growth, market share and profit margins of a company and gives it a competitive advantage.

A typical fibre manufacturing system comprises three stages of process control, as illustrated in Fig. 5.1:

1. inputs,
2. processes,
3. output.

The quality of the input raw materials used for synthetic fibre manufacture is continuously monitored, compared with set standards, and

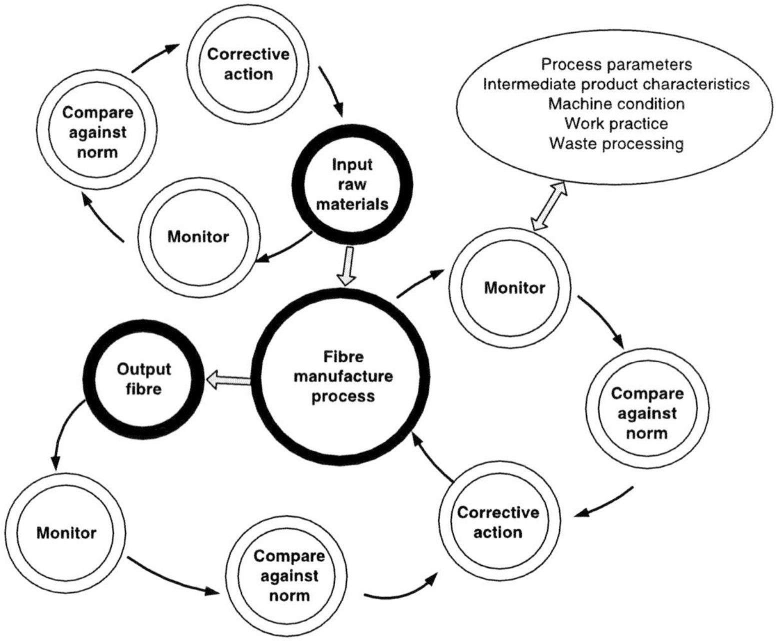

5.1 Process control in fibre manufacture.

corrective actions initiated to ensure consistency in quality. During the manufacturing stage various parameters, such as polymer processing conditions, intermediate dope characteristics, fibre-spinning conditions, by-product and waste processing parameters, are monitored and adjusted to ensure the desired fibre quality. The resulting fibre is also subjected to testing to evaluate various properties and to adjust the manufacturing process if required.

Fibre formation basically involves conversion of polymer into a long, continuous, fine thread-like structure, having properties essential for use as fibre. This conversion process involves three basic steps:

1. spinning/extrusion,
2. drawing/stretching,
3. heat setting.

These steps have their own significance. The process of spinning involves conversion of a polymer into a filament form. The drawing process involves stretching or large deformation to impart molecular orientation and crystallinity to the fibre structure. Heat setting is used to remove the internal stress and stabilise the filament. Like any other process, process control in fibre spinning means control of quality and cost in order to make the process sustainable or viable. In this section, emphasis is given to control of quality rather than cost, although there exists some correlation between the two.

By the term 'quality' an attempt is made to define certain attributes that a material must possess. In the particular case of fibre, 'quality' means certain attributes or properties that the fibre is expected to maintain within certain limits. Some of these properties are fineness, cross-sectional shape, mechanical properties, dyeability, etc. It is recognised that the properties of a fibre depend upon its morphological structure, and hence process control may be considered to control the structure and thereby the properties of the fibre to make it suitable for a particular end use.

In this chapter, process controls, which influence the quality of fibre produced, have been discussed with respect to the polymerisation, fibre-spinning, and post-spinning (drawing and heat-setting) stages. To control a process and hence the properties of fibre, it is also necessary to have a clear idea regarding the structure or morphology development at the above-mentioned stages. The factors which influence the structure development during fibre formation in these stages are further classified into two categories, namely material variables and process variables.

5.2 Process control in polymerisation and fibre spinning

Process control refers to In-Process Quality Control and is aimed at maintaining consistent uniform quality of the material in process at various stages of its manufacture or processing. The process control in the synthetic fibre plant starts with the testing of raw material and should continue with the intermediate products viz. polymers, spun fibre and finished product.

5.2.1 Process control in polymerisation

The polymerisation process variables constitute a vast field, which is discussed in this section. The quality and contents of the raw materials used for polymerisation to manufacture a particular polymer are critically important, along with the polymerisation technique.

Polyester is one of the widely used polymers, not only for textiles but also for many other industrial applications. Polyesters are now the major type of synthetic fibre produced (60%) and consumed worldwide. Polyester is manufactured industrially in two ways: the dimethyl terephthalate (DMT) method and the pure terephthalic acid (PTA) method. The critical parameters involved in the selection of the raw materials DMT, PTA, and ethylene glycol (MEG) for polyester production are detailed in Tables 5.1, 5.2, and 5.3 respectively.

Table 5.1 Properties of dimethyl terephthalate

Melting point	141°C
Boiling point	280°C
Colour	Pure white
Acid number, mg KOH/g	0.03 (max)
Saponification number	573–583
Residue on ignition	0.08%
Ester interchange value	> 90
Aldehyde ester, ppm	30 (max)

Table 5.2 Properties of purified terephthalic acid

Acid number, mg KOH/g	673–678
Ash content, ppm	8 (max)
Total metal content, ppm	6 (max)
Moisture, %	0.2 (max)
p-Touic acid, ppm	130 (max)
Benzoic acid, ppm	30 (max)
4-Carboxy benzaldehyde, ppm	20 (max)
Purity, %	99 (min)

Table 5.3 Properties of MEG (ethylene glycol) (Vaidya, 1988)

Boiling point	195–198°C
Colour	Clear water white liquid
Density	1.1–1.11
Refractive index	1.4330–1.4340
Water content	< 0.1%
Hydroxyl number	> 175
Ester interchange value	> 90

The given polymer type (polyester, nylon, etc.) manufactured from the polymerisation process is the starting material used for formation of a fibre. The polymers used as raw materials for fibre formation must satisfy certain criteria, such as linearity of molecules, stereoregularity, molecular weight, thermal and chemical stability, etc. Molecular weight or the size of the molecules plays an important role in deciding the final properties of the fibre (Mcintyre, 1998). As fibres are thin, hair-like structures, to impart sufficient strength or structural integrity to the fibre the length of the polymer molecules plays an important role. The linearity and stereoregularity of the molecules influences the packing of the molecules and hence structural integrity. The quality of the polymers (such as molecular weight and its distribution) used for fibre formation is extremely important and is taken care of during the polymerisation stage.

Different types of polymers, for instance thermoplastic, thermoset, thermotropic, lyotropic, etc., are used for fibre formation. The nature of the polymers influences the type of polymer-to-fibre conversion process. Although

Table 5.4 PET polyester polymer property (Fourne, 1999)

Properties	Range/values
Intrinsic viscosity, dl/gm	0.4–0.7 (0.63 optimum)
TiO_2 (%)	0.03
Density (g/cc)	1.37–1.4
Melting temperature-T_m (°C)	250–265
Melt spinning temperature (°C)	290
Melt density (g/cc)	0.98
Melt viscosity at 290°C (Pa.s)	50–60
Refractive index, nD	1.57–1.58

Table 5.5 Nylon 66 polymer property

Properties	Range/values
η_{rel} (in n-H_2SO_4)	2.5
TiO_2 (%)	0.03
Density (g/cc)	1.14
T_m (°C)	255–260
Extractables (%)	≈ 0.1
Melt spinning range (°C)	280–295
Melt density (g/cc)	0.98
Melt viscosity at 290°C (Pa.s)	50–60

Table 5.6 Properties of PAN powder

Colour	White
Bulk density (g/l)	200–250
Particle size (μm)	20–30
Mol. wt. (g/mol)	80000–83000
Intrinsic viscosity	1.61
Ash content (%)	< 0.12
Nitrogen content (%)	23–24
Sulphur content (%)	0.27–0.54
Carbon content (%)	65.7–67.4
Hydrogen content (%)	5.45–5.90
Comonomer content (%)	5–14
Acid number	< 0.25 mg alkali /g PAN
Water content (%)	≤ 0.7

different types of polymers are used for fibre formation, the common synthetic fibre-forming polymers are thermoplastics. These common thermoplastic polymers can further be classified into two categories: polymers obtained through the condensation polymerisation route and polymers obtained through the addition polymerisation route.

Polyester (polyethylene terepthalate) and nylon are the two common fibre-forming polymers obtained through the condensation polymerisation route, whereas vinyl polymers such as poly(acrylonitrile) (PAN) and polypropylene are polymers obtained through the addition polymerisation route (see Tables 5.4, 5.5, and 5.6 for typical properties). These two classes of polymers differ considerably in the molecular weight needed for fibre formation. The polymers obtained through condensation polymerisation have relatively lower molecular weights when compared to the polymers obtained through addition polymerisation. There are large differences in the molecular weights used for fibre formation and these differences are responsible for the differences in rheological properties of the two classes of polymers. The fibre grade polyester polymers with a molecular weight of 15 000–20 000 from polymerisation are formed into chips and fed into the fibre melt spinning process.

5.2.2 Process control in fibre spinning

Based on the type of polymer, different polymer-to-fibre manufacturing processes are available. These manufacturing processes can be broadly classified into two types:

1. melt spinning,
2. solution spinning.

Melt spinning is used for polymers that can be melted, resulting in a stable melt. Polymers that cannot be melted, or whose melt is not stable, are dissolved in a suitable solvent and processed via the solution spinning route. Depending on the type of solvent or solvent–polymer system, solution spinning can be further classified into:

1. dry spinning,
2. wet spinning,
3. dry-jet wet spinning.

A fluid is spinnable under given deformation conditions, provided that steady-state continuous deformation or elongation of the fluid jet is possible without any break. In spinning, it is essential that elongational viscosity increases with strain rate, because stress on the filament increases as the filament moves away from the spinneret. In the absence of such an increase in viscosity, a localised defect could get drawn endlessly, leading to catastrophic failure. In a visco-elastic fluid, part of the deformation energy is stored, and when it reaches a critical value cohesive or brittle fracture occurs. Deformation rate and relaxation time of the visco-elastic fluid are the two important parameters influencing the cohesive fracture. Relaxation of stored elastic energy reduces the chances of failure. Relaxation times of spinning fluids range from milliseconds for a solution used in wet spinning to several seconds for high molecular polyolefin melt. Thus cohesive fracture is more frequent with fibre grade polyolefins. Such effects are much less frequently seen in polyester and polyamides, hence a relatively high speed of spinning is possible for these polymer types.

The melt spinning process has been extensively studied and various material and process variables that influence the structure and property development of the filament have been identified (Kothari, 2000; Ludwig, 1964; Mark et al. 1967). In melt spinning, the key parameters for process control are polymer type, molecular weight, molecular weight distribution, throughput rate, quenching condition and spinning speed. The polymer type influences several important variables, for example die-swell, glass transition (T_g), and crystallisation rate. The die-swell in fibre formation has several aspects: die-swell is governed by the same visco-elastic factors that are responsible for instability in the exit zone, known as melt fracture. In extreme conditions, die-swell itself may be the source of unstable or irregular spinning.

Die-swell dictates the design of spinneret, that is the size, length and spacing of holes in the spinneret (see Fig. 5.2). Die-swell also depends on process variables such as throughput rate, extrusion temperature and spinning speed.

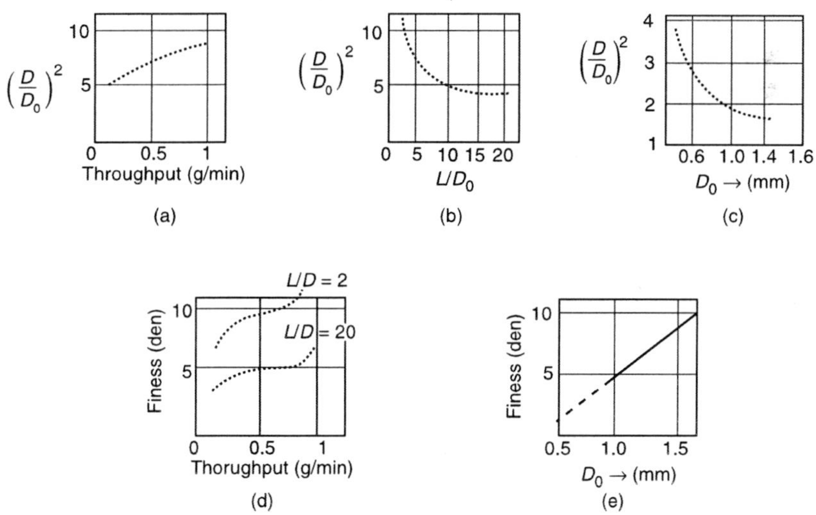

5.2 Influence of spinneret geometry on die-swell during polypropylene spinning (Fourne, 1999). (a) Influence of the throughput Q at T = 190°C, D = 1 mm, L = 8 mm, $[\eta]$ = 1.9, (b) Influence of the L/D ratio at T = 190 °C, throughput Q = 0.3 g/min × hole, other data as before, (c) Influence of the capillary diameter for T = 280°C, throughput Q = 0.3 g/min × hole, L/D = 20, $[\eta]$ = 1.9, (d) Influence of the throughput Q and L/D on the smallest filament titre, (e) Influence of the capillary diameter D_0 on the smallest filament titre.

Thus, for a particular polymer having a molecular weight range, the process parameters should be optimised to control the resultant die-swell ratio. In case of polypropylene, which is prone to large die-swell, a die-swell ratio of 2–3 has been reported. The die-swell depends on the production process of polypropylene, that is the melt viscosity and temperature, throughput rate, and spinneret hole dimensions. The die-swell ratio decreases with increasing temperature as a result of lowering of the viscosity combined with a narrower distribution of molecular weight. Die-swell increases with the throughput rate and decreases as the length-to-diameter (*L/D*) ratio or the spinneret hole diameter increase. This limits the diameter of the spinneret hole and thus the finest possible individual filament titre there by limiting the choice of spinneret hole diameter. The die-swell ratios in melt spinning of polyester and polyamides are not large, except when the temperature is close to solidification point, when the relaxation time increases rapidly. At normal extrusion temperatures, the die-swell ratio does not exceed 1.2–1.5.

Polypropylene has a relatively fast crystallisation rate. It is assumed that crystallisation starts almost immediately after extrusion; it increases rapidly with take-up speed or take-up tension. The orientation of molecules is also influenced by the crystallisation and increases significantly as the crystallisation is almost complete (Fourne, 1999). In case of polyester, however, the spinning-induced crystallisation sets in at a take-up speed of above 3500 m/min and increases further with take-up speed. At higher speed, the induced crystallinity results in a sudden change in filament structure from viscous to semicrystalline solid. With increased speed, the necking point of the filament also moves upward, closer to the spinneret, as the spin-line stress-induced crystallisation shifts the crystallisation to higher temperatures.

Similar behaviour has been observed for nylon as well. In nylon, the crystallisation during spinning does not reach completion at speeds below 3000 m/min. The nylon, owing to its low glass transition temperature and high moisture regain, permits crystallisation in winding room conditions. At higher winding speeds, the crystallisation reaches completion. However, an important difference between polyester and nylon can be observed, which is that the T_g and moisture regain values of these two polymers are significantly different. It follows that the structure present in conditioned as-spun yarn at lower speed is mainly due to crystallisation after moisture pick-up, whereas for the yarns spun at higher speeds, crystals have already been generated in the spinning process (Gupta and Kothari, 1997).

Among the various synthetic fibres, polyester fibres and filaments represent a very important product group in commercial terms, hence a number of studies related to process simulation based on polyester have been reported. These studies help us to understand the sensitivity of the as-spun filament quality to any change in operating or process conditions. This knowledge has considerable commercial significance in terms of control of

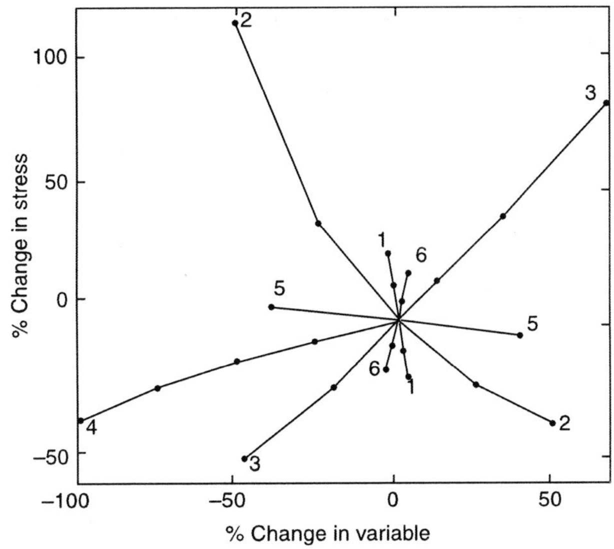

5.3 Sensitivity in respect of stress at the freeze line. The various curves are for changes in 1 Extrusion temperature, 2 Melt flow rate, 3 Take-up velocity, 4 Quench air velocity, 5 Quench air temperature, 6 Melt intrinsic velocity.

critical variables. Failure to control these variables may result in a number of downstream operational problems such as poor drawability, a high level of broken filaments, increased product non-uniformity, unacceptable processability, and so on.

It has been suggested (Nadkarni, 1997) that the incremental mechanical energy input required for a specific increase in the orientation level is unique for all glassy polymers, although the absolute stress needed to achieve a specific orientation is governed by the molecular weight. This implies that for a given molecular weight, there exists a unique relationship between applied stress and the resulting molecular orientation in the absence of crystallisation. Thus spin-line stress (σ_L) controls the orientation of the as-spun filament. The sensitivity of the as-spun filament properties is therefore directly related to the sensitivity of σ_L to changes in the process variables. The critical process variables are those that will influence σ_L. Figure 5.3 illustrates the sensitivity of σ_L towards different process parameters.

From Fig. 5.3, it is clear that to control as-spun filament orientation, the extrusion temperature, intrinsic viscosity of the polymer melt, the take-up speed and the melt flow rate can be identified as the critical process variables. It is also clear from the figure that σ_L is highly sensitive to change in temperature and melt viscosity, and relatively less sensitive to melt flow rate and take-up velocity. The angle and direction of the slope decide the nature of the influence of a parameter on σ_L. The high sensitivity of spin-line stress to extrusion temperature and intrinsic viscosity has important implications for maintaining product quality in a multiposition commercial plant.

As temperature and viscosity are related, a variation in temperature of the spinnerets will result in variation in spin-line stress, thereby increasing the CV of elongation. Accurate control and monitoring of the temperature of the spin pack are absolutely essential to deliver a thermally homogeneous melt to the spinnerets. The viscosity of melt is also influenced by any possible thermal degradation of the melt during extrusion. In the case of polyester, a variation in moisture level in the chips in a batch process could lead to a variation in intrinsic viscosity of the melt, and hence in spin-line stress. In plants with continuous polymerisation, variation in polymer intrinsic viscosity may also result from reactor perturbation. A sudden drop in polymer intrinsic viscosity may cause the melt to stick to the spinneret, thereby disturbing process continuity. Therefore, online monitoring of viscosity is extremely important.

The spin-line stress is also influenced by melt flow rate and take-up speed, although the sensitivity of the response is not as great as that related to temperature and viscosity. For a particular denier of filament the throughput rate and the take-up speed are coupled. Thus for increasing production, the throughput rate and take-up speed must be increased proportionately. This in turn will increase the spin-line stress due to increased deformation gradient. Increased spin-line stress will lead to higher orientation and a different fine structure, and hence a different property of the product. Thus, the fibre line conditions must be modified to ensure and maintain product quality. If possible, options such as increasing the spinning position or increasing the number of holes in the spinneret within certain limits should be considered.

Figure 5.4 provides a sensitivity plot of the freeze line location in melt spinning. As can be seen, the freeze line location is highly sensitive to melt flow rate and quench air temperature. Although the freeze line location does not influence the fine structure of the filament greatly, it has critical practical significance in terms of spin-line stability. Any air turbulence near the freeze line will lead to fused filaments or filament breakage, resulting in process interruption or poor quality of product. Shifting of the freeze line closer to the spinneret can minimise the possibility of air turbulence. This should preferably be achieved by reducing the quench air temperature, rather than by increasing flow rate.

The volume and temperature of quench air is also polymer-specific owing to the differences in glass transition temperatures of different polymers. The glass transition temperature of polyester is about 50°C higher than room temperature, while that of nylon is about 20°C higher. During the air quench process, each individual filament drags a cylindrical layer of air

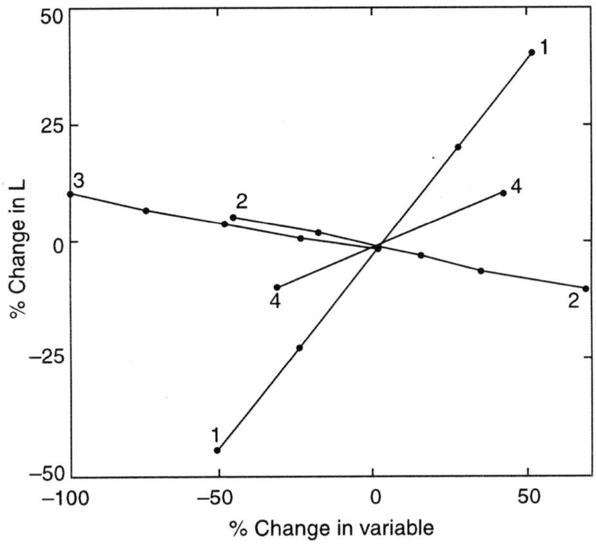

5.4 Sensitivity plot of freeze line location in melt spinning. The various curves are for changes in 1 Melt flow rate, 2 Take-up velocity, 3 Quench air velocity, 4 Quench air temperature.

around itself, which is responsible for cooling. Air flowing through the open spaces between the filaments does not participate in cooling. Therefore, the quantity and temperature of quench air depends on the density of filament and the type of polymer. In general, polyester requires a minimum quantity of quench air and the temperature needed is also high, at around 20°C, whereas for nylon and polypropylene, the quantity of quench air required must be increased and the temperature can be lowered down to around 14°C. The heat transfer coefficient of the polymer also influences the quenching process; polypropylene requires the maximum quenching length and polyester requires the minimum quenching length. The quench length also depends on the flow rate of quench air and is limited by the turbulence factor. Turbulence will lead to variation in yarn titre. It is possible to use slightly turbulent air flow while spinning filaments for a staple fibre because of their relatively large titre; moreover the staple length of the fibres is much shorter than the wave length of the filaments.

Spin finish does not influence the structure of the fibre, but it does influence many of its properties, as well as processes, from fibre formation to further processing and finishing. Although spin finish is only a minor transient part of the fibre production system, it plays an important role in processing, performance, and quality of the final product. It acts as an interface between the fibre and any other surface with which the fibre is in contact. Without the application of the correct type and level of spin finish, the filament or staple fibre formation process is incomplete. Improper type or level of spin finish increases the possibility of yarn break rate or overthrown ends at winding during fibre formation. A good spin finish must offer following features:

- good lubricity to reduce friction,
- reduction of static charge,
- good filament-to-filament cohesion,
- good, stable solubility or dispersability,
- non-stickiness,
- no migration from the yarn surface,
- easily removal by washing,
- effectiveness at low levels of application,
- non-corrosive, non-toxic, and biodegradable properties.

The three primary functions of spin finish are to provide lubrication (frictional behaviour), anti-static property, and cohesion between fibres.

The frictional behaviour of a fibre after application of the spin finish depends on a number of factors such as speed of processing, viscosity of the spin finish, pressure between the spin finish film and the body with which it is in contact, and surface roughness of the frictional surface and temperature. Effect of processing speed and viscosity on the frictional behaviour of fibres coated with spin finish is shown in Fig. 5.5, divided into three zones, namely the boundary region, semi-boundary region, and hydrodynamic region (Behary *et al.*, 2003).

The boundary region occurs at a relatively low speed, between 10^{-4} and 0.1 m/min, or at low pressure, where there is intermittent contact between the spin finish film and the surface. The semi-boundary region occurs at intermediate speed, between 0.1 and 5 m/min, or intermediate pressure, where there

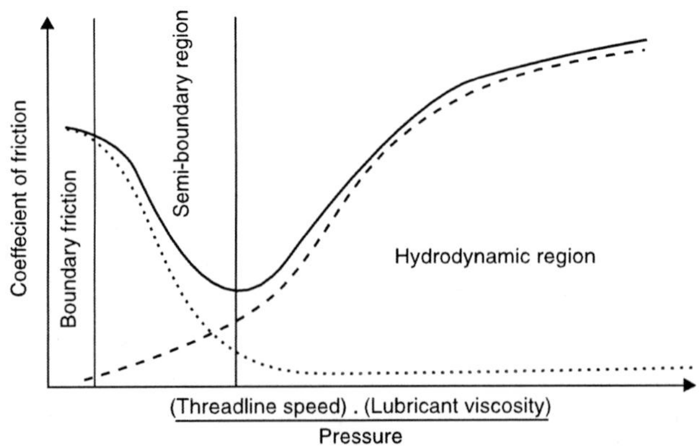

5.5 Frictional behaviour of liquid-lubricated fibres (Behary *et al.*, 2003).

is a mixture of contact and fluid friction. In this region, the friction decreases with increase in speed and fluid friction. The hydrodynamic region occurs at high speed (>5 m/min) or high pressure, and friction increases both with increase of speed and viscosity. The hydrodynamic region is mainly governed by the shearing of the fluid film. A limited increase in the surface roughness leads to a reduction in friction owing to a point contact type situation; for example, a bright, hard chrome has a higher friction than a matte, chrome friction surface. The friction coefficient of spin finish coated fibre mainly depends on the viscosity of the spin finish and it can exceed the friction coefficient of the uncoated yarn. The temperature dependency of the friction coefficient of spin finish coated fibres is inconsistent; in general, however, a reduction in friction with increase in temperature is the expected trend.

Anti-static behaviour relates to the prevention of build-up of static charge. Spin finish can only reduce the static charge that has developed on the fibre; as freshly spun fibres acquire a static charge, the spin finish plays an important role in dissipating that charge. Ionic substances are better anti-static agents than non-ionic substances; however, because non-ionic substances are less moisture sensitive, they can work at relatively low relative humidity of only 40%. The anti-static effect can be increased by increasing the concentration of anti-static agent or spin finish pick-up, but that in turn will increase the hydrodynamic friction.

Cohesion between the filaments attributable to the application of a spin finish is basically the force necessary to shear the fluid in a plane parallel to the tow. Hence, a factor that influences cohesion between the filaments will also influence the hydrodynamic friction. Thus, any attempt to increase fibre cohesion by application of a spin finish will also increase the hydrodynamic friction.

Spin finish can be applied by either a centrifugal pump or a gear pump. A schematic diagram of different types of spin finish applicators is given in Fig. 5.6. For continuous filaments, the spin finish level on yarn lies between 0.8 and 1.5 w/w%, while staple fibre tow requires about 4%. The spin finish level for a particular product depends on the final requirement of the spin finish

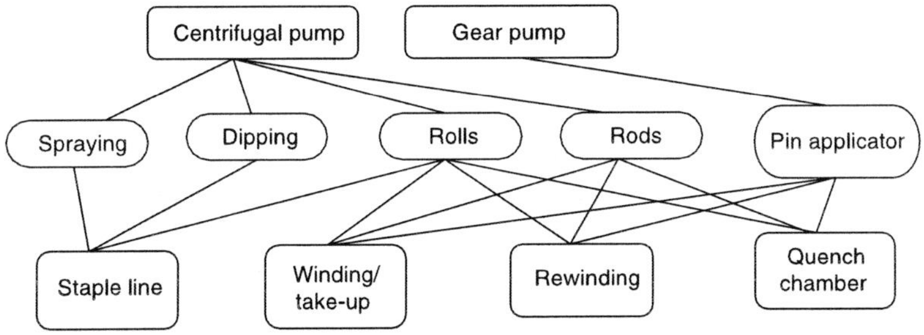

5.6 Schematic diagram of different types of spin finish applicators.

for that product, which varies from about 0.6% to 0.8%. A higher level spin finish application during the fibre formation stage is necessary to compensate for losses elsewhere. Such losses occur during application of finish, during drawing of yarn, or simply as a result of drying. As more spin finish is applied at various later processing stages, the level of spin finish during fibre formation may be decided based on the immediate next process requirement.

5.3 Post-spinning process control: drawing and heat setting

Post-spinning operation is a key process in deciding the structural parameters such as molecular orientation, level of crystallinity and sometimes the types of crystal structure. As a consequence, the post-spinning operation also influences the important end-use properties of the product such as, mechanical, thermal and sorptional properties, as well as the dimensional stability.

5.3.1 Process control during drawing stage

Drawing or stretching is an important post-spinning operation for fibres. The as-spun filaments have a certain fine structure (orientation and crystallinity). The level of fine structure development depends on the polymer characteristics, spinning conditions, and the age of the spun filament. In most cases, the level of fine structure developed is adequate in terms of the required properties of a fibre to be used in most textile applications.

The undrawn filament can be stretched to several times its length on application of a load near or above T_g, and this extension is irreversible. A typical load-elongation curve for the undrawn filament depicts three distinctly different zones during deformation under stress, as shown in Fig. 5.7.

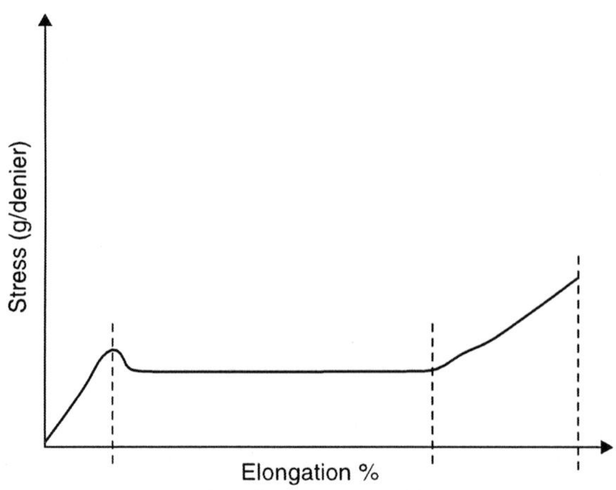

5.7 Typical load-elongation curves of an undrawn filament.

1. The first zone shows a linear relationship between stress and strain.
2. The second zone is associated with development of neck and plastic deformation.
3. The third zone is associated with strain hardening.

The second and third zones are responsible for development of the fine structure (orientation and crystallinity) in the drawn filament.

Semicrystalline filaments with a lamellar morphology after spinning are transformed into a fibrillar structure after drawing. The factors controlling the drawability are limited to the density of entanglements and to the number of tie molecules in the undrawn material. The drawing performance and the results depend on the chain morphology, as well as the crystalline morphology, of the undrawn material. The amount of molecular orientation and the amount of crystallinity in the predrawn material are equally important. Low-crystalline material is easier to draw; more extensive draw ratios are possible and higher strength may be obtained. Also, the degree of crystallinity in such cases increases. By contrast, if the initial degree of crystallinity is high, then it often decreases upon drawing.

In addition to the degree of crystallinity, the size of the crystalline blocks is also important. The size of the paracrystalline blocks determines, to a large extent, how many link molecules will be formed. The link molecules are formed mostly from polymer chains on the peripheries of the crystalline blocks; for smaller blocks, the ratio of the circumference to the volume of the block increases. Moreover, smaller units are easier to relocate and lead to a better distribution of tensile force. With increasing number of link molecules, the strength, and possibly the modulus as well, increases. Drawing performance in relation to fibre properties is given in Table 5.7.

The usual draw ratios for polyester, polyamides and polypropylene vary within the range 3.5–4.5, but for high-tenacity yarns they can reach up to 1:7. The drawing behaviour of different polymers varies. In the case of polyester, which is predominantly amorphous in nature after spinning, two deformation regions have been identified during drawing: the flow deformation region and the stress-induced crystallisation region. Little crystallinity and orientation is produced in the flow deformation region, but a rapid increase in these features is seen in the stress-induced crystallisation region. Stress-induced crystallisation occurs only when the strain rate is sufficiently high to generate a critical stress. With increase in drawing temperature, to achieve the critical stress level, the strain rate has to be increased to counteract the enhanced mobility of the molecules.

Owing to its relatively low glass transition and moderate sensitivity towards moisture, humidity plays an important role in the drawing and storage behaviour of Nylon 6. Like polyester, nylon also draws through a neck.

Table 5.7 Drawing performance in relation to fibre properties (Walzack, 1997)

Properties of undrawn fibre	Drawing performance				Properties of drawn fibre				
	Drawing tension	Maximum Draw ratio	Temperature	Drawing rate	Crystallinity	Orientation	Modulus	Creep	Thermal stability
Low crystallinity	L	H	L	H	↑	H	H	H	L
High crystallinity	H	L	H	L	↓	↓	L	L	H
High pre-orientation	H	L	↑	L	↑↓	H	L	L	H
Small crystallites	L	H	L	H	↑	H	H	H	L
Large crystallites	H	L	H	L	↑↓	L	L	↑↓	↑↓

H - Higher or High; L - Lower or Low; ↓ -Tendency to be higher;
↓ -Tendency to be lower; ↑↓ - Ill defined result or complex relationship

Drawability of as-spun polypropylene filament depends on:

- polymer molecular weight,
- polydispersity,
- morphology of the as-spun filament,
- drawing temperature,
- deformation rate.

With increase in molecular weight, drawability increases initially and then decreases. Drawability increases with decrease in polydispersity. Initial morphology strongly influences the drawability of polypropylene filament. Rapidly quenched polypropylene filaments show pseudocrystalline smectic structure, whereas gradual cooling leads to α-monoclinic structure.

The pseudocrystalline smectic structure can be deformed to a higher degree than the α-monoclinic structure, leading to higher orientation and crystallinity, and thus better mechanical properties. Drawability of the smectic structure is at a maximum at 60°C, but the α-monoclinic structure needs a higher draw temperature to attain the peak draw ratio. Evidence of the existence of smectic structure in samples drawn at temperatures lower than 60°C has also been reported. It is observed that both crystalline orientation function (f_c) and amorphous orientation function (f_a) increase with draw ratio, but the rate of increase of f_c is greater than that of f_a. Tenacity, modulus, and yield stress also increase with increasing molecular orientation. The chances of the development of longitudinal defects or voids several micrometers in length increase with an increase in draw ratio. These voids are essentially reduced density zones. Further drawing leads to the filaments turning white owing to the presence of very large voids. Length-to-breadth ratio of voids decreases with increase of draw ratio.

5.3.2 Process control during heat setting stage

The fundamental objective of heat setting is to render dimensional stability to the fibre and hence to the textile structures made using those fibres. Without heat setting, the fibres are prone to thermal shrinkage, yarns exhibit a snarling tendency, and fabrics tend to crease. To prevent or minimise these unwanted behaviours, heat setting is necessary. The root cause of dimensional instability in the fibre is the internal stress that develops during the process of drawing. The completely random or high-entropy state and the completely ordered or low-energy state are the two most stable states of polymer molecules. For the process of fibre formation, initially the polymer is in the form of a fluid, either melt or solution. In the fluid state the molecules are in a high-entropy state and highly energetically stable. During the conversion of the fluid to fibre, the molecules are transformed into an oriented, semicrystalline state which is thermodynamically

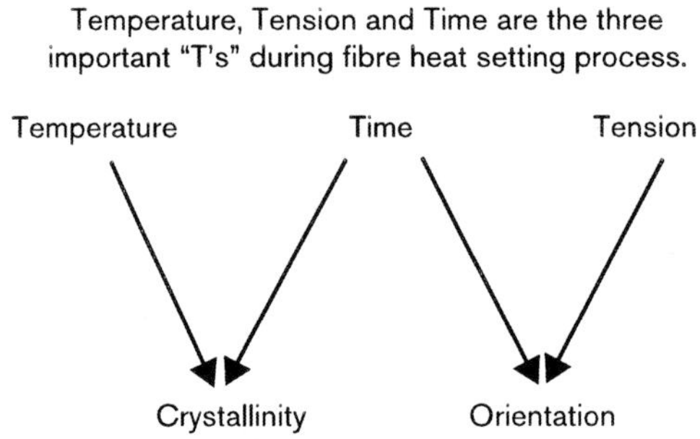

5.8 Dependence of heat-setting parameters on crystallinity and orientation of the fibres.

unstable, leading to development of internal stress in the fibre system. To make the fibre stable, the internal stress has to be reduced or removed.

Heat setting is a process of exposing the fibre to a temperature higher than the use temperature of the fibre, thereby allowing the molecules to rearrange themselves to acquire energy minima depending upon the condition (i.e. temperature, tension, and time), as illustrated in Fig. 5.8. As previously mentioned, the two most stable arrangements of molecules are either completely random or completely ordered, so to minimise internal stress, the molecules either tend to randomise or to crystallise. Given the large size of the molecules and the conditions of heat setting (i.e. temperature and time), randomisation is relatively easier than crystallisation. Thus, after heat setting, there will be a definite change in the fine structure of the fibre leading to a decrease in orientation and there may be some enhancement of crystallinity or removal of some crystal defects.

Oriented structures may be annealed in two different ways: with or without a dimensional restrain. If the annealing process is conducted without the dimensional restrain, the fibre will shrink; this shrinkage will increase with increasing draw ratio and annealing temperature, but will decrease with increasing drawing temperature. If the shrinkage is prevented by mechanical restrain, there are still changes in the fibre structure that take place, although the changes will be somewhat different and their influence on the material properties will also be different. A fibre annealed with restrain, obviously, will not shrink, but crystalline orientation will decrease, and amorphous orientation usually decreases even more. Without restrain, all other conditions being equal, the disorientations are much larger, particularly the crystalline disorientation. The fibre's response to annealing depends strongly on the relationship between the temperature of original crystallisation and temperature of cold drawing on the one hand, and on the temperature and time of annealing on the other.

Annealing may be treated as a kind of 'preshrinking' – a way to prevent fibre shrinkage during its actual use. Such treatment, however, is accompanied by change of other fibre properties, such as tenacity, modulus (especially the initial modulus), orientation, density, dying characteristic, and so on. A slight extension during annealing, which is sometimes treated as an additional drawing step, often prevents an excessive decrease of properties with a simultaneous gain in thermal stability of the fibre. Every annealing process, irrespective of restrain or lack of it, leads to an increase in crystalline melting point of the polymer.

5.4 Key control points in synthetic fibre manufacture

Important issues during synthetic fibre manufacture are the raw material quality used for polymer synthesis, polymer molecular weight and purity during polymerization process, fibre quality, and machinery conditions during the whole process. The control points in synthetic fibre manufacture are categorized into material testing and maintenance which are discussed in this section.

5.4.1 Material testing

The quality of all the incoming material (raw material), intermediate product (polymer chips/granules/powder), and final product (fibre/filament) should be tested and only materials satisfying the norms/standards should be accepted. Some examples are described below.

Fibre/filament

Testing of fibres is a necessary and integral part of the production process for the following reasons:

- quality control and quality assurance,
- process optimisation,
- product development,
- control and monitoring specification.

Methods of investigation of the *fibre micro structure* are summarised in Table 5.8.

The *fibre macro structure* (cross-section, length, fineness, etc.) can be investigated in various ways:

- Cross-sections can be assessed by cutting the sections using a microtome and subsequent analysis by optical microscope or scanning electron microscopy.

Table 5.8 Methods of investigation of fibre micro structure (Fourne, 1999)

Element	Method	Dimensions
Fibre	Optical microscopy	> 0.2 μm
Fbrils/lamellae	Electron microscopy	> 100 A
Voids	SAXS	10–1000 A
Crystalline regions	WASX	1–10 A
Non-crystalline regions	SAXS	10–100 A
	IR, NMR, DMA etc.	Molecular group and arrangement
Orientation	X-ray, IR, birefringence, sonic velocity	Lamellae, crystallites, molecular segments or groups
Density	Density gradient column	

- Fibre length or staple length is a very important characteristic for further processing. Important information/characteristic values may be observed such as:
  - average fibre length (mm);
  - coefficient of variation of fibre length (%);
  - most frequently occurring fibre length (%);
  - proportion of most frequently occurring fibre length (%).

Measurement can be based on either weight or fibre frequency. The measurement technique is based on the formation of a 'fibre beard' and analysis of the data by computer.

- Fibre fineness in terms of nominal fineness, actual fineness, target fineness, or commercial fineness can be evaluated using gravimetric, optical, vibration or air permeability methods.

Mechanical properties

Mechanical properties are evaluated by tensile testing; the various aspects involved in tensile testing are demonstrated in Fig. 5.9. The stress–strain curve gives important information about the fibre for process control, further processing, and end product properties.

Uniformity

Non-uniformity may be internal or external in nature. Figure 5.10 shows the influence of yarn uniformity on various properties. External uniformity can

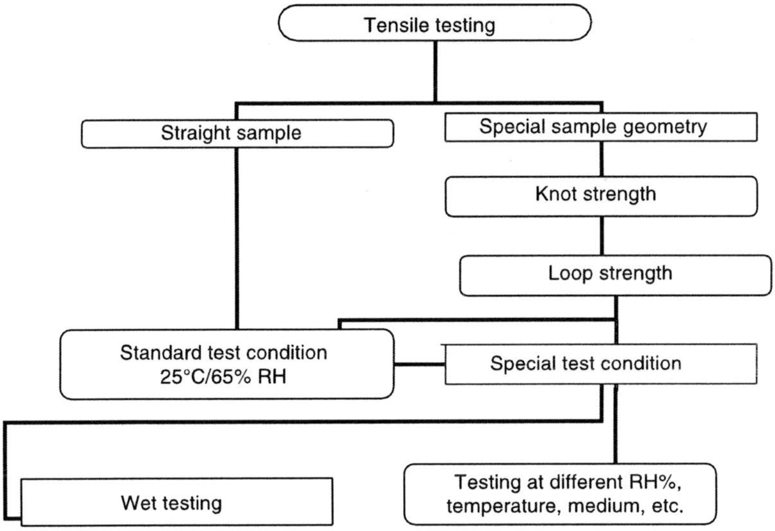

5.9 Tensile testing of fibres.

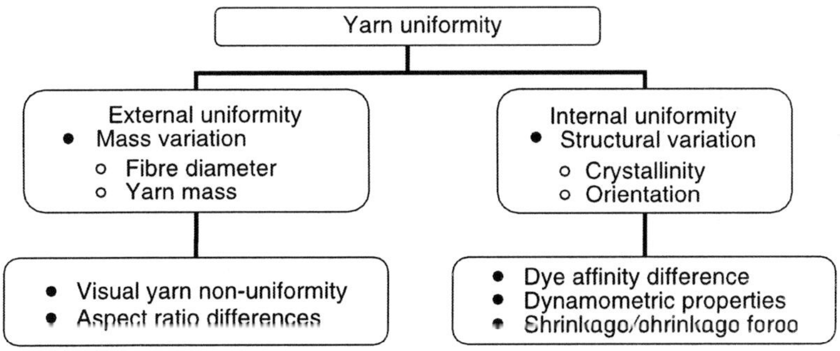

5.10 Yarn uniformity.

be judged visually or with the help of test equipment, as in the Uster test. Internal uniformity cannot be judged visually and is principally characterised by the properties of fibre structure through certain tests, such as dye uptake, shrinkage force, and so on.

5.4.2 Maintenance

Maintenance is necessary to delay the wear and tear of various machine parts and to maintain the working potential of the plant. Cost of breakdown in a fibre plant is significantly high due to wastage of material and loss of production, including cleaning and restarting time. Thus continuous

Table 5.9 Maintenance schedule (Fourne, 1999)

Parts	Clean after	Renew/exchange after
Extruder screw, cylinder	Year	2–3 years
Static mixer	2–6 weeks	–
Spin pack and filter	1–8 weeks	Replace filter
Spinnerets	1–8 weeks	6–24 months
Quench air rectifier, front mesh	2–4 weeks	According to test result
Yarn guides, oil applicators	Daily	2–6 weeks
Traverse guide	Daily	1–4 weeks
High speed bearings in winders, godets		2–6 months at ≥5500 m/min 3 months at ≈5500 m/min 1–4 weeks at 8000 m/min
Stuffer box component	Daily	< 1 week
Yarn cutting knives	Daily	3–8 h using steel knife steel 8–30 h using hardened steel
Hot plates for drawing	2–3 days	1–2 months (hard chromed) 4–6 months (hard coated)
Aspirator mouth piece		6–8 weeks

cleaning, checking, and maintenance, as well as routine preventive maintenance, are extremely important. Machines and equipment are also constructed to be maintenance-free, repair friendly, and capable of running continuously without breakdown. Adequately trained maintenance personnel and an adequate stock of spare parts are also important. These features are shown in a typical maintenance schedule, as provided in Table 5.9.

5.5 Future trends

Integration of back processes in manufacturing stages, the gradual advancement of technology from batch process to continuous production process, and automated process control are the major changes to have taken place in the fibre manufacturing sector during the past decade. The integration of the raw material synthesis and polymerisation process has ensured high input material quality at reduced cost for the fibre formation process. Continuous manufacture ranges from polymerisation to fibre formation, drawing, and heat setting and this has resulted in reduced batch variations, increased fibre production rate, better control over fibre quality, and reduced cost of manufacture. Rapid innovation in the synthetic fibre domain has ensured the delivery of niche and profitable fibres. Imparting special functionality to the synthetic fibre has captured the attention of the textile industry. Continued attempts to automate the fibre manufacturing process and to engineer material properties have paved the way for the production of fibres with desired qualities at affordable cost.

5.6 References

Behary, N., Campagne, C., Caze, C. and Perwuelz, A. *Journal of Applied Polymer Science* 2003, **89**, 645–54.

Fourne, F., 'Synthetic fibres machines and equipment, manufacture, properties', in *Handbook for Plant Engineering, Machine Design and Operation*, Hanser Publisher, Munich, 1999.

Gupta, V. B. and Kothari, V. K. (ed.), *Manufactured Fibre Technology*, Chapman and Hall, 1997.

Kothari, V. K. (ed.), *Progress in Textiles: Science and Technology*, Vol. 2, IAFL Publication, India, 2000.

Ludewig, H., *Polyester Fibres: Chemistry and Technology*, Wiley Interscience, New York, 1964.

Mark, H. F., Atlas, S. M. and Cernia, E. *Man-made Fibres: Science and Technology*, Vol. I, Interscience Publishers, New York, 1967.

Mcintyre, J.E., Lewin, M. and Pearce, E. M. (eds.), 'Polyester Fibre', in *Handbook of Fibre Chemistry*, Marcel Dekker Inc., 1998.

Vaidya, A. A., *Production of Synthetic Fibres*, Prentice Hall of India Pvt. Ltd., New Delhi, 1988.

Walzack, Z. K., *Formation of Synthetic Fibres*, Gordon and Breach, New York, 1997.

6
Process control in blowroom and carding operations

R. ALAGIRUSAMY, Indian Institute of Technology Delhi, India

DOI: 10.1533/9780857095633.2.132

Abstract: The chapter provides insight into the process control aspects related to the blowroom and carding operations in the staple fibre yarn manufacturing process. The main functions involved in these two operations are opening, cleaning and mixing of the fibres. Process control aspects related to these functions are discussed providing quality and productivity related issues and the control activities needed at different zones of these two operations. The control of process parameters in these operations, which will ultimately affect the yarn properties, are also discussed in detail in addition to the control excises needed on raw material selection.

Key words: process control, blowroom, carding, waste control, yarn properties.

6.1 Introduction to blowroom operations

Modern blowroom processes have been shortened considerably by employing the following equipment sequence:

Flock feeder → Opener → Mixer → Fine cleaner → Card feeder

Laying down a mix properly using a flock feeder is the first process control step. The mix must be laid as randomly as possible, and the mix ratio should be maintained not only across the length of the lay down, but also across the width so that, in a given take-off across the width, the mixing ratio is maintained. It is important for the flock feeder to be set up to ensure a small tuft size; this makes the cleaning process for subsequent machines more efficient. The tuft size achieved with the flock feeder will depend upon:

- take-off depth
- degree of opening, determined by number of beating points acting per unit of time
- production rate

It should be noted that for a given production rate, the idle time (when no take-off occurs) must not exceed 15% of the total time.

The next step is opening. In the first opener, a higher speed is always advisable to reduce tuft size. Because of its gentle opening action, this machine does not cause a substantial increase in the number of neps (knots) and short fibres for a wide variety of cotton mixes. The smaller tuft size delivered by this machine helps to remove contaminants and promotes better blending in the mixer.

The most critical function carried out in the blowroom is mixing. Here, components are mixed as randomly as possible in six or eight chambers, depending on the production rate and the desired number of doublings. It is critical to ensure that the desired column height is achieved in every chamber before the cards are started and that this height is then maintained throughout. Poor mixing can result in significant variation in the process and in the final quality of the yarn quality. In some cases it can also result in a streaky appearance after dyeing.

The next step is fine cleaning. In this part of the process, the beating and opening action of the grip can cause damage to the material, if the feeding distance or speed of the beater/cleaner is not correctly controlled. The greatest increase in the number of neps and short fibres occurs during fine cleaning. For particularly sensitive types of cotton, a reduction in the number of beaters, or a switch from a saw tooth beater to a pin type beater could limit this increase. Depending on the type of cotton involved, there is much scope for process optimisation in terms of establishing the correct feeding distance and speed of the beater/cleaner. Waste removal and dedusting also occur continuously throughout the opening and cleaning processes. In this respect, air (suction) pressure has been found to play a critical role: the manufacturer's instructions for operating the machine should also always be followed.

Depending on the type of cotton being processed and the extent of the cleaning/opening processes that have been carried out, the number of neps can increase during these stages of processing by between 100% and 250%, while the number of short fibres can increase from 2% to 10%. These neps and short fibres, along with other trash and fibre fragments, must all be removed during the carding process. It should be noted at this point that the performance of the card has undergone significant improvements in recent years; hence, while the fibre should be subjected to as little stress as possible during blowroom operations to minimise neps and short fibres, these can be effectively removed by the card.

6.1.1 Control of raw material coming into the blowroom

The cotton industry constantly relies on the correct selection and mixing of cotton. Decisions are subjective, and are made based on experience, taking into account both the cost and the characteristics of the fibre, which are linearly related. Some textile mills substitute a certain percentage of low cost cotton during the cotton mixing process, so that the quality only deteriorates to a small degree, and effective use is made of the available cottons throughout the crop year. Factors that can affect the mixing decisions made are: variation in cotton quality over the course of a crop year or within one bale, fluctuations in the cost of cotton, uncertainty regarding the quality of the cotton in stock, and variation in consumer demand.

It is difficult to use empirical methods to arrive at the correct selection and mixing of cotton based on all of these variables, making it necessary to use subjective judgements mentioned. However, subjective decisions are not always reliable and can lead to costly errors: a scientific approach is therefore required for effective decision making in cotton mixing. For example, cotton bales have traditionally been purchased based on nominal specifications: scientific analysis of the characteristics of the fibres in the bale will produce an accurate evaluation of the variation in the bale, and the generated data can then be used for bale management.

In recent years linear programming (LP) has been used in the textile industry in order to optimise the cotton mixing process. LP models take various fibre mixing quality constraints into account in determining the best selection and mixing of cottons. However, they are not always particularly effective, because their restrictive structures lead to over-simplistic representations of the real system. The optimal solution provided by a LP model is therefore not necessarily accurate in practice.

The decisions involved in cotton mixing are based on multiple requirements or goals (such as optimal strength, optimal cost, and so on), and the management may choose to relax one requirement to examine the impact that is has on a different requirement; for example, the strength requirement may be relaxed by 10% to examine the cost implications. Goal Programming (GP), which allows the inclusion of goals and sub-goals in its calculations in order to achieve a satisfactory level for the individual user, has therefore proven beneficial in the textile industry. The GP approach offers the following advantages over LP:

- Deviations from the target value are minimised.
- Goals are ranked based on their contributions.
- The desired level of satisfaction is achieved based on the ranking assigned to the different goals and the minimisation of deviations from these goals.

The success of the GP model is dependent on the accuracy of the input data, particularly the goal priorities and their weightings. The goal priorities often fluctuate: it may be desirable to relax the target values of some (less important) goals to achieve substantial improvements in another goal.

A procedure for determining goal priorities is required for profitable use of the GP model. The analytical hierarchy process (AHP) is used to aid in prioritising the goals based on the user inputs on a pair-wise comparison of the relative importance of various goals.

The other important input is the number of bales in stock and their fibre characteristics, which vary from bale to bale. In most computer based models designed for cotton mixing, average fibre characteristics, based on a few bales, are used, rather than those of all the bales actually used in mixing. This means that the desired fibre characteristics cannot be guaranteed; however, the alternative, which would be to keep a large inventory of bales that are known to have the required characteristics, is extremely expensive. The solution adopted is the use of a cotton inventory (CI) model, which provides information on the availability of bales and their fibre characteristics during the decision-making process. The qualities available are then used as constraints in the GP model. The average fibre characteristics of the bales actually required in the mix are used in the GP model calculations, which ensures that the actual quality of fibre mixing is close to the desired level. In conclusion, then, an integrated use of GP, AHP and CI provides a rational and effective solution to the problems involved in cotton mixing decisions, and helps to ensure the quality of the raw material coming into the blowroom.

6.2 General process control in the blowroom

The operations in the blowroom that need to be controlled depend upon the desired yarn quality characteristics, and on the extent to which these characteristics are affected by blowroom processes. Some important yarn quality characteristics that are directly or indirectly influenced by blowroom operations are given in Table 6.1.

It is therefore clear that the extent of cleaning and opening carried out, along with the rupture of fibres and large trash particles, is extremely important in determining the quality of the final product. It should also be noted

Table 6.1 Yarn quality characteristics affected by blowroom operations

Characteristic	Affected by
Evenness	Number of short fibres generated
Imperfections	Number of short fibres generated and degree of cleaning and opening
Neps	Additional neps generated and degree of cleaning
Hairiness	Number of short fibres generated
Consistency in count	Delivery of a uniform product
Consistency in properties	Consistency in blowroom process
Cleanliness	Degree of cleaning achieved and possible rupture of large trash particles

that the cleaning equipment currently available cannot adequately clean the stock without sacrificing some spinnable fibres as waste, stressing some fibres to the point of rupture and generating some neps. It is therefore crucial to (i) ensure that the waste generated contains mainly unwanted foreign matter and a minimum of spinnable fibres; (ii) minimise the generation of short fibres by avoiding rupture as far as possible; and (iii) minimise the generation of neps.

Having identified what needs to be controlled, it is then important to set a targeted performance level and to establish procedures to ensure that this level is attained. The targeted performance level depends upon the requirements of the carding process, which takes place after the blowroom operations have been completed. It is important for the card technologist to establish these requirements in terms of a specification of the quality characteristics of the lap or chute-fed mat of fibres received from the blowroom. The specification will depend upon the level of technology and the human resources available, and must be established based on a detailed understanding of the influence of the carding process on the quality of the yarn. The absence of such a specification from the card technologist makes the job of the blowroom operatives extremely difficult. On the other hand, if the required quality specification does exist, then the challenge is in achieving these specifications. Knowledge of the aspects discussed below is required in order to achieve the targeted quality level:

- System capability
- Optimal running conditions
- Technological parameters.

System capability: Every type of opening, cleaning and mixing equipment has a limit to its capability. Any attempt to surpass these limits without updating the design may have disastrous consequences. For example, the maximum cleaning efficiency of a blowroom cannot be more than 70% and even this level may not be achievable for all fibres. Therefore, attempting to achieve a 90% cleaning efficiency will result in unnecessary loss of fibres, fibre breakage and nep generation. Similarly, expecting a fibre tuft to become individualised to a single fibre will also have undesirable results. Coarse cleaning equipment is designed to extract heavy, large, loose trash particles (such as husks, stems, stone chips, sand, leaf particles, etc.) and cannot effectively remove finer trash particles and seed coats.

Optimal running conditions: Any system performs best in the correct conditions: this is not only true for machines but also for people. The performance of the opening, cleaning and mixing equipment can be substantially improved by ensuring that it is run at the correct atmospheric conditions, that the throughput rate is correct, and that it is regularly maintained.

Technological parameters: An understanding of the working of the machines and the parameters that influence their performance is essential. This will help in the selection of the correct options and parameters for any given situation. The workings of the individual machines are complex, and a brief description of the major operations is provided in the following sections.

6.3 Process control in blowroom processes

Blowroom operations involve the functions: opening, cleaning and mixing. Each of these functions has very clear impact on the blow room process as well as the quality of the output. Hence, it is important to understand the factors affecting these three functions in blowroom and improve the process efficiency by controlling these parameters and come up with the correct strategy to operate the blowroom.

6.3.1 Process control in opening

The salient feature of the modern blowroom line is the intensive opening of hard pressed cottons into tufts of uniform size using the minimum number of beating points. The tuft weight and density, and especially the variation in the properties of the tufts, affect the irregularity of material in the blowroom in the short term: small tufts can occur in open and even arrangements. Greater variation in tuft density and weight results in a higher number of neps and a larger amount of foreign matter in the card web, while greater variation in tuft properties results in undesirable yarn characteristics such as variation in count and strength, and other imperfections. Opening is therefore a very important first stage in the blowroom process.

Opening involves the following four main processes:

1. Plucking, often with automatic bale pluckers
2. Interaction between tufts and opposing spikes – this involves hopper feeders, and axi-flow, CVT and CNT cleaners
3. Impact of strikers on tufts, using axi-flow and mono cylinder cleaners
4. Teasing of tufts in nipped state by needles, pins or sawtooth elements

Of these four, the first is a relatively gentle process, while second and third are of moderate intensity and the fourth is most intensive. Machines arranged in such a way that the higher intensity processes occur at a later stage, since opening becomes progressively difficult with decreasing tuft size. Tuft weight and density can be effectively standardised by the use of sawtooth elements at the final beating point, while other adjustments may be necessary to ensure even tuft size when cottons with different openness values are used together in the same mix.

Table 6.2 Factors affecting opening intensity

Action	Factors affecting intensity
Plucking	Depth of penetration. (NB reducing penetration has a negative effect on productivity unless the speed of plucking is increased at the same time.)
Interaction between opposing spikes	Speed; distance between interacting surfaces (setting); spike density; throughput rate.
Impact	Striker rotation speed; density of strikers; distance between feed roller and striking element (setting); speed of suction fan; through put rate.
Teasing	Speed of needle or sawtooth roller; density of needles or saw teeth; angle of inclination of needle/saw tooth; distance between feed roller nip and sawtooth roller; through put rate.

It is important to note at this point that blowroom material should not be reprocessed on the opening equipment, especially if single process lines are used, since this is likely to lead to fibre rupture and nep generation as a result of double beating. The number of neps has been shown to increase by more than 50% when materials are reprocessed in the blowroom.

The intensity of the opening action of the various opening devices is affected by a number of factors, as shown in Table 6.2. If the opening action of one process has to be intensified, the appropriate parameters must be adjusted. However, manipulation of these parameters can have adverse consequences, including:

(i) Possible damage to fibres
(ii) Shattering of large trash particles
(iii) Increased nep generation
(iv) Increased waste.

Some general guidelines for optimising opening intensity are given below:

(i) The parameters of the devices that treat fibres gently (i.e. those at the beginning of the line) should be manipulated in the first instance. This will reduce the possibility of fibre damage and nep generation. Only once the intensity potential of these machines is reached should changes in the parameters of the more intensive machines be attempted.
(ii) A reduction in the throughput rate and/or thickness of feed improves the opening capability of any machine.
(iii) Speed is the most convenient parameter to change, and should thus be adjusted first, before any changes are made to setting or other machine parameters such as needle density.

(iv) Atmospheric conditions also play an important part in opening. A damp atmosphere does not facilitate fibre separation.
(v) Blunt opening elements are detrimental to good opening.

6.3.2 Process control in cleaning

Cotton cleanliness is a key factor that determines the quality of the yarn. Efficient cleaning in the blowroom, along with effective processing of cotton containing a low level of trash during blowroom and carding processes, are essential in producing yarn free from foreign matter. In conventional blowroom lines, a cleaning efficiency of 80% in cotton with high trash and 70% in cotton with low trash is achieved, while in modern blowroom lines the cleaning efficiency is much lower, at 50–60%. However, the modern lines achieve a high degree of opening, which allows better cleaning during the carding process. Less lint is lost as waste in modern lines: only 20–30%, depending on the trash level in the cotton, compared to 25–50% in conventional lines. The overall cleaning efficiency for the blowroom and carding processes combined should be about 95%.

The opening process discussed in Section 6.3.1 leads to trash and other foreign material being released from the cotton as a result of the various forces acting on the fibres. In order to effectively clean the cotton, therefore, the trash and other contaminants must then be separated from the fibres. Since the trash particles present in cotton are of different sizes and are not homogeneous in nature, different separation principles are used in the various cleaning machines. The factors that affect the liberation of trash and other contaminants and its separation during cleaning are given in Table 6.3.

As cleaning requires both liberation and separation of trash and other contaminants, effective cleaning can only be achieved if both of these properties are carried out correctly and efficiently, and in a manner appropriate to the nature of trash/foreign matter to be extracted. For example, the procedures that should be adopted (and intensified if necessary) for liberating and extracting large and heavy trash particles would be quite different from that used for removing seed coat fragments. The cleaning equipment in the blowroom has been designed keeping with this in mind, as follows:

- Coarse cleaning equipment
- Effective for large and heavy loose trash particles
- Cleaning equipment of moderate intensity
- Effective for medium sized trash particles
- Fine cleaning equipment
- Effective for small and lightweight trash particles that are strongly adhered to the fibre.

Table 6.3 Factors affecting intensity of trash liberation and separation

Action	Factors affecting intensity
Liberation by:	
Centrifugal force	Rotational speed of opening element; diameter of drum or roller; velocity of air flow; radius of curvature of bend in duct.
Impact	Speed; setting between feed nip and line of action of opening equipment.
Pneumatic force	Speed of suction fan.
Frictional force	Sharpness of grid bars; angle of inclination; distance between interacting surfaces.
Separation by:	
Gravity	Size of slot; setting between grid bars.
Suction	Size of screen perforations; air discharge rate.
Buoyancy	Velocity of air flow; location of separation edge.
Magnetic force	Magnetic power; location of magnets.

Fine cleaning equipment is much more aggressive than coarse equipment in its action on cotton tufts; there is therefore a risk of damaging fibres, crushing trash particles and generating neps the opening ability or trash liberating ability is intensified without due care. A further important point is that the design of the machines means that any cleaning process also involves associated fibre loss, which causes costs to rise.

6.3.3 Process control in mixing

It is impossible to overstate the importance of the correct mixing of fibres that have come from different bales. It is well established that fibre characteristics are subject to significant variation within one bale. In some cases even after bale management techniques are employed, some bales have too high a degree of variation with respect to maturity, fineness, and length percentage of short fibres that they are best discarded. Even then, the characteristics of the fibres in the bales are not completely identical, simply because of naturally occurring variations. Moreover, it is sometimes necessary to blend two or three bale groups for cost optimisation. In either case, a homogeneous mix is absolutely crucial for ensuring consistency in yarn characteristics. The selection of bales and their segregation, blending practices and mixing equipment all affect the homogeneity of the mix. The following points should be kept in mind:

- As fibres from different bales get mixed in tuft form, smaller tuft sizes lead to greater homogeneity.
- The use of a larger number of reserve chutes in the mixing equipment allows greater homogeneity.
- The use of two blenders working in series or in tandem produces a more homogeneous result that one blender of the same capacity working alone.

6.3.4 Correct strategy for blowroom processes

The points below provide guidelines on the adoption of the correct strategy for the three main blowroom processes (opening, cleaning and mixing) described above.

1. Trash should be checked both on a quantitative and qualitative basis, as it is necessary to establish not only the percentage of trash present but also the type of trash, that is, whether it is large or small trash, seed coat fragments, fibre fragments or dust. The right equipment should be selected for the predominant trash type in order to improve performance. For example, if most of the trash present is in the form of large particles, the coarse cleaning equipment should be adjusted. If, on the other hand, the main type of trash is dust, then the suction system should be adjusted instead.
2. If the opening and cleaning capability of the machines needs to be increased, the coarse cleaning equipment should be adjusted first, followed by the fine cleaning machines. This will cause the heavier particles to be extracted, thereby reducing the risk of crush, which would otherwise make subsequent extraction more difficult. The possibility of fibre damage is also decreased.
3. It should be remembered that the card (discussed in detail in Section 6.4) is a more effective cleaning tool than any of the equipment in the blowroom, especially with respect to the extraction of smaller trash, seed coat fragments and dust. Thus the extraction of finer particles should for the most part be accomplished by the card rather than in the blowroom. The blowroom and card should be seen as a integrated unit, each with a distinct role in the cleaning process.
4. Fine cleaning equipment (such as a cleaner with a clamp feed arrangement) can remove seed coat fragments to some extent. Increasing the speed of the opening equipment to an excessive extent will not improve cleaning but will cause more fibre loss, fibre damage and nep generation.

5. A reduction in the production rate will improve the cleaning capability of all machines.
6. For a given production rate, the flow of material should be made as uniform as possible without too many stoppages of individual machines in the line. This can be achieved through the correctly adjusting the throughput rate.
7. The cleaning efficiency of the blowroom should be fixed at around 50%. The total nep increase should not be allowed to exceed 80–100%. A typical permissible increase in neps for various machines is as follows:

   - Uniclean 10–12%
   - Unimix 20%
   - ERM 30%
   - Chute feed 20%

   The 2.5% span length or upper quartile length of the fibre should not fall below 1 mm.
8. Atmospheric conditions should be maintained at the appropriate level; a damp atmosphere, for example, does not facilitate cleaning. The bales should be conditioned in the blowroom for 12 h, preferably in an opened state.

6.4 Process control in carding: control of process parameters

Both process-related and machine-related parameters must be taken into account for the purposes of process control in carding, with wire maintenance and control of card waste also significant. The process parameters can be further divided into three categories: those relating to speed, those relating to the gauge or settings between the various interacting elements, and those relating to the properties of the material. The machine parameters relate to the clothing of various parts, as the dimensions of the components remain the same. A list of important parameters is given in Table 6.4.

The optimisation of process parameters is essential in making the best use of the equipment currently available. The following sections describe the major parameters that can be controlled in the various different parts of the carding equipment. It should be noted, however, that a change in a single parameter may not bring about the desired result, unless other parameters are also adjusted appropriately. Moreover, the processes employed during carding to separate out liberated trash particles and dust from fibres inevitably lead to some fibre losses, but it is important that any parameter adjustments allow such losses to remain at minimum levels.

Table 6.4 Important carding parameters

Process parameters			Machine parameters
Speeds	Settings	Material	
Production rate	Licker-in to feed plate or roller	Thickness of lap	Licker-in clothing
Licker-in speed	Taker-in to feed plate or roller	Sliver count	Cylinder clothing
Cylinder speed	Taker-in to cylinder		Doffer clothing
Flat speed	Cylinder to flat		Flat clothing
	Cylinder to doffer		
	Cylinder to front plate		
	Cylinder to back plate		
	Pre-carding segment to flats		
	Flats to post-carding segment		
	Cylinder to suction hood		
	Mote knife/carding bar to licker-in		

6.4.1 Licker-in zone

The role of the licker-in is to open the fibres as they are fed onto the card and then to transfer them to the main cylinder. Major parameters that can be adjusted in this zone include the speed of the licker-in, the setting of the licker-in to the feed plate or feed roller, and the setting of the mote knife and carding bar to the licker-in. Some representative measurements from the licker-in zone are as follows:

- When the thick lap or mat of fibres is fed at around 0.13 m/s, the licker-in teeth impinge the fibres at 13 m/s (i.e. around 50 km/h).
- Approximately 600 000 licker-in wire points pass per second through the lap fringe.
- The typical draft between the feed plate or roller and the licker-in is in the order of 1000, which causes the lap to become 1000 times thinner upon reaching the surface of the licker-in.

Licker-in speed

Speed is the easiest parameter to regulate, and is therefore probably the most frequently adjusted parameter. Table 6.5 provides an overview of the consequences of increasing the speed of the licker-in.

Table 6.5 Influence of licker-in speed

Action	Consequences
Intensive opening of fibre tufts	More trash may be exposed
Reduction in tuft size on licker-in	Smaller tufts are passed for carding
Greater force exerted on fibres as they are teased out from the feed nip	Increased liberation of trash particles with possible risk of fibre damage (especially for long and fine fibres)
Higher centrifugal force experienced by tufts on the licker-in surface	More trash may escape, and immature fibre clusters may become trapped within tufts, leading to risk of fibre loss
Lessening of draft between licker-in and cylinder at excessively high speed	Difficulties in transferring fibres to cylinder
Aggressive opening action between combing bar segment and licker-in	More trash may be liberated, with possible risk of fibre rupture

Increasing the speed of the licker-in leads to better cleaning and carding, but also causes more waste generation and fibre damage. Difficulties in transferring the fibres to the cylinder also can be experienced at excessively high speeds. For adequate transfer of fibres, the draft between the cylinder and the licker-in should be around 1.5–1.7 for cotton and 2.5–2.9 for synthetic fibres. Poor transfer of fibres may cause unopened fibre tufts to be passed on to the cylinder in an erratic manner, as well as leading to the loss of good fibres as licker-in waste.

Setting of feed plate or feed roller to licker-in

This setting mainly affects the distance between the point of release of fibres from the grip of the feed plate or roller and the line of action of the teeth of licker-in. The fibres must be gently teased away from the nip as they are released: fibre tufts should not be plucked as a whole. The fibres arrive at the licker-in teeth in the form of a tapered fringe, the length of which depends upon the length of the fibres.

If the feed plate or roller is brought closer to the licker-in, a large part of this fringe then comes into contact with the licker-in teeth. As a consequence, more fibres with trailing ends still lying at the nip or beyond will be pulled out, leading to a more aggressive action. If, on the other hand, the distance between the two is increased, when the tip of the fringe reaches the teeth, many fibres with trailing ends will be released from the feed nip. Since nothing will obstruct their removal from the fringe, they will be simply be plucked away without being teased, causing a decrease in opening intensity.

Setting of mote knife and carding bar to licker-in

The setting of the mote knife plays a role in determining the quantity and quality of waste generated: the closer the mote knife is to the licker-in, the more waste is generated. This waste is also richer in lint. The setting is usually kept at 0.018 inches.

The carding bar further opens out tufts before they are transferred to the cylinder. The aggressiveness of this opening action is determined by the distance between the carding bar and the licker-in and the relative speed between them: a closer setting leads to greater opening intensity. However, if the distance is too small, fibre rupture can occur; on the other hand, if the distance is too great, the tufts will not be adequately opened. This means that the opening must be carried out by the cylinder, which may lead to nep generation. The setting between carding bar and licker-in is usually 0.019 inches.

6.4.2 Cylinder region

In the cylinder region of the card, there are a number of process parameters to be controlled and optimised. These include the speed of the cylinder, the speed of the flat, the setting of the cylinder to the flat, and the setting of the pre- and post-carding segments.

Cylinder speed

In any card, the highest cylinder speed that can be achieved is limited by mechanical design considerations. Today, though a speed of 600 rpm is attainable on a modern card, it is more usual to run the cylinder at 300–500 rpm. The speed of the cylinder plays a particularly significant role in the transfer of fibres onto the doffer. The consequences of increasing the cylinder speed are given in Table 6.6.

In terms of quality control, increasing the speed of the cylinder has many advantages, including improving cleaning efficiency (especially fine dust removal), and reducing neps and fibre clusters. However, it can lead to the generation of short fibres, particularly when the fibres used are long and fine.

Flat speed

Flats work in close partnership with the cylinder in carrying out carding actions. Because of the carding action between the cylinder and the flat, fibres are distributed on both surfaces. The substantial speed difference between the two elements leads to very fast loading of fibres onto flats, causing the flats to lose their opening capacity. Flats must therefore be removed from the carding zone for cleaning, with loaded flats replaced by fresh ones. Although fibre loading occurs very quickly as the flat enters the carding

Table 6.6 Influence of cylinder speed

Action	Consequences
Intensive opening action between cylinder and flat, and between cylinder and carding segments.	More individualisation; higher stress exerted on fibres; more liberation of trash particles, seed coats and short fibres.
Increased centrifugal force on fibres and trash particles.	More liberation of trash particles, seed coats and short fibre; increase in transfer efficiency with concomitant decrease in cylinder load and improved opening.

zone, the flat then continues to absorb dust, neps and trash particles thrown by the cylinder throughout the rest of the process. With the increase in production rate brought about through the use of modern cards, it is imperative to use a higher flat speed in order to effectively clean the stock by removing more waste in absolute terms. Moreover, the less intense waste extraction carried out in the blowroom is primarily aimed at avoiding nep generation and shattering of the seed coat and other trash particles; hence, more flat waste must be extracted at the card.

Cylinder–flat setting

The aggressiveness of the opening action depends upon the setting of the cylinder to the flat. There is an optimum setting for each type of fibre, depending on its fineness, dust level and tenacity. Over the entire flat zone, the setting is gradually reduced in the direction of the material flow in order to gradually increase the opening intensity.

If the setting is too close, intensive opening of fibre clusters results, with liberation of dust and trash, but the nep and short fibre content may increase due to the high level of stress acting on the fibres. A wide setting will cause insufficient opening and nep disentanglement. This in turn can cause an increase in the nep level and in the short thick areas of the yarn. The nep level should therefore determine the optimum cylinder–flat setting.

Setting of pre- and post-carding segments

Modern cards with high production rates are equipped with several (1–5) stationary flats or carding bars both in the region between the licker-in and the flats and in the region between the flats and the doffer. These additional opening devices were introduced in order to compensate for the loss in opening ability caused by a high throughput rate. The pre-carding segments handle tufts, while the post-carding segments handle opened fibres.

The setting should be close enough to allow the fibre tufts to be thoroughly pre-carded thoroughly before reaching the flats. This leads to the liberation of dust and short fibres, which are immediately sucked away. The post-carding segments similarly release some dust and husk particles which are also immediately extracted.

6.4.3 Sliver forming zone

The main parameter to be controlled in this region is the linear density of the sliver, which is connected to the speed of delivery. It is often impossible to achieve a very high delivery rate and a fine sliver, due to the mechanical limitations of the machine and technological difficulties associated with the subsequent processing of very fine sliver. On the other hand, production of very coarse sliver at a slow delivery rate though is not mechanically restricted but can lead to different processing difficulties. For example, a fine sliver will lead to creel breaks whereas a coarse sliver causes problems in drafting. A wide range of sliver densities is possible between these two extremes, allowing the correct density to be selected for a particular production rate. The choice is narrowed down by the attenuation capability of the machines used downstream, principally the speed frame and ring frame.

The range is further narrowed down by the quality of carding. At a given production rate, the production of coarse sliver increases the load on the cylinder, which in turn increases the nep level and allows unopened fibre clusters to pass through, as well as leading to an increase in the majority hook. Hence it is advisable to produce the finest sliver possible when other considerations are taken into account such as quality, processability and productivity of the yarn. This optimum value of sliver fineness varies between different mixes and between different fibres.

6.5 Process control in carding: control of card clothing, wire maintenance and card waste

The clothing of various parts of the card is the principal machine parameter that can be controlled, with the most important aspects being cylinder clothing and to a lesser extent doffer clothing.

6.5.1 Cylinder clothing

The three most important parameters of cylinder clothing to be controlled are:

1. point density
2. inclination angle
3. height.

Clothing manufacturers offer a variety of options to suit a particular type of fibre on a particular generation of machine.

Point density

The wire point density of the cylinder clothing principally influences carding intensity. However, the most significant aspect is the number of wire points available per unit of time. A combination of low speed with high point density or high speed with low point density can be expected to provide the same result. The development of cards with a high production rate and hence increased throughput rate necessitated an increase in point density in order to compensate for the resultant loss in combing (or number of available wire points per fibre). For a given fibre and production rate, an increased point density is expected to give a better result, but only up to a certain level: beyond this the opposite effect may occur due to the clogging of inter-wire point spaces by fibres.

The selection of the appropriate wire point density is dependent on:

- Coarseness of fibre: coarse fibres are easy to separate out and hence do not require high density.
- Type of fibre: synthetic fibres require lower point density than cotton, principally due to their long length and increased bulk.
- Production level.
- Trash level: a high percentage of trash may cause the wire points to become loaded with particles very quickly rather than releasing them to the flats.

Inclination angle of wire points

The inclination angle of the cylinder wire points is extremely important in determining both carding intensity and transfer efficiency. When the inclination angle increases, fibre transfer to the doffer and trash ejection from the cylinder surface both show improvement, but the carding intensity is reduced.

Height of wire teeth

The height of the wire teeth also has a significant impact on carding intensity and transfer efficiency. Shorter teeth make it more difficult for the fibres to escape the carding action by moving towards the base of the teeth. Fibre transfer onto the doffer is also facilitated when the fibres are positioned on cylinder surface near the tip of the wire points. Moreover, if the height of the teeth is reduced, more teeth can be accommodated per unit length of wire, as there is a geometric relationship between height, angle of front and back edges, and pitch. Due to their increased throughput rate, modern high production cards require a shorter tooth height of around 2 mm.

6.5.2 Doffer clothing

The selection of the correct clothing for the doffer is also extremely important. The point density here should be low for coarse cotton and high for fine cotton. No advantage is gained by using a high point density with coarse cotton; in fact, it presents a risk of fibres being damaged by becoming too tightly pressed onto the surface of the cylinder.

6.5.3 Wire maintenance

Wire maintenance is another important aspect of process control in the carding region. For a modern 2 mm high cylinder wire, the use of a normal grinding stone is not recommended; instead, a TSG grinder from GRAF should be employed. In fact, if a TSG grinder is unavailable, wire grinding should not be attempted.

The wire should be ground every second or third month, in order to maintain the necessary sharpness. The number of traverse points should increase according to the age of the wire: for successive grindings, the number of traverse points should follow in the sequence 3, 5, 10, 17, etc. However, a microscope should be used to confirm the quality of the grinding: if it is found to be insufficient, the number of traverse points should be increased.

For doffer grinding, a normal grinding machine will suffice. It is important to ensure that all the wire points are touched by the grinding stone, and that a slow and gradual grinding is carried out, in order to achieve the best formation. If the grinding is too harsh, burr formation will result, which will in turn increase the number of hooks in the fibre. The effective length of the fibre from this card will thus be reduced.

The grinding of flat tops is also important: yarn quality improves with each grinding, and frequent grinding using a machine with an emery fillet will result in fewer neps and more consistent yarn. The flat tops should be changed at the same time as the cylinder to ensure good quality and consistency.

Before changing the wire, the individual card quality should first be checked. The licker-in wire must be changed after every 150 000 kg of fibre processed; if changed earlier, yarn quality will be further improved.

Stationary flats must similarly be changed after every 150 000 kg. As with the licker-in wire, however, an earlier change is recommended for the first three or six stationary flats at the licker-in side (every 100 000 kg). This helps to maximise the carding effect between the cylinder and doffer, which is critical for improved yarn quality.

6.5.4 Control of card waste

As raw material constitutes the bulk of the cost involved in yarn manufacture, the adequate control of waste at different stages of manufacture is of considerable importance in process control. Waste that is not reusable should be monitored closely in order to achieve the maximum yarn yield with minimum waste. Card droppings cannot generally be reused, so waste control measures should be concentrated in this region. However, waste reduction should not affect cleaning efficiency and yarn quality: the aim must be to minimise the loss of good fibres while ensuring that the required level of trash removal takes place.

The optimisation of process parameters such as card settings/speeds, modifications/attachments to the cards, along with other developments in modern cards, have resulted in a reduction in lint losses. The cleaning carried out by a card has a significant influence on the yarn and fabric quality and also on the performance of the material in spinning and subsequent processes. Seed coats, motes, fuzz, immature fibre clusters and trash are usually removed during carding. A number of factors affect the level of waste generated during carding; the most important of these are discussed below.

Influence of air currents

Air currents play a critical role in the separation of trash from lint in the licker-in region. These air currents are generated by the rotation of the licker-in; however, due to the licker-in design, the air pressure around the device is variable. The region above the licker-in is at relatively high pressure, as the air currents generated by the licker-in are forced through the limited space between the wire surface of the licker-in and the bonnet. The mote box region is again at lower pressure compared to the undercasing region. The air currents generated can be used to effectively remove unwanted trash particles.

Card modification or attachments

Modifying cards through the use of attachments is often a means of improving quality. One example of such a modification is the use of double combing segments at the licker-in, with stationary flats at the back and front of the card. The double combing segment with mote knives is fitted in place of the normal licker-in undercasing and double mote knives; a set of two stationary flats are fitted at the rear of the card before the back plate. Another set of three stationary flats are fixed at the front of the card where the stripping plate is normally located. The amount of waste generated by the licker-in is found to be slightly higher when the combing segment was used, but the total waste generated over the whole carding process shows hardly any change. The overall cleaning efficiency is also generally comparable, but the nep level at the end of the process is found to be lower when above attachments are employed.

Another type of attachment that may be used is the fibre retriever. This consists of a pair of baffle plates (replacing the mote knives), which partially enclose a section of the licker-in between the feed plate and the licker-in undercasing. One of the two plates is kept under the licker-in undercasing so that it separates the mote box region from the undercasing region, while the other plate is placed below the feed plate. The latter leaves a small gap from the ground level as an inlet for air. A close setting between the two plates causes an upward air draft between them as the licker-in rotates. This helps in the retrieval of the light lint that falls into the mote box, without detrimentally affecting trash removal. The air and some short fibres are then discharged through the undercasing perforations. The fibre retriever is extremely useful in waste control as it improves trash extraction in the licker-in region and also reduces losses in spinnable fibres.

Speeds and settings of the licker-in

A higher licker-in speed results in an increased amount of droppings in the licker-in zone, with marginal improvement in cleaning efficiency. The setting of the undercasing to the licker-in at the entry point is an important factor in licker-in waste control and trash removal. A closer setting between the licker-in and the undercasing leads to an increase not only in licker-in waste but also in the trash content of the waste; the air currents generated by the licker-in are also stronger, leading to improved trash removal.

Flat speed and direction of movement

In addition to individualising the fibres, as discussed above, the flats also contribute significantly to the cleaning achieved in the card. While larger trash is found in the licker-in region, lighter trash is removed by the flat strips. This trash is composed of lead-bits, motes, broken seed coats and immature fibre clusters. The cleaning action of the flat is dependent upon the type of flat used, the flat speed, the flat to cylinder setting, the front plate settings and the condition of the flats. Yarn quality increases with the amount of flat waste generated, but only up to a point: beyond this, increasing the amount of waste generated has no observable effect on quality. Hence, flat waste should be kept at the optimum level through the selection of the correct flat speed. A flat speed of 4 in/min is generally adequate for semi-high production cards while for high production cards, the flat speed can be increased up to 5 in/min. Flat waste should also be checked for uniformity of thickness across the width of a card. The settings of any cards with thicker flat strips should be carefully monitored.

If the flats move in the same direction as the cylinder, the removal of strippings is an easy procedure. If the flats move in the opposite direction, the cylinder carries the material to be cleaned by the flats just above the licker-in. The flats take up the trash but do not transport it through the machine as they do in normal forward-movement systems; instead the trash is removed from the machine immediately.

Setting of cylinder undercasing to cylinder

The setting between the cylinder undercasing and the cylinder affects the cylinder fly waste, which is composed mainly of good fibres. A waste level of 0.2–0.3% in this region is usually considered satisfactory. A setting of 0.034″–0.056″–0.068″–0.102″ should be maintained from entry to exit, and the undercasing should be kept smooth by regular cleaning and polishing. The two halves of the undercasing should fit closely together with no gap between them.

Atmospheric conditions

It is essential to maintain the correct atmospheric conditions in the card room. A lower relative humidity (RH) can cause frequent web breakages, excessive fly liberation and lapping on the doffer. On the other hand, if the RH is too high, the web can sag and the licker-in and cylinder can become overloaded. The ideal temperature for polyester and acrylic fibres should be around 85–90°F, with an RH of 55–60%.

6.5.5 Other process control considerations in carding

Some other issues must also be taken into account when considering process control in carding. These are briefly outlined below:

- Feed variations should be minimised wherever possible, since lower feed variation leads to lower draft deviation, and thus improved carding quality and consistency in yarn quality. Even if the card is used with an autoleveller, feed variations should still be kept as low as possible (plus or minus 10%). With the latest chute-feed systems, the feed variation can easily be controlled to within 5%.

- The selvedge of the feeding bat should be good, and should not be folded or doubled, as this leads to an increase in neps, and sometimes to cylinder loading.
- The lap fed to the carding machine should be narrower than the nominal width of the machine.
- For processing cotton, a minimum suction pressure of 800 Pa should be maintained at the trash master (at knife) for effective removal of trash and dust particles.
- Scraper blades should be correctly maintained: worn or damaged blades will cause the card web to stick to the crush rollers. The correct pressure should be maintained between scraper blade and crush roller; insufficient pressure will also result in the card web sticking. Similarly, if the calendar roller pressure is too high, card web sticking will increase along with it.

6.6 Yarn count issues and other common process control problems for blowroom and carding operations

The yarn count and its variations are influenced to a large extend by the blowroom and the carding operations. Any short-term variations introduced in these processes, if not controlled in the subsequent processes also, can become elongated and manifest as long-term yarn count variations as they reach the yarn stage. Hence, it is important to analyse and understand the issues related to this aspect and take remedial actions in blowroom and card.

6.6.1 Yarn count issues in the blowroom

Yarn count and count variation are two important aspects of yarn production that need to be controlled throughout the blowroom and carding processes. Yarn count control must begin in the blowroom: it is important to maintain a steady average weight of the feed material in order to achieve good consistency in yarn count at the ring frame stage. Some of the main factors that cause variation in the blowroom material, and hence inconsistent final yarn count, are as follows:

- insufficient opening of cotton and variation in tuft size
- ineffective functioning of the evener motion
- irregular air-flow on screen

- variation in cotton level in the blended reserve
- uneven mixing of waste and cottons
- uneven screen surfaces
- faulty air currents
- uneven or worn fluted rolls and calendar rolls
- dirty screens or screens with rough surfaces
- incorrectly set or chocked dampers
- too large a draft between calendar section and lap winder
- types and speeds of beaters unsuitable for the quality of cotton

6.6.2 Yarn count issues at the card

During carding, in the stripping cycle, any difference in the amount of waste extracted by different cards, along with short-term variation in sliver, plays a major role in causing variation in yarn count. Exceptionally large variations in card sliver, with a wave length greater than the total length of the sliver in the draw frame creel, cannot be evened out by doubling (i.e. by combining two or more lengths of yarn into a single thread). To some extent, this type of variation can be controlled by feeding cans from the first passage draw frame deliveries over a number of second passage draw frame deliveries.

The evening out of long-term variations can be achieved, since these irregularities are at somewhat different stages of development as they emerge from the first passage draw frame. During doubling, slivers which are not related to the possible extent should be brought together.

6.6.3 Control of sliver uniformity percentage (U%)

Short-term variation in card sliver contributes to 10% of the total, provided the silver is regular. The card sliver U% thus plays an important role at this point. As the corresponding length of card sliver to a lea of yarn is about 3.5–12 mm, the card sliver U% mainly affects variation within the bobbin. In the case of combed counts, however, the contribution of card sliver U% to yarn count variation would be insignificant.

High card sliver U% can be caused by a number of factors, as follows:

- Worn gear wheels
- narrow setting of the undercasing
- large waste particles reaching the doffer
- uneven or worn doffer or worn feed roller bearings
- draft at excessively high tension
- jerky motion of calendar roller

- too fine a web
- faults in the coiler
- incorrect setting of the coiler base
- incorrectly sized coiler trumpet not suitable to the hank feed
- poor functioning of roller weighting
- uneven feed to the nose of the feed plate
- worn clothing and improper clothing settings
- uneven heights in the wire clothing
- excessive humidity
- poor feed roller grip, caused by failure of feed roll weighing
- bent feed roller and bent surface of feed plate

6.6.4 Common process control problems

In order to take quick control measures, it always helps to list out and analyse the common process control problems that occur on day to day basis. High variation in output material, nep formation, poor cleaning efficiency are some of the common problems associated with blowroom operation and nep formation and holes or patches in the card web are common problems associated with carding operation. These aspects are discussed one by one.

High variation in blowroom material

Variation in blowroom material causes significant problems at later stages in the process (such as in the yarn count, as discussed in Section 6.6.1 above), and can be caused by a wide variety of factors. Firstly, the condition of the cotton itself can be an issue: insufficient opening of cotton and wide variation in tuft size leads to inconsistent material. It is also important to select the correct equipment for the type of fibre being processed; for example, if the wrong type of beaters are selected, the material produced will be variable. Other human errors include the use of excessive soft waste in the mixing or uneven mixing of soft waste with cotton; unequal feeding of the mixing in hopper bale breaker; improper synchronisation in the amount of material fed and delivered from the beaters in the processing sequence; incorrect settings of dampers, leading to irregular air flow; and unnecessary and frequent adjustment of lap feed-regulating motion.

Mechanical problems can also cause inconsistency in blowroom material. These can include intermittent failures of the feed/delivery mechanism; inadequate sensitivity of the piano-feed regulating motion; worn links in the feed-regulating motion; worn pedal link, knife and cone drum bearings; malfunctioning of the length measuring motion; dust and dirt clogging the cage; insufficient pressure in the condensers after the beaters; defective functioning of solenoids and air filters in the modern scutchers; and extreme variation in atmospheric conditions.

Nep formation in the blowroom

Blowroom processes invariably introduce neps into the fibres, but the number of neps can and should be regulated by adequate process control. If the moisture level of the cotton is either too high or too low, an unacceptable degree of nep formation will occur; similarly, extremely fine cottons with high trash content will usually result in increased neps.

As with the control of variation in blowroom material, the remaining factors causing excessive nep formation can be broadly divided into two types: those caused by human error (e.g. poor setting), and those caused by machine failures. The first category covers issues such as excessively narrow settings between the feed roller or pedal and the beater; inappropriate ratio of fan to beater speed; excessively wide setting between stripping rail and beater; beater speeds set too high or too low; fan belts set too slack or too tight; and the use of more beaters than required. Potential machine failures include rough or blunt blades and bent pins on beaters; damaged and rusty grid bars; bent conveyor pipe lines; worn out stripping rails; and air leakage and obstruction of cotton flow through the pipe line.

Poor cleaning efficiency in the blowroom

The last of the principal problems occurring in the blowroom is poor cleaning efficiency. This can be caused by incorrect settings and speeds of the various types of equipment in the blowroom, such as improper adjustment of the angle of grids; too close a setting between the grids/grid bars and the beater; excessively wide settings between the evener roller and the inclined lattice; too high a fan speed causing back draft from the gutter flue; and excessive feed to the beaters. Other factors that can contribute to poor cleaning efficiency include beaters with blunt striking edges; air leakage in the beater chamber and dust receptacle; high variation in trash content in the different cottons used in the mix; and excessive waste accumulation at the air passage.

Nep formation in cards

As well as in the blowroom, neps can also be formed as a result of incorrect settings and other issues at the card. The main factors that lead to unacceptable formation of neps at this point include: setting of lap selvedge guides too close; too wide a setting between back plate and cylinder; high licker-in speeds; too wide a setting between licker-in and feed plate; blunt licker-in wire or dull flats; too wide a setting between cylinder and flats or doffer; too much space between licker-in cover and feed plate; ends of front and back plates damaged; undercasing dirty or choked with fly waste; cylinder/doffer not stripped properly after lapping; rough surface on the front and back plate; use of blunt grinding stone; insufficient stripping; and higher doffer speeds.

Holes or patches in the card web

Holes and patches can form in the card web for a number of reasons: clusters of cotton embedded on cylinder wires and an excessive number of unopened cotton tufts can both cause problems with the card web. Similarly, an incorrect (too wide) setting between the licker-in and cylinder can lead to holes or patches developing. Finally, equipment faults and damage can also affect the card web: these include damaged cylinder/doffer flat wires; cylinder/doffer wires embedded with excessive seed coats and grease; and faulty air current under the licker-in, cylinder and doffer.

6.7 Bibliography

1. Chattopadhyay, R. 2001. *Process Control in Spinning*, Department of Textile Technology, Indian Institute of Technology, Delhi.
2. Garde, A.R. and T.A. Subramanian, 1987. *Process Control in Spinning*, ATIRA, Ahmedabad.
3. Ratnam, T.V. and K.P. Chellamani, 1999. *Quality Control in Spinning*, SITRA, Coimbatore.
4. Artzt, P. 1985. *Melliand Textilberichte, Influence of Various Card Clothing Parameters on the Results Obtained in High-speed Carding on Cotton*, E789-E796 / 701-712, English Edition.

7
Process control in drawing, combing and speed frame operations

A. GHOSH, Government College of Engineering & Textile Technology, Berhampore, India and A. MAJUMDAR, Indian Institute of Technology Delhi, India

DOI: 10.1533/9780857095633.2.158

Abstract: The chapter commences with a discussion of the need for process control in the intermediate machines of a spinning line; the draw frame, comber and speed frame. The process control of each of these machines has been discussed under separate sections. Each section includes a discussion of the key issues related to each machine, including machine productivity, material handling, contribution to yarn quality, common defects and their causes, as well as a variety of other issues pertinent to process control.

Key words: comber, draw frame, drafting, speed frame, yarn quality.

7.1 Introduction

Process control is an integral part of yarn manufacturing. As the material flow in spinning process is sequential, the output of one machine becomes the input for the next. Each of the machines in a spinning line performs specific and unique tasks. This makes process control critical, as inferior performance by just one machine can spoil the good performance of all the other machines. The key processes leading up to spinning are as follows:

- The blow room performs opening and cleaning functions.
- Carding individualises and cleans the fibre tufts.
- The draw frame removes irregularities from the slivers.
- Combing primarily removes the short fibres.
- The speed frame converts the slivers into rovings so that they can be used as a suitable feed material for ring spinning.

If an analogy is drawn with soccer, then blow room operations and carding form the 'Defence' and the ring frame becomes the 'Forward' as it spins the yarn as the final outcome of the process. The draw frame, comber and speed frame play the role of 'Midfield' as they bridge the activities of the 'Defence' and 'Forward' operations.

The objective of this chapter is to discuss process control activities related to the use of the draw frame, comber and speed frame. Process control is primarily aimed at controlling machine or process parameters such as speed, temperature, humidity, pressure and pH. In contrast, quality control is focused on the measurement and control of product characteristics. Within the spinning process, there is a significant overlap between process control and quality control. The quality of the intermediate and final products is monitored both online and offline, ensuring that any deviation is identified quickly and allowing the swift implementation of process control steps designed to counteract the quality problem. The process control activities related to the use of the draw frame, comber and speed frame are discussed in detail in the following sections.

7.2 Process control in drawing: key elements

Drawing has a strong influence on subsequent processing, with the performance of the draw frame affecting blend and count variation of yarn as well as potentially exacerbating yarn faults. The minor alteration of key elements can be employed to produce a variety of yarns with a range of different attributes. In the production of carded yarn, for example, sliver is drawn twice to achieve the desired doubling, parallelisation and evenness. Similarly, in combed yarn production, the slivers are drawn twice pre-combing, with an additional draw post-combing. The key issues concerning draw frame process control will now be discussed.

7.2.1 Roller setting

The optimum roller settings are governed by a number of factors including fibre type, length, fineness and bulk, along with sliver weight and selected back zone draft. The roller settings for synthetic fibres such as acrylic and polyester are usually wider than those for cotton, as such fibres usually have greater bulk and, as such, require a wider roller setting due to the longer friction field they produce (Salhotra, 2004). However, higher roller settings produce more floating fibres in the drafting zone, leading to an increase in yarn irregularity. This problem is further exacerbated if the short fibre content is excessively high.

The front draft zone setting is a decisive factor for final yarn evenness. A front zone setting 2–3 mm higher than the 5% fibre length (by number) given by the USTER AFIS instrument produces the most efficient result (Chattopadhyay, 2002). Depending on the type of fibre, fibre bulk and manufacturer's recommendation, the back zone setting should be 3–6 mm wider than that of the front. A lower back-zone draft and wider back zone setting are generally used in combination to counteract the effects of the drafting force.

7.2.2 The top rollers

The top rollers are a key element in the drawing process as, when pressed with relatively high force against the lower rollers, they guide the fibre. Increasing the top roller pressure narrows the gap between the pressure fields at the back and front of the fibres, and exerts controlled floating fibre movement in the drafting zone. This may result in improved sliver evenness. However, if the top roller pressure is too high, there may be significant overlap between the back and front pressure fields in the main drafting zone, hindering smooth fibre motion and thereby resulting in higher sliver irregularity.

The top rollers, in general, should be of an equal diameter, with cots buffing at regular intervals to ensure good performance. Softer top roller cots enclose the fibre strand to a greater extent than harder cots and thus provide better guidance to the fibres. However, this increased contact means they wear out more rapidly. Normally 83° shore hardness cots are used for top rollers (Chattopadhyay, 2002) for processing of carded slivers. However, lower shore hardness can give better performance for processing of combed slivers.

7.2.3 The bottom rollers

The bottom roller types most commonly used in draw frame processing are axial and spiral flutes. Spiral flutes produce more efficient running and clamping of the fibres than axial flutes (Klein, 1987), and rotation of the top rollers against the spiral flutes takes place more smoothly, facilitating even production. In order to achieve both smooth processing and enhanced grip, spiral flutes should be used in the front rollers, and parallel flutes in the middle and back.

7.2.4 Draft distribution

Most of the modern draw frames employ a two-zone drafting system. The main function of the back zone is to prepare the material for the main drafting process, which takes place in the front zone. In order to achieve effective drafting, the fibres should enter the front zone in a taut, straight condition. The back zone draft varies between values of 1.3 and 1.7. The back zone draft should be set above the critical value at which it exhibits the stick-slip phenomenon (Klein, 1987). This critical value also depends on the fibre orientation in the sliver, as poor fibre orientation necessitates a higher back zone draft. The back zone draft in the breaker draw frame should therefore be slightly higher than that of finisher, and the total draft in the breaker draw frame should be kept lower than that of finisher (Salhotra, 2004).

7.2.5 Number, hank and disposition of slivers

Ensuring good blend homogeneity is an important function of the drawing process. A simple yet effective method to achieve this is complete randomisation of the feed cans. However, careful control of the components is even more important. For two blend components, A and B, the hank of sliver of component A can be estimated from the following expression:

$$H_A = \frac{N_A(100 - P_A)}{N_B P_A} H_B \qquad [7.1]$$

where H_A and H_B are the hanks, N_A and N_B are the number of slivers and P_A and P_B are the blend percentage of components A and B respectively, with $P_B = (100 - P_A)$. The sliver thickness of each component should not differ significantly as this may lead to an insufficient grip on the finer sliver, adversely affecting the drafting process and producing an irregularly blended sliver (Salhotra, 2004). Controlling the feed arrangement can further reduce the risk of irregular blending. In the production of a polyester-cotton blend, the polyester slivers should be fed at the furthest feed positions. For example, in six or eight sliver feeds, the arrangement of cotton (C) and polyester (P) slivers should be PCPCCP and PCPCPCCP respectively (Salhotra, 2004).

7.2.6 Dust removal and waste

Dust is a key problem for both the personnel and machinery involved in processing, especially in the production of rotor yarns, where an accumulation of dust and debris in the rotor grooves can negatively affect the quality of the final product. More than 80% of dust is removed in draw frames equipped with appropriate suction devices (Klein, 1987). In addition, draw frame processing removes very little waste, and this waste in the form of clean draw frame sliver may be recycled by feeding it back to the blowroom mixing process.

7.2.7 The autoleveller

When a premium quality product is required, an autoleveller system is employed on the finisher draw frame in order to control short and medium term fibre variation. Generally open loop autoleveller is used in the drawframe as the measuring or sensing is done on the feed side (six or eight slivers) and correction is done in the process. Figure 7.1 depicts the measuring system (scanning or sensing rollers) of autoleveller in Rieter RSB 851 drawframe. The group of slivers coming out from the trumpet is passed

7.1 Scanning rollers of Rieter RSB 851 drawframe.

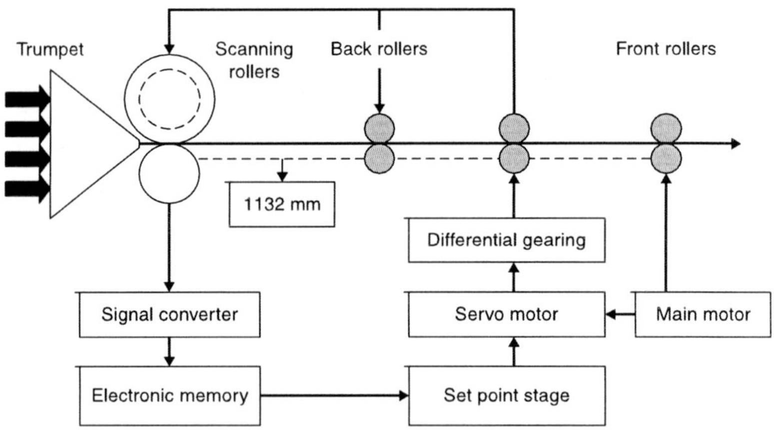

7.2 Autolevelling system of Rieter RSB 851 drawframe.

through the nip of tongue and groove roller combination. A high amount of pressure is created at the nip by using spring and lever system. Some movement occurs for the tongue roller when the mass per unit length of the feed slivers deviates from the nominal value. This movement is converted to electrical signal (voltage) by signal converter (distance transducer). The signal is stored in an electronic memory (Fig. 7.2). This memory ensures that the draft is changed exactly when the irregular portion of the silver reaches the main draft zone. The electronic memory transmits the measured voltage to the setpoint stage with a certain delay. The setpoint stage uses the measured voltage and machine speed to calculate the number of revolutions needed for the servomotor. The servomotor transmits the additional speed to the middle roller of the drafting system through a planetary gear mechanism. This ensures that the draft is adjusted accordingly only in the main draft zone. The speed of the back roller, sensing (scanning) rollers and creel rollers is also varied, ensuring a constant delivery speed without requiring any alteration to the back zone draft (break draft) and tension draft.

When an autoleveller encounters a sudden mass variation in the sliver which continues over a certain length, the machine gradually rectifies the problem, with the inertia of the system preventing an instantaneous adaptation of the fault back to the desired set value. The partially corrected section, thus produced, is known as the correction length, and the time that elapses to reach the set value is the correction time. The correction time depends on the following factors (Chattopadhyay, 2002):

- inertia of the regulating system,
- delivery speed,
- draft,
- extent of mass variation of the sliver from the set value,
- change of mass level, for example, normal level to light or heavier level to normal.

The development and application of sensor and actuator systems has reduced the correction length in the modern draw frame to 3 cm, depending on the operating speed. The autoleveller helps to reduce both count variation in the yarn, and ring frame end breaks. With the use of an autoleveller system, the lea count CV% has also been brought down to less than 1.0. However, improper setting of the autoleveller may be counterproductive and prove detrimental to the quality, so adjustment of the settings must be done in accordance with the manufacturer's instructions.

7.2.8 Machine productivity

To ensure the most efficient production, the ratio of draft to number of doublings can be adjusted. Increasing the draft to a higher or lower value than the number of doublings, for example, allows speed and quality to be balanced to best effect. The effect of such changes on yarn quality is negligible, and increasing the speed of the draw frame does not affect the quality of the yarn appreciably, meaning that, if breaks are within specified limits, drawing speeds can be increased whenever required (Garde and Subramanian, 1978).

7.2.9 Material handling

The storage and handling of slivers in subsequent operations are further variables in the drawing process. The following factors must be reviewed regularly:

- the condition and adequacy of the can spring,
- the condition of castors,
- the gap between the sliver column and the inside of the can.

Sliver storage capacity is related to the can diameter and height and, as a general rule, the standard spring pressure of sliver storage capacity is 80%. Ideally, a gap of 8–10 mm is left between the sliver and the inside of the can. Where the gap is larger, the eccentricity between the can and the tube wheel must be corrected.

Piecing of the sliver should be avoided, as separated portions can never be properly drafted, and thus produce a potential fault when subsequently processed (Garde and Subramanian, 1978). Batch creeling and minimum draw frame stoppage both help to ensure that the possibility of such faults progressing to the next process is minimised. Additionally, removal of fixed lengths of sliver from the delivery can whilst batch creeling can further reduce the production of potential faults.

7.3 The impact of drawing on yarn quality

Yarn quality is greatly influenced by the performance of the draw frame. Count variations both within and between individual bobbins, inappropriate roller settings, and fibre conditions are all key elements affecting the quality of the yarn.

7.3.1 Count variation within bobbins

Roller lapping and fibre shedding are two phenomena responsible for the production of yarn count variation within a bobbin, and as such must be closely monitored. In the past, the use of a three-zone drafting system in the breaker draw frame increased incidences of roller slip, giving rise to very high within lea count variation. However, this problem has been alleviated in modern draw frame processing with the incorporation of a two-zone drafting system.

The selection of optimum roller settings and appropriate top roller pressure also constitute key factors. In addition, the use of 1.03 creel draft between the lifter and back rollers, and the placement of web draft between the front and coiler calendar rollers should be taken into consideration, as they have significant bearing upon sliver wrapping and irregularity.

An inherent drawback of the roller drafting system is the formation of a drafting wave, with wavelength varying from 2 to 3 times the fibre length. High sliver U% is symptomatic of the drafting wave, and it is therefore imperative that the U% of both breaker and finisher sliver are controlled to restrict bobbin count variation to the specified range.

7.3.2 Count variation between bobbins

The control of count variation between individual bobbins presents a greater difficulty. Count variation between bobbins is attributable to improper control of draw frame wrapping. Checking 5 m of wrapped draw frame sliver quickly alerts operators to any serious deviation, allowing reparative steps to be taken (Garde and Subramanian, 1978). In addition, the use of autolevelling systems on modern draw frames ensure proper control of the hank of drawn sliver, reducing irregularity and thus bringing down the between bobbin count variation.

7.3.3 Yarn irregularity

Imperfect roller drafting aggravates the uncontrolled motion of floating fibres, increasing yarn irregularity, whilst excessive short fibre content further exacerbates the problem. An essential aim of drawing is thus to parallelise the fibres, facilitating effective drafting at the subsequent stages of the process. Irregularity inherited from upstream processes is also minimised by means of doubling. Improved drafting systems and reduced irregularity of the drawn sliver are therefore key to the minimisation of yarn unevenness (Slater, 1986).

7.3.4 Yarn imperfections

Poor drafting conditions in the draw frame can lead to increased imperfections in the yarn. Investigation of the quantity of neps produced during drawing reveals that neps in carded yarn outnumber those in carded web, suggesting that certain aspects of the drafting process, a lack of fibre orientation and presence of fibre hooks in the sliver for example, contribute such additional neps (van der Sluijs and Hunter, 1999).

7.3.5 Yarn hairiness

Yarn hairiness is directly related to fibre preparation. An increase in the number of draw frame passages enhances fibre parallelisation, partially alleviating the problem. In addition, Bar *et al.* (1990) suggest that use of an autoleveller system at the draw frame further reduces yarn hairiness.

7.3.6 Yarn strength

Poor condition of the draw frame drafting system affects the strength of the yarn. Inappropriate roller settings and pressure selection, poor surface conditions and eccentricity of the rollers all lead to a reduction in yarn strength (Garde and Subramanian, 1978). However, yarn can be strengthened by increasing the number of draw frame passages, due to the enhanced fibre parallelisation.

7.4 Process control in drawing: common problems

Some of the common defects in the drawing process and their causes are discussed below.

7.4.1 Roller lapping

Roller lapping may restrict the flow of material through the production line, whilst additionally leading to the wastage of raw material. Static accumulation on the top rollers and fibre tackiness are the key causes of roller lapping (Chattopadhyay, 2002). Low humidity hinders the dissipation of static charge, which, in combination with inadequate fibre finish (in case of synthetic fibres), causes static accumulation on the top rollers. In contrast, fibre tackiness results from high temperature and humidity. Cotton wax, for example, can smear and lead to roller lapping if the room temperature and relative humidity are too high. Lapping of the top or bottom rollers is equally likely. However, a higher occurrence of lapping on the top roller is a likely indication of high relative humidity, whilst lapping of the bottom roller suggests low relative humidity.

Depending on the type of cotton and the processing speed selected, roller varnish may be used to prevent lapping. The quality of the varnish and uniformity of application are critical, otherwise this action may prove counterproductive. The roller lapping can be kept at a low level via the following methods:

- proper control of relative humidity and temperature in the department,
- avoiding sticky and honey dew infested cottons,
- using an appropriate amount of antistatic spray on manmade fibres,
- applying a berkolising treatment to cots at regular intervals,
- restricting the addition of soft waste in the mixing to within 2–3%.

7.4.2 Coiler tube choking

Choking of the coiler tube is caused by a deposition of dust, fibre, tint or fibre finish, particularly in the production of synthetic fibre. When fibre finish is reduced, low humidity encourages static generation and inter-fibre repulsion, leading to increased sliver bulk and eventual choking of the coiler tube. Such restriction of the coiler tube diameter can prove very problematic. However, the following corrective measures may be prescribed to eliminate the problem of choking (Salhotra, 2004):

- reducing the sliver weight per unit length,
- using cans having greater spring pressure,
- maintaining optimum relative humidity in the department,
- cleaning the sliver passing section at regular intervals,
- using trumpets, guides and plates made of either steel or a chromium-plated matted surface.

7.4.3 Over-correction and under-correction in the autoleveller

Regular monitoring of the electronic autoleveller system is required to ensure the optimum performance. Over and under-corrections are the two frequently encountered problems in the use of the autolevelling system. If an over-correction occurs, a thicker portion of fibrous strand is converted to a thinner portion, or a thinner portion is converted to a thicker portion, due to an inappropriate (over-) adjustment of the draft. This may be caused by either incorrect signal generation or incorrect signal processing by the electronic system. In the case of an under-correction, an inappropriately thick or thin portion of the fibrous strand remains as a faulty portion of lesser magnitude, even after autolevelling. This happens due to a change in the draft that is insufficient to rectify the problem. In an RSB 851 draw frame, a potentiometer (R38) is adjusted (between 0 and 10) to tackle the problem of over and under-corrections. The schematic representation of over and under-correction is shown in Fig. 7.3.

Sliver testing has been recommended by Rieter to determine whether the RSB 851 draw frame is over or under-correcting. The scheme of the sliver test is shown in Fig. 7.4. If nominal doubling (N) is 6, then the machine is operated with (N – 1) and (N + 1) doubling that is 5 and 7 slivers, processed with the autoleveller switched on. The count of the delivered sliver is checked, and the deviation percentage is calculated to establish the extent of over-correction or under-correction. If an over-correction is revealed, the

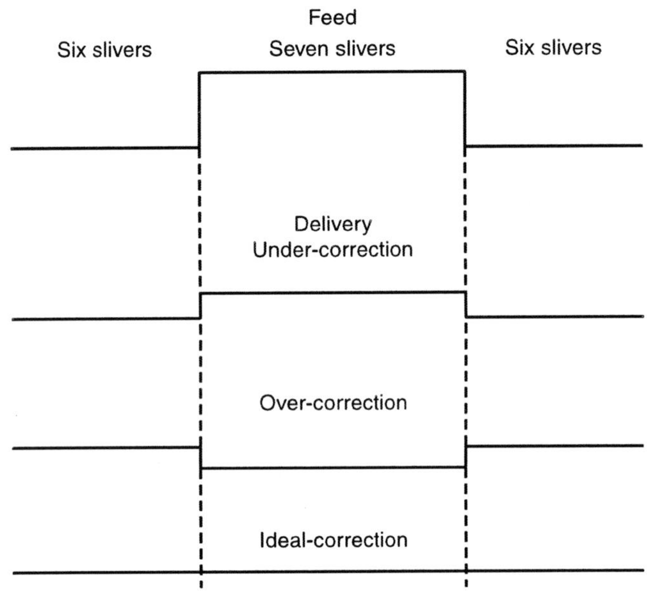

7.3 Over- and under-correction in autoleveller.

R38 potentiometer is turned towards zero, while if an under-correction is revealed, it is contrastingly turned away from zero.

The autolevelling system facilitates correction by altering the draft in the main drafting zone. The position of the scanning device (measuring device) and the position of correction zone are thus different. Therefore, the measured signal (thickness or thinness of the input slivers) has to be stored in electronic memory and released, after some delay, to actuate the servomotor as soon as the faulty region of the sliver reaches the correction zone. An electronic pulse generator is used to measure this delay time.

7.4.4 Early and late levelling in the autoleveller

Early and late levelling occur when the timing of the draft alteration and the arrival of the faulty portion in the correction zone do not match. In the case of early levelling, the timing of the draft alteration precedes the arrival of the faulty portion in the correction zone, as shown in Fig. 7.5. In contrast, late levelling occurs if the timing of the faulty portion's arrival in the correction zone precedes the alteration of the draft (Fig. 7.5). Early and late levelling lead to the production of thick and thin places in the sliver.

The timing of levelling is optimised by changing the setting of a numeral switch. In the RSB 851 draw frame, the setting of the numeral switch (S93)

SLIVER TEST			Date:
Machine No.			
Nominal sliver count (ktex)			
Doubling (N):			
Setting of numeral switch (S 93):			
Setting of R38 potentiometer:			
Test number	(N-1) slivers	N slivers	(N+1) slivers
1			
2			
3			
4			
5			
6			
7			
8			
9			
10			
Average			
Deviation %			

$$\textbf{Deviation \%} = \frac{\text{ktex}_{(n-1)} - \text{ktex}_n}{\text{ktex}_n} \times 100$$

Deviation %: Positive means over-correction

Deviation %: Negative means under-correction

$$\textbf{Deviation \%} = \frac{\text{ktex}_{(n+1)} - \text{ktex}_n}{\text{ktex}_n} \times 100$$

Deviation %: Positive means under-correction

Deviation %: Negative means over-correction

7.4 Sliver test for over- and under-correction.

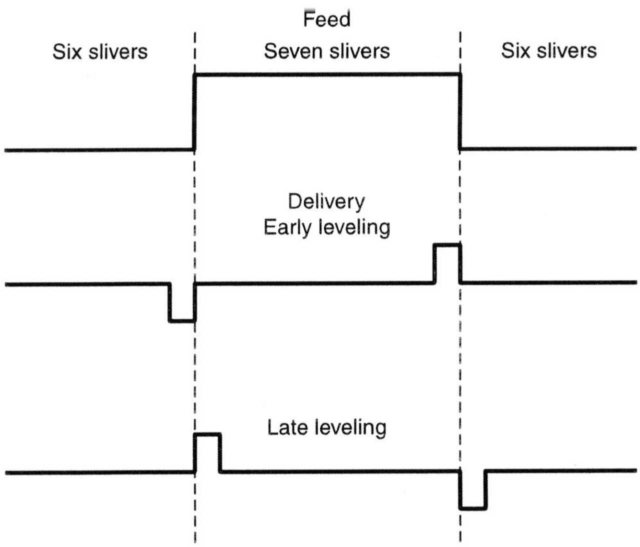

7.5 Early and late levelling in autoleveller.

Table 7.1 Optimum timing of RSB 851 drawframe autoleveller

S93 Setting	Pulses	Front zone setting: 36–40				Front zone setting: 40–50			
		Break draft				Break draft			
		1.16	1.28	1.41	1.70	1.16	1.28	1.41	1.70
0	192								
1	191								
2	190								
3	189								
4	188	▒							
5	187	■	▒			▒			
6	186	▒	■	▒		■			
7	185		▒	■	▒	▒	▒		
8	184			▒	■		■	▒	
9	183				▒		▒	■	
A	182							▒	
B	181								▒
C	180								■
D	179								▒
E	178								
F	177								

can be varied between 177 and 192 pulses, to cover the wide range of front zone and break draft settings. Table 7.1 depicts the optimised setting of the numeral switch for various front zone and break draft selections. The distance between the scanning and correction points in the RSB 851 draw frame is 1132 mm. The correction point is at a distance of $(1132 - 0.9 \times A)$ mm from the scanning point, where A is the front zone setting in millimetres. As the front zone setting increases, the distance between the scanning point and correction point reduces, decreasing the time required for the sliver to travel from the scanning to the correction point. Thus, the number of pulses is reduced (Table 7.1). Similarly, an increased break draft in a front zone with the same setting facilitates faster travel of the sliver between the middle and front roller nips, leading to a lower number of pulses for optimum timing.

7.4.5 Common defects in drawing and their causes

Some of the common quality problems in draw frame processing and their causes are discussed below (Ratnam *et al.* 1994).

Causes of roller lapping in drawing are:

- channelled or worn out roller coverings,
- inaccurate setting of top roller clearers or worn clearers,

- worn out flutes in the bottom rollers,
- unacceptable humidity,
- extreme top roller pressure,
- inappropriate use of varnishes on the top roller cots,
- insufficient suction in the pneumafil,
- inappropriate roller settings for the type of material processed.

Causes of irregular selvedge in drawing are:

- inadequate top roller pressure,
- undue spread of sliver at the feed,
- worn flutes on the bottom rollers,
- extreme pneumatic pressure.

Causes of end breaks in drawing are:

- inappropriate sliver piecing,
- doubled sliver in the feed,
- thin card sliver resulting from web falling at cards,
- inaccurate trumpet size,
- use of cottons with excessive honeydew content,
- insufficient top roller pressure or break draft,
- damaged surfaces in drafting or calender rollers,
- overfilling of cans and poor material handling practice,
- inappropriately wide setting between drafting rollers,
- improper ambient conditions in the department.

7.5 Process control in combing: key elements

Combing reduces short fibres, neps and trash, improving the mean fibre length. This ensures that the resultant yarn has better evenness, with fewer imperfections and greater strength. The performance of a comber is influenced by both the production rate and the extent of waste extraction. For a given production rate and the waste level, an optimum choice of batt weight, feed length per nip and nips per minute should be made. Backward feed produces a better quality than forward feed, but at a cost of higher waste extraction with the same parameter setting. In modern spinning, a forward feed is predominantly used for medium and coarse count yarns, whilst a backward feed is reserved for fine count yarns. The key factors relating to process control in combing are discussed in the following section.

7.5.1 Pre-combing process

There are two types of pre-combing process:

1. Lap doubling process (classical method): Card → sliver lap machine → ribbon lap machine → comber
2. Sliver doubling process (modern method): Card → draw frame → doubling machine → comber

For the processing of medium staple cotton producing a medium yarn count, no differences exist between the lap doubling and sliver doubling processes (Klein, 1987). However, for the processing of long to extra-long staple cotton producing a fine or very fine yarn count, use of the sliver doubling process is more advantageous than the lap doubling process. This is attributed to the higher cohesion and parallelisation of fibre produced by the sliver doubling process.

As the carding process generates over 50% of trailing hooks, an even number of drawing passages (preferably two) must be provided between the card and comber. This ensures that the fibres are presented to the comber with the majority of hooks arranged in a leading direction.

Morton and Nield (1953) have observed that by interposing a third drawing operation between the card and comber stages, the amount of comber waste produced increased from 13.1% to 16.8%. However, if the extra process was accompanied by an additional reversal of sliver at some point in the process, ensuring that the majority hooks were still presented in a leading direction at the comb, the waste produced was only 11.3%. Hence, whilst an additional drawing process prior to combing does give rise to more parallel fibres and fewer hooks, this must be accompanied by a sliver reversal arrangement to offset the presentation of hooks in wrong direction.

7.5.2 Thickness of the lap sheet

A thin lap sheet facilitates smooth penetration of the circular comb. However, it exerts a poor retention of neps and impurities at the sheet, decreasing the possibility of their removal in the subsequent combing cycle (Klein, 1987). Therefore the thin lap sheet gives rise to good cleaning of the circular comb, but a poor self-cleaning effect. In contrast, a thick lap sheet exerts a higher retaining power and more effectual bite, but applies a strong load to the circular comb. In addition, if the lap sheet is excessively thick, the circular comb may no longer be able to penetrate effectively, allowing the fibres in the top layer of the sheet to escape. Hence, selection of an optimum lap thickness is essential in effective combing (Fig. 7.6). In some of the modern combers, the linear density of feed lap is around 80 g/m.

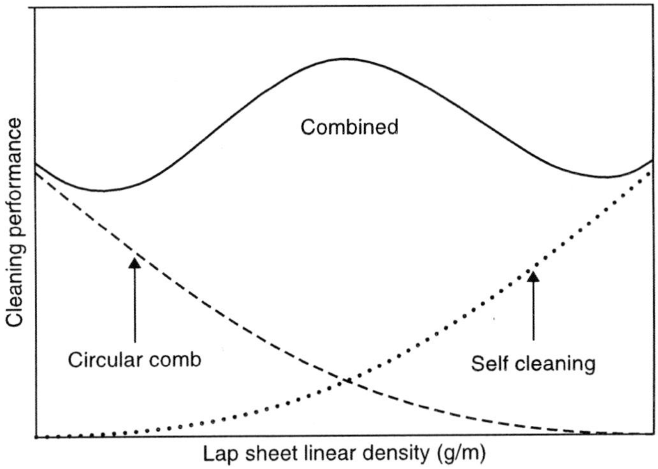

7.6 Effect of lap sheet thickness on cleaning performance.

7.5.3 Parallelisation of the fibres in the lap sheet

The degree of orientation of the fibres in the lap sheet has considerable influence on the performance of a comber. An increase in pre-comber draft increases the fibre parallelisation of the lap sheet. A poor degree of orientation not only increases the loss of good fibres, but also increases the load on the circular comb. In contrast, a high degree of fibre parallelisation significantly reduces the holding power of the sheet so that it is no longer able to hold back the neps and impurities as it usually does (Klein, 1987). As a consequence, some of the neps and impurities pass through the top comb. It is now accepted that a certain level of fibre disorientation associated with lower pre-comber draft is essential for removal of neps and impurities. Higher linear density of lap and higher amount of fibre disorientation actually increases the strength of the lap, which is considered to be an important factor influencing the overall cleaning performance of the comber.

7.5.4 Type of feed

In general, there are two types of feed:

1. Forward feed: The material is fed whilst the nipper is rocking towards the detaching rollers. The major operational sequence is:

   combing → feed → detachment → combing

2. Backward feed: The material is fed during the return of the nippers. The operational sequence is:

   combing → detachment → feed → combing

Table 7.2 Threshold length of fibres in the sliver and noil for forward and backward feeds

Fibre length	Forward feed	Backward feed
Length of shortest fibres in the combed sliver	$d - f$	d
Length of longest fibre in the noil	d	$d + f$
Boundary length	$d - \frac{f}{2}$	$d + \frac{f}{2}$

For fibre of a particular length, whether it would go into noil or retain itself in the combed sliver is dependent on the modes of feed (Chattopadhyay, 2002). This is depicted in Table 7.2 where d and f are the detachment setting and feed length per nip, respectively. The boundary length in Table 7.2 is defined as the average value of the shortest fibre length in the sliver, and the longest fibre length in the noil. It can generally be assumed that a fibre shorter than the boundary length is an addition to the noil, and longer than boundary length is an addition to the combed sliver. An increase in boundary length causes the noil to increase, and noil extraction increases for both type of feed with the increase of the detachment setting. Nevertheless, an increase in the feed length per nip can result in differing responses. In the processing of a forward feed the noil reduces, but in a backward feed it increases with the increased length of feed length per nip.

A higher value of feed length per nip reduces the number of combing cycles needed before fibres are seized by the detaching roller. Generally, a lower feed length per nip is chosen for longer, finer fibres, because they require a higher number of combing cycles.

Usually, a forward feed is selected for high production rates when quality demand is moderate, with the noil percentage kept between 5% and 14%. However, when a higher quality is necessary, a backward feed must be used, with a noil percentage in the range 14–25% (Klein, 1987).

The cleanliness of the combed sliver is also dependent on the feed mode. The backward feed always produces better sliver cleanliness than the forward feed (Klein, 1987). This can be ascribed to the increased combing of fibres before they are seized by the detaching roller.

7.5.5 Top comb

The needle density of the top comb varies from 26 to 30 per cm depending on cotton micronaire. A higher needle density results in better nep removal, but with the additional risk of strong loading and possible damage, which can impair the combed sliver quality. Periodic cleaning of the top comb is necessary to ensure an even and consistent quality of the fleece. The cleaning interval depends upon the cotton quality, production rate, density and top comb condition.

The depth of top comb penetration also affects the noil elimination. Lowering the top comb by about 0.5 mm leads to a 2% increase of noil, and a higher elimination of neps (Chattopadhyay, 2002). However, excessive penetration of the top comb disturbs fibre movement during piecing, resulting in a deterioration in quality.

7.5.6 The setting of the combing elements

Efficient selection of the nipper to half lap and top comb penetration settings are essential factors in producing the most effective combing operation. In a modern high speed comber, a setting of 0.5 mm between nipper and half lap with a top comb penetration of 0.5 give a good performance (Chattopadhyay, 2002). A uniform setting exerts better quality values from all heads.

7.5.7 The density of points on the circular comb

Older machines had comb segments made of needles. More recently, sawtooth clothing is used as, in contrast to the needles, it is more robust, needs less maintenance and is universally applicable. However, the uniform density of needles or saw teeth can cause problems, as the difficulty of penetration for the front row of needles may cause the material to bounce back. A gradual increase of the density from the front to the back rows gives a more efficient combing performance.

7.5.8 Piecing wave

Combing generates a piecing wave which exhibits periodic variation, as visible in the spectrogram in the form of peaks at about 30–75 mm wave length (Klein, 1987). An eccentric withdrawal of the web from the web pan reduces the piecing wave to some extent. The piecing wave can be minimised by the use of an autoleveller in the drawbox. Removal of the piecing wave can be facilitated by the use of a post-comb drawing. However, more than one post-comb drawing can cause the problem of sliver licking, so multiple drawings are therefore avoided.

7.5.9 Control of comber waste

The principal function of the comber is the removal of short fibres of a predetermined length as waste. Significant numbers of neps, along with some foreign matter such as seed coats, are also removed. A higher percentage

Table 7.3 Guidelines for optimum levels of comber waste for different length distributions

Nature of fibre length distribution of mixing	Short fibres % by number		Mean length	Approximate level of optimum comber waste % by weight in relation to yarn quality
	Baer sorter	Fibres <15 mm		
Triangular	>20%	>20%	< 26 mm	No optimum: increasing waste continuously improves quality
		< 18 %	≥ 26 mm	0.5 × short fibre (%)
Flat	<18%	< 18 %	Any	0.5 × short fibre (%)

of comber waste may not always ensure better removal of short fibres and neps, and therefore will not enhance yarn quality or performance beyond a certain level. To ascertain the level of required waste is therefore a case of optimisation, balancing the improvement of quality against the higher cost involved in the removal of comber waste.

It therefore follows that the removal of short fibres and neps alongside retention of the longer fibres holds the key to efficient combing. The following points are worth noting (Garde and Subramanian, 1978):

- Card trailing hooks usually prevail over leading ones. They are removed when fed as leading hooks by keeping an even number of passages between the card and the comber.
- A higher level of pre-comber draft can ensure lower level of comber waste without affecting the resultant yarn quality. However, there are risks associated with high pre-comber draft which undermine the inter-fibre cohesion, including problems such as lap licking and frequent sliver breaks.
- The strategic place to counteract neps is during the carding rather than the combing process. It is more economical to operate cards at a low production rate than to extract high comber waste. Since neps escape easily through needle spacing, their elimination is only possible with fibre removal, which is expensive.

The nature of fibre length distribution in the cotton or mix dictates the optimum level of comber waste to give the desired yarn quality, yarn appearance and end breaks at ring frames. A study conducted at ATIRA regarding this optimisation is summarised in Table 7.3.

Steady improvement in yarn quality can be achieved with an increase in comber waste to a point where most fibres below 15 mm length are removed (Garde and Subramanian, 1978). As a rule of thumb, one can assess the comber waste level as 0.5 × short fibres (%). For large scale production,

Table 7.4 Norms for improvement in mean length

Count (Ne)	Fibre length (mm)		Waste (%)	Expected increase in length (mm)
	Effective length	Mean length		
28–34	26–27	21–23	7–9	1.0–1.2
	28–30	21–23	10–12	1.2–1.8
35–44	31–34	24–26	11–13	2.0–2.5
46–60	34–36	25–27	12–14	2.5–3.0
61–90	36–39	26–28	13–15	2.5–3.0
91–120	36–39	27–30	14–16	3.0–4.0

Source: ATIRA norms.

however, the minimum level of comber waste that gives the desired yarn quality with minimum ring spinning end breakage should be chosen. High quality yarns necessitate a higher level of comber waste removal, in order to ensure a low level of neps and short-length thick fault (A and B types).

The level of waste removal should always be judged from the perspective of short fibre reduction or fibre length improvement. The comber is not a perfect fractionating device; some short fibres are mixed with the comber sliver, while some longer fibres escape into waste. This fractionating deficiency, however, cannot be allowed beyond certain level. The fractionating efficiency can be expressed by two methods. The first method estimates the relative short fibre content in the comber lap and comber sliver, whilst the second method evaluates the improvement in mean length. The first index lacks reliability on account of the disagreement between the different methodologies used to ascertain short fibre content. Therefore, the method of expressing fractionating efficiency in terms of improvement in mean length has been found to be most reliable. Table 7.4 gives norms for improvement in mean length for various mixings and levels of comber waste. The waste at combers needs to be checked and controlled due to the following reasons:

- excessively high waste means financial loss,
- a lower level of waste than normal could lead to unacceptable yarn quality and performance,
- between–comber waste variation could contribute to the between lea count variation.

7.5.10 Machine productivity

Beyond a certain productivity level combing quality is affected, with the type of raw material and quality demand both influencing the resultant product. Usually the comber production rate is reduced by 30–40% when processing fibres producing fine counts, in order to realise desired quality levels. An endeavour to raise the production beyond the standard rate leads to a deterioration in the quality of the combing. Hence, there is not much scope to increase comber production.

7.6 Process control in combing: the impact of combing on yarn quality and common problems arising from the process

Combing plays a significant role in improving yarn quality. This is discussed in the following section.

7.6.1 Count variation within bobbins

For combed materials, the piecing wave irregularity is one source of count variation within bobbins. However, the offset arrangement of web collection leads to partial compensation of the piecing wave, whilst the piecing wave can be minimised by using an autoleveller in the comber drawbox. In addition, a further post-comber drawing considerably reduces the piecing wave irregularity.

7.6.2 Count variation between bobbins

The differences in draft and waste between combers can lead to count variation between the bobbins. Regulation of between bobbin count variation can be exercised through uniformity of waste levels and drafts between combers (Garde and Subramanian, 1978).

7.6.3 Yarn irregularity and imperfections

Combing exerts a strong influence over short fibre content, facilitating individualisation and parallelisation. This has a significant bearing on the drafting operation, and any compromise on combing quality can lead to poor yarn evenness (Slater, 1986).

Combing is the second strategic point used to combat the incidence of neps after carding. This is dependent on both the noil level and cotton type, with combed cotton yarns usually containing about one-tenth the number of neps found in carded yarns. The combing process appears to be more efficient at removing fibrous neps than seed-coat fragments (van der Sluijs and Hunter, 1999). Neps removal is governed by various parameters, such as the needling of the cylinder and top comb, the setting of the half lap and top comb, and the overall maintenance. The factors causing neps are as follows:

- low needle density in top comb,
- inappropriate top comb setting (with late entry restricting the depth of penetration),
- poor mechanical condition of the comber.

Combing reduces the occurrence of thick and thin places in the yarn by improving the mean fibre length and parallelisation.

7.6.4 Yarn faults

Combing removes undesirable short fibres and approximately 50% of the fibre trash content, whilst simultaneously improving fibre parallelisation. It is thus essential to reduce yarn faults, and can overcome some of the drawbacks of blowroom and card. The following factors must be optimised to control the yarn faults (Garde and Subramanian, 1978):

- level of pre-comber draft,
- feed length per nip,
- half lap to nipper setting,
- density, type and arrangement of wires on half lap and top comb,
- penetration of top comb,
- level of comber noil.

7.6.5 Yarn hairiness

Understandably, combed yarns are less hairy than carded yarns as the short fibres are removed during combing. The greater the amount of waste, the less hairiness the yarn exhibits.

7.6.6 Yarn strength

The improvement in yarn strength brought about by combing strongly depends on the level of comber waste, mechanical condition and setting of the combing elements. The greater the amount of waste, the higher the yarn strength will be. However, poor mechanical condition of cylinder and top comb needles and non-uniformity of settings can all lead to loss in yarn strength.

7.6.7 Common defects and their causes during the combing process

Some of the common defects in the combing process and their causes are discussed below (Ratnam *et al.* 1994).

Causes of high comber sliver variation are:

- variation in the waste extraction between heads,
- difference in the setting between the back detaching roller and nipper,
- chocking of half lap with seed coats or immature cotton,
- excessively wide setting between the half lap and comb cleaning brush,
- eccentric top and bottom rollers in detaching and draw box rollers,
- misaligned and bent nippers,
- inappropriate needle spacing and broken or bent needles,
- difference in detaching roller diameter and improper timing of top comb,
- fluctuation in pneumatic pressure for draw box top roller weighting,
- variation in the piecing distance of the combed fleece,
- poor condition of saddles and top detaching roller brackets.

Causes of differences in noil between heads are:

- difference in top comb penetration between the heads of the same comber,
- difference in the setting between half lap and bottom nipper,
- uneven and insufficient nipper grip,
- variation in diameter and pressure of the top detaching rollers,
- difference in suction pressure between the heads.

In some of the modern combers, the waste is collected centrally making it difficult to get the headwise noil value.

Causes of poor combing efficiency are:

- presence of fibre hooks or disorderly fibres as a result of non-standard preparatory processing,
- timing of the combing cycles not adjusted correctly at the proper indices,
- chocking of top comb with short or immature fibres,
- insufficient penetration of the top comb,
- half lap cleaning brush loose on shaft or set too far from the cylinder,
- extreme variation in short fibre content in the mixing,
- inappropriately wide setting between half lap and nipper,
- incompatible mixing of soft waste.

Causes of lap running slack are:

- inappropriate feed ratchet movement,
- lap loose on the shaft,
- inadequate tension draft between the lap and feed rollers.

Causes of poor nep removal efficiency in combers are:

- ineffectively wide setting between half lap and bottom nipper,
- ineffectively wide setting between comb cleaning brush and stripping rail,
- low penetration of the brush with the half lap,
- insufficient penetration of the top comb,
- uneven nipper grip,
- damaged needles in top comb or half lap,
- damaged or absent top roller clearer cloth,
- incorrect atmospheric condition,
- incorrect choice of top comb needle density.

Causes of long fibres in the waste are:

- a lap containing numerous trailing hooks, which can give rise to excessive waste containing a high proportion of long fibres,
- disorderly arrangement of the fibres in the lap,
- improper timing of the comber such that the already detached fringe is fed back into the last few rows of cylinder needles during backing off,
- uncontrolled disturbance of the fibres in the lap during detaching.

Causes of short fibres in the slivers are:

- high fibre-to-fibre friction,
- improper timing of top comb penetration,
- presence of more hooked fibres in the lap.

7.7 Process control in speed frame operations: key elements

The speed frame is an intermediate machine, used between the draw frame and the ring frame. Enhanced roving quality is essential to meet the high quality standard demanded in many yarns, and the quality of the rove depends on both fibre quality and processing parameters. Process control in speed frame use is mainly centered on the factors discussed below.

7.7.1 Roving twist

Appropriate selection of the roving twist is imperative for the quality of yarn, as well as key to the smooth processing of material in the ring frame.

Table 7.5 Typical values of roving twist multipliers

Material	Twist multiplier
Carded cotton	1.40
Combed cotton	1.0–1.2
Synthetic fibres (polyester 32 mm)	0.95

Typical values of roving twist multipliers for different materials are given in Table 7.5. High roving twist may cause drafting problems in the ring frame, leading to the formation of thick spots in the yarn. To prevent this and improve the opening up of the twist, a higher break draft can be used at the ring frame, but this is at the cost of yarn evenness. Similarly, inadequate roving twist greatly increases the roving breakage rate at the speed frame. The optimum roving twist can be chosen by checking the roving strength at a gauge length of 5 cm. For a roving made out of medium staple combed cotton, for example, a strength between 0.55 and 0.60 cN/tex is optimum. The roving strength depends upon the following factors (Chattopadhyay, 2002):

- fibre length and fineness: longer and finer fibres give greater roving strength,
- parallelisation of the fibre: increased fibre parallelisation reduces roving strength,
- roving linear density: a coarser density produces roving with greater strength,
- twist per inch: roving strength is directly proportional to the twist per inch,
- uniformity of twist: roving strength increases with the twist uniformity.

7.7.2 Roving tension

Roving tension should be maintained at an optimum level to encourage uniformity. Low roving tension leads to the production of soft package, which frequently collapses, whilst high roving tension produces a hard, compact package. However, excessive roving tension can cause false draft, roving stretch and roving breakage (Klein, 1987). Roving tension depends on the number of wraps on the pressure arm, with a greater number of wraps, producing a higher roving tension. Similarly, a bigger roving package operating at a higher speed also increases roving tension. In this case, the roving is sufficiently consolidated by use of a higher twist, to arrest roving stretch.

7.7.3 Condenser

A wide ribbon width in the drafting zone causes more fly liberation, irregularity and hairiness of the roving. Use of condenser in the drafting zone restricts the ribbon width at the front roller nip, making the roving more compact and reducing breakage. Furthermore, the presence of the condenser increases inter-fibre friction, improving the roving quality by reducing fly liberation (Chattopadhyay, 2002).

However, use of the condenser can also cause irregularities in the draft. The fibres at the edge of the ribbon, which touch the surface of the condenser, decelerate during drafting, leading to a deterioration of roving evenness.

7.7.4 Spacer

The hank of the material, total draft and fibre bulk determine the spacer size. For example, 5–6 mm spacer size is used for producing roving of 1.4–1.6 Ne from the sliver of 0.16 Ne. The optimum selection of spacer size not only improves the yarn strength and evenness, but also reduces long thin and thick faults in the yarn (Klein, 1987).

7.7.5 Top roller pressure and hardness

The key considerations regarding top roller pressure and hardness on the speed frame are similar to those for the draw frame. Excessive pressure on the back top roller increases the resistance to roller movement, enhancing the risk of torsional vibration (Chattopadhyay, 2002). Such disturbances cause greater short term variation and roving irregularity. This can be counteracted by the use of softer cots, which give better speed frame performance. Cots with 80–850 shore hardness are generally used, and it is advisable to use softer cots for the front top roller and medium hardness cots at the back top roller.

7.7.6 Roller setting

Selection of the optimum roller setting ensures greater roving evenness by minimising the drafting difficulty. The back zone roller setting should be slightly wider than the front zone setting, and care should be taken that the selected front zone setting is no wider than recommended, as this can increase the occurrence of long thin and thick faults in the yarn. Similarly, a lower back zone setting increases the back zone drafting force, which may lead to torsional vibration, thereby increases roving unevenness. For combed cotton yarns, saddle gauges of 50 and 52 mm, for the front and back zones respectively, are commonly used. The bottom roller setting of 44 and 53 mm, for the front and back zone respectively, are used to allow the overhang. Synthetic fibres require a wider setting than cotton due to the absence of short fibres (Salhotra, 2004).

7.7.7 False twister

The false twister is a grooved insert that increases the roving twist in the unsupported length between the flyer top and delivery roller of the drafting system (Klein, 1987). As a consequence of false twister use, both the roving breakage rate and fly generation are reduced. Additionally, roving becomes more compact, increasing the capacity of the bobbin. The number of the grooves in the false twister has a significant bearing on the generation of false twist in the unsupported length of roving. Traditionally made from rubber, modern false twisters are usually made from nylon because of the greater wear resistance this offers.

7.7.8 Between row variation

The front row bobbins are slightly finer than those of back row, resulting in higher yarn count variation. Remedial measures can be taken to minimise row-to-row differences. The extension of the flyer top on the back row so that the roving meets the flyer top at the same angle of approach for both the rows is one method, whilst the use of different flyer top designs in each row can also produce a more effective twist in the front row. The improved designs of flyer and flyer top have eliminated the old concept of feeding the rovings of front and back rows of the speed frame in different ring frames.

7.7.9 Machine productivity

The productivity of the speed frame suffers due to the end breaks. The control of end breaks is thus of the utmost importance. Most of the breaks occur within the flyer, and although a higher production rate of the speed frame does not affect yarn quality, it does have a significant influence on the rate of end breakage. Speed frame production can be increased by reducing creel and within flyer breaks, speeding up the machine and production of coarser rove. The reduction of breaks can be achieved as follows:

a) Reducing creel breaks

1. Carded process

- check diameter of condensing trumpet at the draw frame;
- test for correct coiling and can spring plates, as they can cause breaks through entanglement.

2. Combed process

- eliminate one draw frame passage in the post-comber drawing if more than one is scheduled;
- reduce the speed of the high speed draw frame if only one post-comb drawing is used;
- use six doublings and six draft rather than eight, to reduce excessive parallelisation;
- add one card sliver to five combed slivers at the draw frame as a last resort.

b) Reducing breaks within the flyer: End breaks within the flyer are mainly dependent on mechanical conditions, such as tendency to vibrate, lack of smoothness of the flyers and unevenness of the rove. Use of false twisters or twist-masters on flyer-tops increases the twist in the rove between the flyer top and front roller nip, and thus reduces breaks. The false twister with a small hole and greater depth is most efficient.
c) Speeding up the machine: The machine can be run at the highest speed until the end breaks are no longer excessive. This increase does not affect the yarn quality.
d) Production of coarser rove: Draft can be easily increased on the speed frame by 20% from the normal level, and increasing it at the ring frame does not affect the yarn quality to any appreciable extent.

7.7.10 Material handling

During doffing, doffers normally keep the doffed bobbins on the top arms and carry 8–10 bobbins by hand to the storage place. This practice is not only laborious but also sometimes results in bobbins falling on the floor and the rove being spoiled. Suitable trolleys should thus be used for handling the roving bobbins. Furthermore, the roving bobbins should not be held up or stored for an extended length of time, as the inner layer pressure increases the fibre coherence in roving, present problems during drafting at the ring frame. This precaution should be particularly taken when spinning 100% polyester or its blends (Salhotra, 2004).

7.8 Process control in speed frame operations: the impact of speed frame operations on yarn quality and common defects related to the process

The stretching of roving, count variation within and between bobbins, and the roving doff size etc. all have a significant impact on the final yarn quality. The contribution of speed frame process on yarn quality is discussed below.

7.8.1 Count variation within bobbins

A major source count variation within bobbins is the stretching of roving resulting from undue bobbin speed fluctuation (Garde and Subramanian, 1978). The count variation in the layers of a roving package draws it down to the yarn stage in the form of within count variation. Lower inter-fibre cohesion, due to a high degree of fibre parallelisation, may also lead to stretching at creels. Furthermore, faulty building mechanisms often results in roving tension fluctuations during a complete build, with the front row bobbins most significantly affected by the uncontrolled roving stretch, due to the extended unsupported roving length between the front nip and flyer top.

7.8.2 Count variation between bobbins

The count variation between bobbins is mainly ascribed to the difference in average roving hank. The difference in average roving hank arises due to the following facts (Garde and Subramanian, 1978):

- draft differences between draw frames and speed frames,
- excessive hank difference between front and back row of speed frame bobbins,
- significant trends in hank over a roving bobbin caused by irregular control of bobbin speed.

These factors should be properly controlled in order to minimise the count variation between bobbins. If the differences are consistent but under control, such as hanks of front and back row bobbins, the bobbins can be creeled on separate ring frames with suitable drafts.

7.8.3 Yarn irregularity and imperfections

The speed frame drafting system has substantial effect on yarn evenness and imperfections (Slater, 1986). The improved drafting system with optimum roller settings and top roller pressure reduces roving irregularity, helping to improve yarn unevenness and imperfections.

7.8.4 Yarn faults

Speed frame processing can generate yarn faults in different ways (Garde and Subramanian, 1978). Uneven stretching of the rove is one source of the generation of long thin faults of H and I type, whilst a major source of E, F and G faults is the breakage of roving. Every piecing after the break results in either an end breakage at the ring frame, or a long thick fault. It is therefore imperative to control breakage rate at the speed frame. In addition, a lot of fly is generated at the speed frame, and to control short, thick yarn faults it must be ensured that the generated fly does not enter the rove.

7.8.5 Yarn hairiness

Yarn hairiness is influenced by roving twist, package size, location of the roving bobbin and the roving hank (Chattopadhyay, 2002). Increasing the twist increases the compactness of the roving and thereby reduces the fibre spread under the top roller pressure. In addition, yarn hairiness decreases with the reduction of the roving doff size. However, yarn spun from the front row of roving bobbins is more hairy than that spun from back row. For a constant linear density of yarn, hairiness increases with coarser roving due to the fact that the higher front zone draft at the ring frame causes a higher spread of fibres at the front roller nip. Furthermore, the use of condensers on the speed frame drafting unit arrests the marginal fibres, accounting for the reduction in yarn hairiness.

7.8.6 Common defects and their causes

Various types of common defects in speed frame processing and their causes are discussed below (Ratnam *et al.* 1994).

Causes of ratching or stretching of roving are:

- inappropriate choice of winding-on or ratchet wheel,
- erroneous starting position of the cone drum belt,
- incorrect selection of the initial and build bobbin layers,
- variation in the bare bobbin diameter,
- faulty shifting of the cone drum belt.

All the aforesaid points are taken care of in modern speed frame eliminating the possibility of stretching of roving.

Causes of roving breaks are:

- erroneous choice of creel draft,
- sliver entanglement at the feed,
- incorrect piecing during the back process,
- inappropriately wide back zone setting,

- loose or broken top and bottom aprons,
- excessive break draft,
- inadequate top roller pressure,
- vibration in the flyers,
- use of excessively narrow spacers,
- excessively deep meshing of the draft gears,
- broken or damaged teeth in the draft gears.

Causes of roller lapping are:

- excessive spindle speed and inappropriately high drafts,
- damaged surfaces in the top roller cots,
- use of varnishes on the top roller cots,
- damaged aprons or condenser guides,
- improper choice of spacers,
- excessively wide setting of the back zone,
- improper ambient conditions in the department.

Causes of soft bobbins are:

- inaccurate reduction of the bobbin surface speed,
- faulty lifter wheel resulting in fewer layers,
- faulty ratchet wheel or spur wheel at the builder mechanism,
- incorrect positioning of the bobbin wheel.

Causes of slough off are:

- improperly functioning builder mechanism during bobbin rail reversal,
- faulty selection of the density lifter wheel setting,
- incorrect taper formation,
- higher bobbin weight than the manufacturer's recommendation.

Causes of slubs are:

- excessive roving breaks,
- waste accumulation at creels, clearers and flyers,
- incorrect choice of spacer,
- inappropriately low break draft,
- restrictive back zone setting,
- absence of a positively driven top roller clearer,
- excessive lashing during end breaks.

Causes of slack ends are:

- slipping cone drum belt,
- sliding of bobbin rail,
- clogging of long collars with waste and dirt.

7.9 Conclusions and future trends

With the advancement of computer and information technologies, the process control activities applied to spinning are becoming increasingly sophisticated. Some of the modern versions of autoleveller draw frames are equipped with fuzzy speed control, for example, whilst the cone drum-based building mechanism of the speed frame has been replaced with a microprocessor-based control. As well as process control, the very creation of the machinery is benefitting from technological advancement. The Rieter high speed combers for example are designed in accordance with computer aided process development.

In a customer dominated market, the quality requirement of textile products is becoming ever more stringent, making the role of process control increasingly crucial. It is envisaged that the role of computers in spinning will only increase, and that future process control activities will become heavily reliant on online and automated systems. However, the textile process engineer will still have a significant role to play, and therefore a thorough knowledge of process control is of paramount importance to cope with the impending challenges.

7.10 Acknowledgement

The authors are grateful to Indranil Ghosh, R & D Manager, Vardhman Group, for his valuable comments and suggestions during the preparation of this chapter. The authors are also thankful to Rieter AG, Switzerland, for extending the permission to reprint some of the figures (7.1, 7.2 and 7.4) and table (7.1).

7.11 References

Bar H P, Furter R and Harzenmoser I (1990), Influence of autoleveling and on-line quality control on the quality of ring yarns, *Textil Praxis*, 45, 362.

Chattopadhyay R (2002), *Advances in Technology of Yarn Production*, New Delhi, NCUTE Publications.

Garde A R and Subramanian T A (1978), *Process Control in Cotton Spinning*, 2nd Ed., Ahmedabad, ATIRA.

Klein W (1987), *Short Staple Spinning Series, Vol 3: A practical Guide to Combing and Drawing*, Manchester, The Textile Institute.

Morton W E and Nield R (1953), The effect of lap preparation on waste extraction at the cotton comber, *J. Text. Inst.*, 44, T317–T334.

Operating instruction of RSB 851 drawframe, Lakshmi machine works, Coimbatore, India.

Ratnam T V, Seshan K N, Chellamani K P and Karthikeyan S (1994), *Quality Control in Spinning*, Coimbatore, SITRA.

Salhotra K R (2004), *Spinning of Manmade and Blends on Cotton System*, Mumbai, The Textile Association (India).

Slater K (1986), *Textile Progress: Yarn Evenness*, Manchester, The Textile Institute

van der Sluijs M H J and Hunter L (1999), *Textile Progress: Neps in Cotton Lint*, Manchester, The Textile Institute.

8

Process control in ring and rotor spinning

P. K. MAJUMDAR, Government College of Engineering & Textile Technology, Serampore, India

DOI: 10.1533/9780857095633.2.191

Abstract: Control of process parameters is extremely important in ring and rotor spinning to ensure production of yarns with the desired quality at the highest possible speed and with minimum waste. The chapter discusses the importance of different key process parameters and their influence on process performance and quality of yarns. It also discusses the effect of different factors on those key process parameters.

Key words: spinning tension, end breakage, twist loss, spinning triangle.

8.1 Introduction

Ring and rotor spinning are the final stages of yarn formation. Any defect in processing in this final stage may lead to the formation of defective yarns and also higher end breakage. It is therefore extremely important to keep the spinning process under control to deliver yarns with the desired quality at the highest possible speed and with minimum waste. The key parameters which need to be under control for both ring and rotor spinning are:

- spinning tension,
- end breakage,
- minimising waste.

Effective control of these parameters will ensure both yarn quality and the productivity and economy of the spinning process.

8.1.1 Understanding spinning tension in ring spinning

Control of spinning tension on the ring frame is extremely important in controlling yarn characteristics as well as spinning performance. If tension is too high or too low, it will cause higher end breakage leading to poor productivity, higher wastage and ultimately making the process uneconomical. It is important to understand both the theory involved and the practical issues involved in controlling spinning tension.

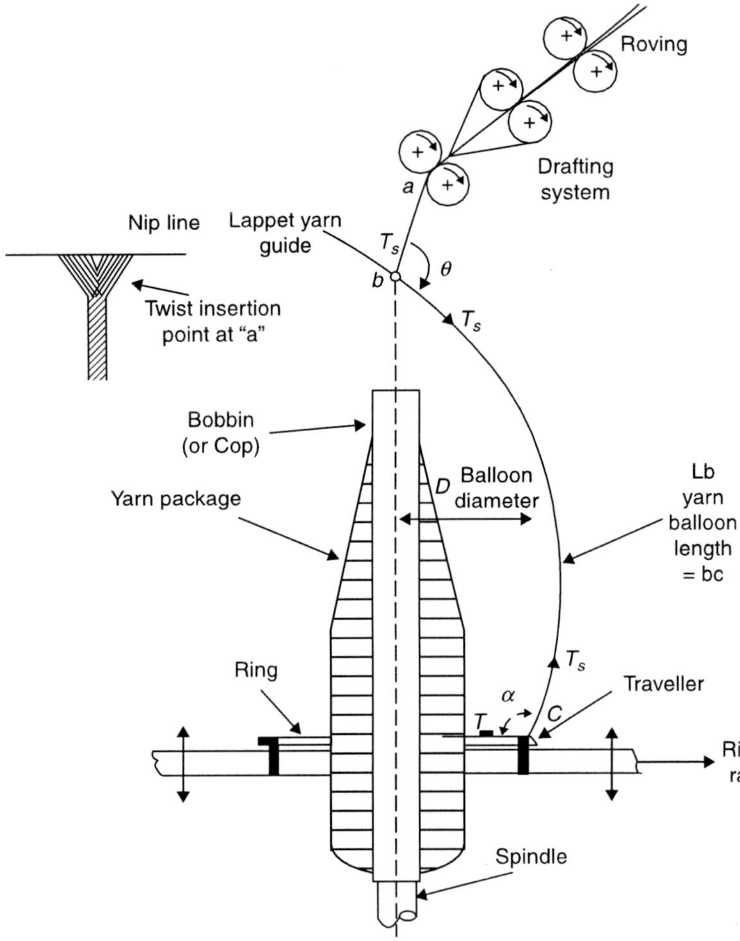

8.1 Passage of yarn in ring spinning.

Tension in different zones

Figure 8.1 illustrates a typical ring spinning system. In ring spinning, the traveller and the balloon length, L_b, must move around a common axis in order to wind the yarn onto the spinning bobbin. Work must therefore be performed against the frictional force of the ring on the traveller, the traveller on the yarn, and against the air drag on the traveller and balloon length. The combination of these forces causes tension in the yarn. This work is additional to the effort needed to overcome the friction of the spindle bearings and air drag on the forming yarn package.

Tension develops in the yarn in three zones:

- the winding zone,
- the balloon zone,
- the yarn formation zone.

In the winding zone, the length of yarn between traveller and forming package develops winding tension, T_W. In the balloon zone, tension occurs between the traveller and lappet guide (often referred to as the *balloon tension*). This tension, at a given point on the balloon length, varies with amplitude (i.e. the radius of the point) measured from the common axis. The yarn formation zone is the area between the pigtail lappet guide and the front rollers of the drafting system, where yarn tension is referred to as *spinning tension*, T_S.

Theoretical models of ring spinning

A proper understanding of the theory of ring spinning is essential for effective control of the process. The yarn rotates around the inner circumference of the lappet guide at almost the same speed as the traveller. The radius of the lappet guide is sufficiently small for any central forces generated to be inconsequential. The motion of yarn between the lappet and front drafting rollers is therefore principally related to the velocity along its length, that is the thread line velocity. Consequently, the forces of interest are the air drag along the yarn length, the tension at the lappet guide, and the resistance to bending around the guide. The air drag is proportional to the square of the thread line velocity, but this velocity is usually small in comparison to the rotational velocity of the yarn. The force caused by air drag along the yarn length is thus assumed to be negligible. The bending resistance, due to the flexural rigidity, of the yarn is many times smaller than T_0 and requires no further consideration. T_0 is therefore the only effective force governing T_s and, as a result, analysis of the forces present in ring spinning is usually concerned with the remaining two zones.

In steady running conditions, the traveller presses against the bottom of the internal flange of the ring. The forces acting on the traveller at the point of contact with the ring are shown in Fig. 8.2, where

C = centripetal force $[MR\ \omega^2]$ needed to keep the traveller circulating around the ring
μ = friction coefficient between ring and traveller
η = the yarn-lappet coefficient of friction
θ and α = the angles shown in Fig. 8.1
P = reaction force of the ring and traveller
R = ring radius
Mg = traveller weight

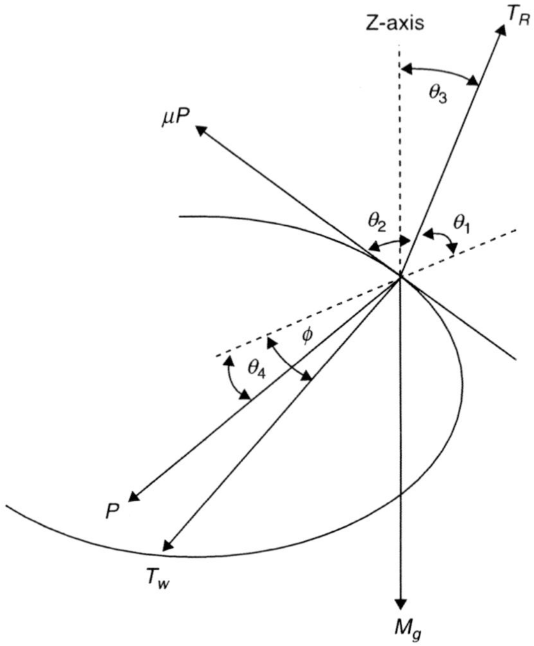

8.2 Forces acting on the traveller.

ω = angular velocity of the traveller
$\theta_1 \ldots. \theta_4$ and ϕ = angles indicated in Fig. 8.2
T_S = the spinning tension
T_0, T_R = the tensions in the balloon length at the lappet guide and at the ring and traveller, respectively
T_w = the winding tension

Strictly, T_w is not the true winding tension. This is because centripetal, Coriolis, and air-drag forces act on the mass of the yarn length from the traveller to the ring bobbin. It can be reasoned that the latter two effects negate each other and are therefore of little relevance. The effect of the centripetal force is to change the path of the yarn; rather than forming a tangent from the package to the traveller, it becomes a curve. The change is a minor one, however, and for the sake of simplicity this centripetal force can also be ignored.

The causes of tension in the balloon zone relate to the forces acting on each elemental length, where r denotes the distance of an element of length from the z-axis. The forces present are as follows:

- centripetal and Coriolis forces associated with yarn motion,
- air-drag forces opposing yarn motion,

- the weight of the element acting vertically downward,
- the resistance to bending due to the flexural rigidity of the element.

Early references to the theory of ring spinning can be found in papers by Ludicke[1] in 1881 and by Escher[2] in 1883. The theory has since been augmented and developed by various researchers[3–14] and, based on these analyses, DeBarr and Catling[15] published a monograph on the subject in 1965. Nerli[16] attempted to develop a computer-based analysis for calculating forces acting on the balloon and the balloon shape, but this attempt was limited by its dimensional formulation. Batra[17–19] summarised theoretical contributions to ring spinning theory and introduced numerical solutions for steady-state balloon shapes.

Lisini[20] proposed a non-stationary mathematical model of the ring spinning process. Fraser[21,22] investigated the characteristics of air drag, such as the effects of air drag magnitude on balloon shape, tension and the bifurcation behaviours of the solution. Stump[23] developed transient solutions of the nonlinear time dependent partial differential equations and studied the stability of representative quasi-stationary balloons. Zhu[24,25] studied the steady-state response and stability of ballooning strings under the influence of air drag and used the Hopf-like bifurcation to predict the limit cycles of ballooning strings. Fan[26] calculated the natural frequencies and mode shapes of a single loop balloon. Tang[27–29] estimated the air-drag coefficients on ballooning cotton and wool yarns. Yarn tension and balloon shape were obtained by experimental data.

Rong Yin and Hong-bo Gu[30] established a more realistic model by incorporating yarn as a linear elastic component, which has been simplified by most investigators. The equations of motion were derived by following Fraser's approach,[31] and numerical solutions to these equations for small elastic parameter values presented. MATLAB software can be used to develop a program corresponding to the model. The effect of yarn elasticity in a quasi-stationary balloon has been investigated. This model can be used to simulate balloon shape and predict yarn tension under given spinning conditions for a small strain value, and to facilitate further research into reducing yarn breakage in ring spinning.

8.2 Factors affecting spinning tension in ring spinning

This section covers the effect of spindle speed, traveller mass, traveller/ring friction, ring diameter and balloon height.

8.2.1 The effect of spindle speed

Yarn tensions are directly proportional to the square of the traveller or spindle speed. When the spindle speed is increased, the friction work between ring and traveller (and hence the build-up) increases as the 3rd power revolutions-per-minute of the spindle. If the spindle speed is too high then the traveller sustains thermal damage and fails. This speed restriction is felt particularly when spinning high-strength cotton yarns. When spinning 100% cotton, fibre dust from the cotton acts as a lubricant, but not all cottons generate the same amount of lubricating film. Traveller speed is not a limiting factor in cotton combed yarn intended for knitting. Because the yarn TPI is lower, the yarn strand is not strong enough, and so the yarn tension becomes the limiting factor. Yarn strength is not a problem for polyester/cotton blends and cotton weaving counts, and the limiting factor is traveller speed. For a ring diameter of 40 mm, spindle speeds of up to 19 500 rpm should not be a problem. Rings like Titan (from Braecker), NCN (bergosesia), etc. are able to meet the requirements.

For spindle speeds greater than 20 000 rpm, ORBIT rings or SU-RINGS should be used. Because the area of contact is larger with these rings, with higher speeds and pressure, the heat produced can be easily dissipated. Normal rings and traveller profiles cannot produce good quality yarn if run at speeds greater than 20 000 rpm.

ORBIT rings are very useful in working 100% polyester at high spindle speeds, where high tension generates a lot of heat between the ring and the traveller. Because the yarn strength of polyester is very high, heat dissipation is the only limiting factor. ORBIT rings with higher area of contact will be able to run well at higher spindle speeds.

The maximum spindle speed for a given yarn tension, or the minimum tension for a given spindle speed, can be determined using the following equation[32]:

$$(nH)_{max} = 94.6b\sqrt{\left(\frac{T_0}{N}\right)}$$

where n is the traveller speed in thousands of revolutions-per-minute, T_0 is in g-wt, N is the count of the yarn in tex, and H is the balloon height in cm.

$$b = \left(\frac{H}{P}\right)_{max} \quad \text{where} \quad P, \text{cm} = 94.6\sqrt{\left(\frac{T_0}{Nn^2}\right)}$$

A single stable balloon cannot form if the tension is less than the value given in the equations. If, at a given spindle speed, the tension is reduced or the balloon height is increased, at a certain point the balloon collapses and forms a neck. The spinning tension decreases as the diameter of the package increases during the build of a bobbin, so balloon collapse is most likely to occur when the yarn is winding onto the full diameter of the package. Collapse is more likely with longer balloons, even though tension increases with balloon height.

The effect of spindle speed on yarn properties has been observed by many researchers. Anbarasan[33] reported that the benefits of advanced speed systems were subsequently extended by the introduction of variable speed systems such as the inverter drive system. This system offers the possibility of selective and continuous speed adjustment for the complete cop build. Optimum spindle speeds can thereby be achieved so that an almost constant spinning tension is maintained throughout the cop build.

8.2.2 The effect of traveller mass

Winding tension increases with traveller mass, and heavier yarns require a greater centripetal force to keep them rotating. Traveller mass is used as a variable to increase tension and generate higher centripetal force for heavier yarns. The traveller mass is usually chosen according to the linear density of the yarn being spun. In fact, all the implications for winding tension apply to tension in the yarn forming zone.

Traveller mass controls winding tension, which in turn produces tension in the lower balloon at the other side of the traveller. Because of yarn/traveller friction, the balloon tension is rather less than half the winding tension. The balloon tension in turn gives rise to the spinning tension between the lappet and the front roller nip; this depends on the winding tension and on the centrifugal force and air drag on the balloon. Within practical limits the coefficient of traveller/ring friction varies inversely with traveller mass, directly with ring diameter, and increases to a maximum at a particular rotational speed.

Heavier travellers are generally used for thicker counts, and the maximum traveller mass consistent with a good spin is chosen, to give an acceptable end breakage rate. As a last resort it may be necessary to reduce the spindle speed in order to control the end breakage rate.

The minimum traveller mass is determined by the need to avoid balloon collapse when winding onto the maximum package diameter and maximum balloon length. The maximum mass is limited by the need to avoid excessive tension, which causes a high end breakage rate. A range of traveller masses can be employed for a given set of conditions, and this range becomes narrower as the spindle speed increases.

The minimum tension required to avoid balloon collapse is proportional to (spindle speed)^2, and the tension produced by a given traveller is proportional to f(spindle speed)^2, where f is a coefficient that depends on the ring/traveller coefficient of friction, yarn/traveller coefficient of friction, and the angle at which yarn is wrapped around the traveller. The minimum traveller mass would be independent of spindle speed if f remained constant but in practice, although f is constant at low spindle speeds, it decreases at higher speeds. A slightly greater mass is therefore required at higher spindle speeds.

Minimum traveller mass is proportional to the yarn count and to (balloon height)^2, but is inversely proportional to ring diameter. If the traveller is too light then the balloon diameter becomes excessive, causing the yarn to beat against the separators. It can also cause traveller 'flutter' or 'chatter', a vibration with audible impacts between the traveller and the ring, which results in higher peak tensions and increased end breakages as well as traveller and ring wear.

If the traveller mass is greater than the minimum then a firmer package of increased capacity can be produced, and the twist differential across the traveller increased. This may increase yarn strength by parallelising the fibres as they are gathered from the delivery roller nip, and may reduce the end-breakage rate, though a heavier traveller will also increase power consumption.

It is generally assumed that in cotton spinning the tension ratio (mean individual thread breaking load)/(mean spinning tension) must not be less than 14 in order to achieve an acceptable end-breakage rate. Higher tensions with ratios of 4 or 5 are acceptable for polyester/cotton blends, and a ratio of 12 for worsted spinning. Yarn emerging from the front roller nip is only partially twisted and its strength may be only about one-third of the ultimate individual thread breaking load.

The minimum traveller mass for a particular ring frame is that which is just sufficient to prevent balloon collapse during winding on the shoulder of the cop bottom, that is, during winding at the maximum balloon height onto the maximum package diameter. Since both the minimum balloon tension and the traveller-ring frictional force are approximately proportional to the square of the spindle speed, the mass of the lightest traveller that can be used is largely independent of the spindle speed, n. The minimum mass is, however:

- inversely proportional to the ring diameter, $2R$,
- proportional to the count, N(tex),
- and proportional to the square of the balloon height, H

where $2R$ and H are in inches and N is in tex. An approximate value of the minimum traveller weight, M, in grammes is given by the expression

$$M = \frac{0.000048H^2N}{2R}$$

8.2.3 The effect of traveller/ring friction

Rings and travellers are the dominant elements in the ring spinning process. The traveller has, among other duties, the function of regulating the spinning tension. The spinning tension must be high enough to keep the thread balloon stable, but not so high as to cause yarn breakage. Winding tension increases with the frictional coefficient between the ring and the traveller (μ_{RT}). It should be noted that the spinning tension in the yarn is directly related to winding tension, so any factor that increases winding tension will also increase tension in the yarn during ring spinning.

Reducing the friction coefficient to the lowest possible level is key to the success of ring spinning. At this point perfectly balanced spinning geometry will have been achieved, meaning that the spinning tension is on a constant balanced level. Friction will be reduced if the rings and travellers establish a symbiosis. The ring/traveller system in short staple spinning operates dry, but at high speeds normal friction systems only work with additional lubricants. The fibres protruding from the yarn body between ring and traveller are crushed and form a steadily regenerating lubricating film in a process known as fibre lubrication. The resulting friction coefficient varies according to the fibre, which may be either dry, wax-containing cotton, or synthetic with added softening agents. The coefficient of friction (μ) can vary in extreme cases from 0.08 to 0.12, explaining why different traveller weights must sometimes be applied in otherwise identical spinning conditions. In order to maintain the same friction or spinning tension with different coefficients of friction, different traveller weights must be used. Dry cotton has a high friction coefficient, while manmade fibres have low to medium coefficients of friction, depending upon the manufacturer.

It has been shown by Catling[34] that the effective coefficient of friction between the ring and the traveller is a very important factor in the balloon stability equation, and largely determines the yarn tension. It was found that traveller speeds are, in practice, limited by the fact that at high speeds an abrupt and unpredictable increase in the friction coefficient is apt to occur. In some circumstances this may merely increase the winding tension (and thereby the number of thread breakages) but more usually the increase in friction is accompanied by overheating of the traveller and very rapid wear of both ring and traveller.

8.2.4 The effect of ring diameter

Package capacity is approximately proportional to (ring diameter)^2, so a large diameter ring may be thought desirable, but increasing the ring diameter also increases yarn tension. When thicker counts are being spun, frames with larger rings are fitted because the yarn is strong enough to withstand the increased tension and a larger package size is desirable. However, the ring diameter is restricted by yarn tension and the minimum angle of lead, as well as other factors such as power consumption, spindle rpm, and traveller speed limitations.

$$\text{Max. spindle speed} \propto \frac{1}{\sqrt{\text{Ring diameter}}}$$

$$\text{Max. linear traveller speed} \propto \sqrt{\text{Ring diameter}}$$

This is why we do not see ring spinning machines with very large rings and very small bobbins (tubes). Large ring radii and yarn packages are used for coarser and stronger yarns, but for finer yarns both the ring radius and package size must be smaller. Most worsted yarns are spun using rings between 45 and 75 mm in diameter at spindle rotational speeds between 7000 and 12 000 rpm. Short staple yarns are spun with smaller rings, but at about twice the spindle speeds used for worsted yarns.

8.2.5 The effect of balloon height

The balloon does not behave exactly like a tensioned string in ring spinning, because of its relatively large radius and the influence of traveller and ring/traveller friction. A stable balloon system can be formed for many combinations of balloon height (H), ring radius (R), package radius (R_1), spindle speed, yarn count and yarn tension. This is because the length of yarn forming the balloon is not fixed even though the axial balloon height is fixed; the yarn length is self-adjusting so that the balloon shape permits the new horizontal force on each element in the yarn to provide the necessary centripetal force on that element.

For a given value of R and h there is a limiting value of P (as defined in 8.2.1) for which a node can be formed. The limiting value of P decreases as the ratio R/H or R_1/H increases; the maximum theoretical value of H/P is π for a ring of infinitely small radius.

To avoid neck formation R/P and H/P must be small and so P must be large. For a given yarn count, this means that either T_0 must be large or ω must be small, both of which are undesirable.

A neck will gradually form in the balloon if unfavourable conditions arise, such as increase in H or a reduction in traveller mass. Balloon collapse occurs when the neck makes significant contact with the package, thereby causing a breakdown in the spinning process. Collapse is most likely to occur when yarn is winding onto the maximum package diameter with the longest balloon length, that is, 58 when H is at a maximum and P is at a minimum.

By restricting the maximum balloon radius, and thereby reducing air drag on the balloon, control rings allow the creation of a longer balloon without a neck, and with a much lower yarn tension than would otherwise be possible. Each section of the total balloon then has a smaller value of H. H/P is thus effectively reduced and balloon collapse avoided even though the overall balloon length has increased. By stabilising tension (T_0), the benefits from the use of control rings permit larger package length, higher spindle speeds and reduced power consumption.

Yarn tension increases as normalised balloon height increases, particularly when the height is greater than 10 cm. When the ratio of yarn-length-in-balloon to balloon height is greater than 1.2, the yarn tension has a 'lower' value and the effect of balloon height on the tension is very weak.

The influence of balloon control rings is quite considerable, especially at long cops. A reduction of the yarn balloon is advantageous and may even be a prerequisite for optimum performance. If balloon control rings are mounted at the correct distance, with the yarn balloon restricted as long as possible during one lift of the ring rail, then a marked increase in performance is possible. Balloon control rings are used to contain the yarn-loop, by reducing the yarn tension and decreasing the balloon flutter instability.[35] Flutter instability here refers to uncontrolled changes in a ballooning yarn under dynamic forces, including the air drag. Due to the significant variation in balloon length and radius during the bobbin filling process, the optimal location for the balloon control ring is difficult to pinpoint. In order to address this difficulty, this study[35] investigates the variation in the radius of a free balloon, and examines the effect of balloon control rings of various diameters at different locations on yarn tension and balloon flutter stability. The results indicate that the maximum radius of a free balloon and its corresponding position depend not only on the yarn length to balloon height ratio, but also on yarn type and count.

A control ring of suitable radius and position can significantly reduce yarn tension and decrease flutter instability of free single loop balloons. The balloon control rings are usually fixed to, and move in sync with, the ring rail. However, the reported results suggest that, theoretically, a balloon control ring that always remains approximately half way between the yarn-guide and the ring rail during spinning could lead to significant reduction in yarn tension.

8.3 Control of end breakage rate in ring spinning

The end breakage is a critical spinning parameter that not only affects the maximum spindle speed but also indicates the quality of yarn, the mechanical condition of the machines and the quality of raw material. Therefore, control of end breakage rate is the prime requirement for getting better ring frame performance and for achieving higher spindle speed.

8.3.1 Theoretical models

The problem of yarn breakage during spinning is serious and difficult to solve due to the complexity of the breakage mechanisms. Yarn breakage phenomena are complicated because the causes are numerous, including the influences of humidity and temperature, but also because the tension and strength of the yarn always vary. In practice, almost all yarn breakages occur at the exit of the front roller, probably because, at that point, the yarn has not yet been twisted and is still weak. Hori[36] has used simple mathematical models of yarn tension and strength variation to calculate the probability of breakages. He concluded that:

1) Variations in yarn strength and tension, and in the average strength and tension, influence the number of the yarn breakages.
2) Above all, the variations in yarn strength are most influential under normal conditions.

Evenness of the yarn is thus more important than lower tension on spinning frames in reducing yarn breakages.

De Barr[37] also discussed the problem of end breakage in ring spinning in terms of a simple mathematical model. Absolute values of end breakage rate could not be calculated, but the treatment enabled the relative importance of such factors as tension, regularity of yarn, angle of lead, etc., to be assessed. The results, some of which would be applicable to any process in which yarn breakages are involved, show that the most important factor affecting end breakage rates is yarn regularity.

The method of calculating end breakage rates was based upon that described by Hori,[36] but the model used was more realistic, and the treatment extended so as to enable the effect of changes in, for example, angle of lead, to be investigated. The variation of T_T/C with angle of lead, α, has been used to calculate the variation of end breakage rate with angle of lead.

From the parameters of the strength and tension distributions, the probability, p, that in any interval, Z_S, the mean tension mT_1, being greater than the mean strength m_S, was calculated by the following equation:

$$N = \frac{100 \times p \times 3600}{Z_s}$$

Ghosh *et al.*[38] proposed a statistical model for predicting the end breakage rate based on the following assumptions:

I. The yarn is composed of a successive chain of links with different lengths.
II. An end break occurs when the spinning tension exceeds the strength of the yarn in the spinning zone.
III. The strengths of the links of fibre bundle just issuing from the front rollers are independent variables and follow a normal distribution, that is, a link of high strength may immediately follow one of low strength. The strength is constant within a link but varies from link to link. The strength of a link depends on the number of fibres in the link and the coefficient of variation of the strength of all the links. The coefficient of variation of the strength of the links is equal to the coefficient of variation of mass per 8 mm length of the yarn (Uster CV%). This is based on the fact that the variance-length curves of fibre assemblies run practically horizontal for cut lengths between 1 and 10 mm.[39]
IV. The strength of the yarn in the spinning zone is about one-third of the mean strength of the yarn.

The probability that the tension exceeded the strength of the links just passing from the front roller's nip was found by the relationship

$$P = 1 - \Phi(Z)$$

where Z is the standard normal deviation and $\Phi(Z)$ can be expressed

$$\Phi(Z) = \frac{1}{\sqrt{2\pi}} \int_{-Z}^{\infty} e^{-\frac{Z^2}{2}} dz$$

8.3.2 Measuring the end breakage rate

Accurate measurement of the end breakage rate can be used in determining operatives' work allocation, and provide valuable data for quality-control purposes.[15] Unfortunately, end breakage rates are particularly difficult to measure with enough accuracy for the results to be useful. Not only are there systematic variations between apparently similar frames at the same time, and for the same frames at different times, but there are also many chance variations resulting from imperfectly understood and randomly occurring phenomena. Useful results can, however, be obtained if care is taken, and the procedure outlined below will be found adequate where the major systematic variations have been eliminated.

Observation of all end breakages occurring on one side of one ring frame during the build of a complete bobbin may be regarded as a 'unit breakage test'. It is usually convenient for one observer to make several unit tests concurrently by observing all the sides attended by one operative, but it is not practicable to extend this by watching more than one operative at a time. Ends gaited after doffing, and those accidentally gaited by the spinner, should be ignored. Multiple breaks on adjacent spindles often have a single cause and should therefore be noted separately.

To reveal trends with time of day, ambient conditions, part of building cycle, etc., records should be made in consecutive constant time intervals of, say, 15 minutes, and ambient atmospheric conditions should be recorded at not less than hourly intervals. After each complete build-cycle, the results should be summarised for each unit test. Trends relative to ambient conditions or stage of build may be revealed by plotting the mean rate for a number of unit tests during consecutive 15 min intervals.

Before making use of the observations, however, it is desirable to determine the reliability of the results in view of chance variations that inevitably occur. Clearly, the greater the number of unit tests comprising the series of observations, the more reliable the results. However, a quantitative procedure is required for determining the number of unit tests needed to ensure that the average is within a particular range of the true breakage rate for each set of conditions.

Suppose the end breakage rates obtained from h unit tests are $b_1, b_2, b_3, \ldots, b_h$; then the average rate is

$$B = \frac{b_1 + b_2 + b_3 + \cdots + b_h}{h}$$

and the mean of the squares is

$$S^2 = \frac{b_1^2 + b_2^2 + b_3^2 + \cdots + b_h^2}{h}$$

Then, if a further quantity

$$E^2 = \frac{S^2 - B^2}{h - 1}$$

is determined, the true breakage rate lies within the range $B - 2E$ and $B + 2E$, and the number of unit tests required to give a breakage rate reliable to $\pm q\%$ is given by

$$x = 40000\frac{hE^2}{B^2q^2}$$

8.4 Factors affecting end breakage rates in ring spinning

The mechanism of end breakage in ring spinning is a complex phenomenon involving different factors. The major factors influencing end breakage in ring spinning are yarn strength in relation to mean value of peak spinning tensions, yarn irregularity, yarn twist and relative humidity.

8.4.1 The effects of spinning tension and yarn strength

Most end breaks occur close to the front roller nip, where the fibres are being twisted, and so the strength and tension in this region are of the greatest importance. It is important to note that increasing the spinning tension does not necessarily increase the end breakage rate, because the corresponding increase in twist density in the balloon leads to an increase in yarn strength in the critical nip zone which, in some circumstances, may be sufficient to reduce the end breakage rate. The effect is particularly noticeable when spinning tension is increased without any increase in spinning speeds. The balloon diameter is reduced, and thus the angle at which yarn is wrapped around the traveller is increased. As a result, both the twist differential and the tension differential across the traveller are increased, and, over a substantial range, the end breakage rate is reduced by increasing the spinning tension.

At high speeds and with large packages, however, the interaction of these factors gives rise to operating difficulties. To avoid balloon collapse, limiting spinning speed and use of proper traveller weight are extremely important while winding on the largest diameter and with the largest balloon, that is, just on completion of the cop bottom, particularly when spinning is carried out without a balloon control ring.

At this point in the build, when the balloon is at its largest and the air drag (and therefore the balloon tension) is maximised, the angle of wrap of the yarn round the traveller is minimised, so that both the tension and the twist differentials across the traveller are very small. Maximum tension is transmitted to the nip zone at a time when minimum twist potential is available for consolidation of the newly formed yarn, and consequently end breakages are common.

8.4.2 The effect of yarn irregularities

The weakest part of a forming yarn, where most end breakages occur during ring spinning, is at the point of twist insertion in the spinning triangle. Three factors are therefore of importance[40]:

(1) The number of fibres in the triangle and the variation of this number
(2) The propagation of twist to the apex of the triangle
(3) The mean tension and tension fluctuation.

Clearly, where there are a large number of fibres in the cross section of the forming yarn, the yarn will be more able to withstand the spinning tension and tension fluctuations, provided that the mean spinning tension is kept well below the breaking load of the yarn (typically 30% below mean yarn strength). End breakage problems will arise when the number of fibres in the cross section of the fibre ribbon varies significantly and/or the peak value of tension fluctuation is too high. The variation of the number of fibres in the cross section causes thin and thick places in the fibre ribbon. As these pass through the twist insertion point at the apex of the spinning triangle, the thin places are more easily twisted than thick places, so thin parts of the ribbon tend to have more twist than thicker parts. If a very thin part of the ribbon becomes over twisted and weak, the yarn will be susceptible to peak tension fluctuations.

It is a well-established fact that yarn irregularity cannot be eliminated even under the best processing conditions. But there are many conditions that increase the variability of lengths of yarn, and which may lead to excessive spinning breakage rates.

The importance of irregularities in yarn strength in determining the end breakage rate has been noted by Hori,[36] and it is clear from findings by De Barr[37] that the end breakage rate is more sensitive to variations in irregularity of yarn strength than to any other factor. The effect is so marked that the range between the lowest attainable irregularity and the highest irregularity that can be tolerated is very small.

8.4.3 The effect of twist in the yarn

Most end breaks in ring spinning occur in the material just issuing from the rollers, and the end breakage rate is very dependent upon the twist in the yarn between thread-guide and the roller nip.[41] The twist in this zone is always be greater than the equilibrium twist in the yarn on the package, so that the strength of the yarn here is normally greater than might be expected from the known spindle and delivery speeds. Furthermore, although the twist in the yarn in this zone decreases with the twist factor, this reduction will be less than the decrease in twist in the yarn on the package. The end breakage rate in soft-twisted yarns should not be as much greater than that in hard-twisted yarns as might otherwise have been expected. Indeed, the increased twist in the roller nip – thread-guide zone may well be very important in the spinning of soft-twisted yarns.

A reduction in end breakage rate can be achieved by using a traveller that presents a considerable obstacle to the passage of twist across it, but this reduction may be at the expense of yarn quality.

The angle at which yarn is wrapped around the thread-guide is also a factor determining the amount of twist in the yarn issuing from the roller nip. In the usual arrangement, the thread-guide is positioned substantially on the axis of rotation of the spindle, and the spindle-roller geometry is such that the yarn from the roller nip to the thread-guide makes an angle of about 30° with the spindle-axis. This suggests an average angle of wrap of 30° but, under running conditions, the actual angle of wrap will vary cyclically during each revolution of the balloon from $30° + \theta_0$ to $30° - \theta_0$. θ_0 is the angle of inclination to the spindle-axis of a yarn element leaving the thread-guide and entering the balloon.

Under most spinning conditions, the instantaneous angle of wrap becomes negligibly small once per revolution and there is effectively a direct path for twist and tension to be transmitted from the yarn balloon to the nip zone. In a system known as vertical spinning, the roller nip is directly above the thread-guide, so that the angle of wrap of the yarn around the thread-guide does not vary appreciably during each revolution of the balloon. It might be expected that the difference between normal and vertical spinning would greatly influence the end breakage rate, but this does not seem to be the case. It may therefore be inferred that the capstan (or angle-of-wrap effect) of the passage of the yarn over the thread-guide is equally effective as a twist barrier and as a tension barrier.

8.4.4 The effect of temperature and humidity

The temperature and humidity of the atmosphere in the mill are important factors in controlling breakage rate.[37] While no published data relating to the effect of temperature is available, common experience suggests that low temperatures should be avoided. Some research into humidity has been carried out at the Shirley Institute,[37] where it was found that effects in the card room and spinning room were in one respect dissimilar. In the card room it was found that very low humidity, such as is experienced in frosty weather, led to excessive drafting irregularity, especially at the draw frame, which would give rise to higher spinning breakage rates. In the spinning room the breakage rates per 100 spindles per hour were 18, 29 and 28 for percentage relative humidity values of 20, 45 and 65, indicating that very low humidity may be beneficial in reducing spinning breakage rates.

8.5 Control of fly generation and twist variations in ring spinning

Fly generation and twist variations are the two disturbing but unavoidable occurrences in ring spinning. Since it is impossible to prevent the fly from being released and avoid twist variation as it is inherent to the twisting mechanism, exercising control over them will ensure trouble free spinning.

8.5.1 The causes of fly generation

In conventional spinning, it is impossible to collect all the fibres emerging from the front rollers into the forming yarn. Some fibres escape as they emerge from the front rollers, while others are thrown out later by centrifugal acceleration. Many short fibres are lost as fly during the processing of staple fibres on spinning machines, and a considerable quantity of fibre debris and dust is released. Fly and dust are deposited on machine components or are continuously being whipped up and around by rotating and circulating devices such as spindles, drums and drive wheels. They have therefore always caused significant disturbance to service and maintenance, as well as diminishing quality. This problem has intensified further with high production speeds and high drafts. In ring spinning machines, the most fly and dust are released in the main drafting zone and the spinning triangle (up to 85%), while the balloon and travellers account for most of the remainder.

8.5.2 The factors affecting fly generation

For exercising better control over fly generation it is essential to know about the influencing factors. While drafting zone and the spinning triangle are the major sources of fly generation, relative humidity also plays a major role to control it.

The effect of the spinning triangle

During ring spinning, the fibres on the two edges of the spinning triangle must be strongly deflected so that they are bound into the yarn at the convergence point. The deflection is higher with a smaller triangle. Not all fibres are bound into the yarn, particularly those with high rigidity and low cohesion with neighbouring fibres. The end breakage rate also influences the waste and fly. The spinning triangle and spinning angle significantly affect the yarn breakage, because the long and narrow width implies a long weak point and hence causes more end breakages. However, a resulting advantage of the small triangle width is that the edge fibres are better bound in the yarn, which gives smoother (less hairy) yarns and hence less fly generation. According to Klein,[42] when the spinning triangle is short, the fibres from the edges must be strongly deflected if they are to be bound into the yarn structure. The quantity of fly waste can be minimised by ensuring that the fibres emerge from the delivery rollers in as narrow a ribbon as possible, and spinning with higher drafts from thicker rovings tends to increase the amount of fly waste.[43] With the development of compact spinning systems, where the spinning triangle is practically eliminated, the problem of fly generation is greatly reduced.

The effect of humidity

Atmospheric conditions influence the rate at which machines can be worked. In a very dry atmosphere, the electrical charges generated by friction, if not conducted away, make controlling the yarn more difficult. Dry conditions create static electricity, and unfavourable fibre repulsions occur, causing the fibres to stray from the normal passage. Low humidity also causes fibres to be stiff, resulting in higher fly generation. Correct relative humidity (RH) ensures that yarn can be worked with minimal risk of breakage, reduces static electricity due to the greater electrical conductivity of the yarn, and lowers the amount of fly liberated. If the humidity is too high, static electricity will be doused, causing stickiness. The fibre mass clings together firmly, which makes uniform drafting difficult along with roller laps and piecing conditions. Generally 55–60% RH is maintained for coarser counts (20–40s Ne). For finer counts (60–80 s Ne) a slightly lower range of 50–55%, is maintained.

8.5.3 Mechanisms of twist propagation

Although twist is actually inserted into the region between the thread-guide and the traveller, the yarn between the front roller nip and the thread-guide, and that wound onto the package, is also twisted. Twist must therefore pass both the thread-guide and the traveller. There is, however, an important difference in that the twist passes the traveller in the same direction as the yarn, while across the thread-guide the yarn and twist are moving in opposite directions.

The thread-guide and traveller offer frictional resistance to the passage of yarn, and the tension in the yarn on the downstream side of either must be greater than on the upstream side. In a similar way, the thread-guide and traveller offer resistance to the passage of twist, and the torsional stress in the yarn on the downstream side (in the direction of twist propagation) must be less than on the upstream side. The twist (turns per inch, say) in the yarn in the balloon must therefore be greater than that in the yarn between front roller and thread-guide, or in the yarn as it is wound onto the package. Under steady-state conditions, however, the rate at which the twist passes the traveller must be equal to the rate at which twist is inserted in the balloon zone. It follows that the twist in the yarn in the balloon must be greater than calculated.

The distribution of twist between front roller nip and package was studied by DeBarr and Catling.[44] Grandrelle or mock grandrelle yarns were photographed during spinning using the technique described by Axson,[45] and the observations confirmed the above predictions. The conclusion of De Barr and Catling does not seem to have been substantiated by direct measurement of twist in the roller nip – thread-guide zone.

Kanai and others[46] obtained data on twist in the roller nip–thread-guide zone by direct measurement and concluded that the turns per inch of a yarn between the nip point of rollers and the traveller has been shown to exceed that of a yarn on the bobbin. The twist density of a yarn above the snail wire has been found to be 15% greater than in the yarn below it. On the other hand, Wegener and Landwehrkamp[47] found that there was a reduction of about 22% in the twist at the balloon control ring, and another reduction of about 19% at the thread guide. In worsted spinning, Gessner[48] found that without a balloon control ring, the twist in the balloon yarn is 123% and 107% below and above the control ring, respectively, and the twist in the roller nip – thread-guide yarn is 78% of the bobbin yarn twist, contradicting the findings of De Barr and Catling. Wegener and Landwehrkamp did not, however, mention whether these results applied to all stages of the doff and to the entire chase cycle. Subramanian and others[49] used a simple method for measuring the twist density in the yarn between the roller nip and the thread-guide of a ring spinning frame, which was found to be reliable in comparison to results obtained by high speed photography. The method consisted of plucking the yarn passing between the roller nip and thread guide with the fingers and counting the turns.

Variations in twist flow over a doff

As the yarn package builds up, the traveller adjusts its rotational speed automatically, leading to variation in the twist level within yarn packages. However, this variation is very small and, taking into account the twist added during unwinding, the effect of traveller speed change (as cop build-up) is practically insignificant.

The variation in twist density over a doff has been studied by Subramanian and others,[49] by taking a large number of readings from the start to the finish of the doff with the ring rail alternately at the nose and at the shoulder of the chase. The twist in the spinning zone is seen to be, on the whole, higher than the mechanical twist when the ring rail is at the nose of the chase, and always lower than the mechanical twist when the ring rail is at the shoulder of the chase.

Variations in twist flow within a chase

With conventional ring spinning frames, the balloon height and the length of yarn in the balloon vary cyclically according to the type of build being used. Since the twist in the yarn in balloon is greater than that in the yarn wound onto the package, if the length of the yarn in the balloon is changing, the twist in the yarn being wound onto the package is also changing. The magnitude of this twist variation is not dependent upon the change in balloon height but upon the rate of change.

With a roving built package, where the changes in tension accompanying changes in balloon height are small, the twist in the yarn on the package is actually greater when the ring rail is ascending than when it is descending. With cop-build packages, however, tension changes are significant and cause the yarn twist in the balloon to increase as the ring rail is ascending. Variations in the yarn twist on the package throughout the chase depend upon the relative magnitude of the tension and balloon height effects. In general, however, twist is expected to vary during the chase. If the twist density of the yarn in the spinning zone at the shoulder of the chase is less than the mechanical twist, a question arises as to whether the twist density in this zone is only momentarily low just when the ring rail is at the shoulder.

In order to clarify this point, Subramanian *et al.*[49] measured twist density in the yarn in the spinning zone over a doff at four different chase positions. They observed that, over the chase, a periodic variation occurs in the twist density in the yarn in the spinning zone. Starting with a very low value at the shoulder, the twist gradually increases till the nose is reached. It then falls to a low value at the shoulder, so that at any intermediate position of the chase, the value is higher when the ring rail is coming down than when it is going up.

8.5.4 The factors affecting twist variation

Although twist variation is inherent to the mechanism by which twist is being imparted in ring spinning, proper selection of spindle speed, traveller weight and drafting angle will help to provide effective control over it.

The effect of spindle speed

The effect of spindle speed on twist variation has also been studied by Subramanian and others.[49] The twist density in the spinning zone was measured at spindle speeds of 6000, 7000 and 9000 rpm. For the same traveller number and yarn count, the twist density in the spinning zone, when winding at the shoulder of the chase, increases as the spindle speed decreases. When winding at the nose, the twist density remains almost unaltered, so that the difference between nose and shoulder twist densities is considerably less at a spindle speed of 6000 rpm than that at 9000 rpm.

The effect of traveller weight

A heavier traveller normally ensures better twist flow in the spinning zone. Even with the heaviest traveller, however, the twist in the spinning zone will be much lower than the mechanical twist when winding at the shoulder.

The effect of varying the traveller weight was studied by Subramanian and others[49] at four different stages of the doff. The twist density was always highest when the heaviest traveller was used, at all stages of the doff. The difference in twist densities with a lighter and heavier traveller narrows down towards the end of the doff.

The effect of the drafting angle

The drafting (or roller stand angle) and spinning angles are important parameters controlling twist flow and variation. There are two distinct points of interest for the spinners; the angle at which the yarn is wrapped around the rollers, and the angle of the roller stand. A wrap of a few degrees reduces the tendency of broken ends to lap on the top roller. Too great an angle, however, will impede the twist flow. It is essential that the twist flow run up to the front roller nip to a distance well within the staple length of the raw materials. It is in this context that the angle of the roller stand is important. If the guide eye is vertically below the roller nip, a large roller stand angle is needed to keep the angle of wrap small. When the line thread is vertical, a 60° stand angle results in approximately a 30° wrap, while a 45° stand angle gives angle of wrap of between 10° and 27°.

The best spinning angle is that which gives the best twist flow in the region between the lappet eye and the roller nip. It was once thought that a vertical thread line would give the best results, because it allows twist flow from the bobbin to be absolutely free up to the nip point. This theory appeared to be sound, but in practice it was found not to be entirely correct. The amount of twist generated at the traveller-ring junction is not in fact capable of going beyond the lappet eye in a vertical thread line. When spinning in vertical position, the yarn is in contact with the guide eye at all times, and it tends to hold back the twist in the yarn.

Studies also show that, as the spinning angle increases from the vertical, the pressure between the yarn and the front of the guide eye drops, until at about 35° the yarn is clear of the guide eye for a short time in each revolution of the balloon, and the twist is free to run up to the front roller nip. When the angle increases still further, the yarn remains clear of the guide eye for shorter and shorter periods in each revolution, until finally it is again in constant contact.

8.6 Process control in rotor spinning

As the yarn formation mechanism in rotor spinning is force-restrained rather than position-restrained, proper process control is extremely important to ensure production of yarns with the desired quality at the highest possible speed and with minimum waste.

8.6.1 Control of yarn tension

Yarn tension is much lower in rotor spinning as compared to ring spinning, but controlling yarn tension is still an important factor in avoiding end breakage. This is because the yarn formation mechanism is force-restrained rather than position-restrained in rotor spinning.

Theoretical models

Mack[50] originally established the differential equations governing yarn shape and tension inside rotors. The tension, T_r, at a radius r is given by:

$$T_r = T_0 + \frac{1}{2} m\omega^2 \left(b^2 - r^2\right)$$

where m is the yarn linear density, ω the yarn rotational speed, and T_0 the tension at the rotor wall. Many authors have followed Stalder[51] who showed that T_0 is small compared with $m\,\omega^2 b^2$ (where b is the rotor radius), and have consequently ignored T_0. This is convenient because T_0 can only be obtained by finding the boundary conditions for the differential equation given by Mack. Arinc[52] was the first to determine T_0, and found that it is decided by the actual air flow inside the rotor. Arinc also showed that, as the yarn linear density increases, T_0 tends towards zero and spinning becomes impossible owing to the change in yarn rotation from forward to backward motion with respect to the rotor. Krause and Soliman[53] derived a theoretical relation between yarn-mass variation and the variation of yarn-spinning tension for open-end rotor spinning, and demonstrated experimentally that spinning-tension measurement represents a useful means of assessing yarn irregularity. For an 'ideal yarn', the relation between the force variability and mass irregularity was calculated numerically as a function of rotor diameter and mean fibre length. The hypothesis agreed well with experimental results. Practical spinning trials gave proof of the correlation between yarn-mass variability, yarn-spinning tension and yarn strength.

Sampling inspection of the yarn-spinning tension and its variability on open-end spinning machinery could well serve as a useful test for quality-control purposes. Chao and Oxenham[54] continued with Arinc's analysis and were able to show that, as the yarn becomes finer, T_0 becomes relatively large compared with the holding-back force against the fibre band at the wall of the rotor. Consequently, as the yarn becomes finer, more and more end breaks will occur owing to the rupture of the fibre band on the rotor wall. Clearly, T_0 is important in determining the ultimate linear density limits in rotor spinning, and needs to be calculated as a routine matter, despite the fact that it is much smaller than the yarn tension at the doffing tube. This is rarely done because a computer solution for T_0 is needed for each doffing tube, air-flow regime, yarn linear density, and twist factor. Grosberg[55] used a simpler method for determining the yarn tension at the rotor wall. He considered that the yarn is situated between the lip of the doffing tube, of radius a, and the rotor wall, of radius b. The tension at b, T_0, can be found by imagining the rotor wall removed, and then T_0 has to work to overcome air drag on the yarn and friction at the navel. However, the yarn loses energy as it travels from b to a, and this is reflected in the work done by the tension through the force required to overcome the Coriolis acceleration. This method offers an approximate solution, which makes it possible to predict changes in internal yarn tension by a simple closed-form solution as follows:

$$T_0 = \frac{1}{4}\omega^2 b^3 E - \frac{1}{2\pi}(m\omega^2 b/t)$$

where t is the twist in the yarn and E is given by:

$$E = \frac{1}{2}C\rho d$$

where C, ρ and d are air drag coefficient, density of air and yarn diameter respectively.

The tension at the rotor wall, as predicted by the above equation, was too large by about 11%. The result is considered useful despite this error because of its simplicity.

8.6.2 The factors affecting yarn tension

Although yarn tension is much lower in rotor spinning as compared to ring spinning, when considering very high speed of rotor and occurrence of twist loss it is extremely important to know in detail about the effect of different factors on yarn tension to keep the end breakage rate within control.

The effect of rotor speed and diameter

In the course of development, rotor speeds have been increased from approximately 30 000 rpm originally to 160 000 rpm today. This has been achieved by reducing rotor diameter. It can be demonstrated that all rotor speeds and diameters introduced in recent decades are governed by the fact that the centrifugal force must be kept within certain limits to avoid end breakage. Assuming that the centrifugal force acting on the thread in the rotor can never exceed yarn tension, this represents a theoretically absolute spinning limit, though this cannot, and is not intended to be reached in practice. Spinning tension must always lie with a sufficient safety margin below normal variations in the inherent strength of the yarn, if the spinning process is to run efficiently.

A minimum possible speed is also allocated to each rotor. There is a risk that the rotor speed (and thus spinning tension) will decline to such an extent that the centrifugal force in the rotor groove is no longer sufficient to generate the twist retention and false-twist effect (between nozzle and rotor groove) on the draw-off nozzle. These twist effects are necessary for spinning stability; if twist integration in the rotor groove is seriously disturbed or interrupted then a thread break occurs. This situation is clearly apparent when calculating the so-called minimum twist multiplier, from which the optimum speed range for each rotor diameter can be derived.

Reducing rotor diameter in the interests of achieving higher rotor speeds and higher output is surprisingly effective. The predicted (lower) limits for rotor diameter have repeatedly been breached by new developments, and nowadays quality yarns are spun (from suitable raw material) with 28 mm diameter rotors operating at speeds of up to 160 000 rpm. The fundamental relationship between rotor diameter and fibre length, although not invalidated, has been decisively modified by such advances in rotor technology.

According to Cheng and Cheng[56] a higher rotor speed leads to higher yarn tension. This is the product of frictional force of the yarn at the doffing tube and the centrifugal force acting on the yarn end in the rotor. The tension in the yarn largely results from centrifugal forces, and so is proportional to the square of the product of the rotor speed and diameter. Derichs[57] suggested that to maintain an optimal tension with increased rotor speed, the following relationship may be used:

$$n = 3.2 \times 10^6/D$$

where n = rotor speed (rpm) and D = rotor diameter.

The effect of twist in the yarn

Krause and Soliman[53] have shown that specific yarn tension increases with increasing twist factor. This is because a higher axial force is necessary at the spinning point in the rotor to enable a higher twisting rate (and hence a higher yarn torque) to be maintained. The false-twist effect that is usually present in open-end spinning also affects the yarn mass in the radial portion through twist contraction, and thus may influence spinning tension in a different manner for different twist factors.

The effect of navel design

The navel and rotor together form the central spinning unit. It acts both as a twist blocker and a false twister element, turning the yarn gently twice around a solid angle of 90°. Navel design affects the local air compression in the rotor and so has a substantial impact on yarn structure and topography. Other than the number of grooves, the radius of the curvature of the navel is the most obvious design criterion, because the area of the shell increases or decreases with it. The length of the yarn guidance on the navel, the friction path, and the yarn tension, can all be manipulated by this means. It is a far too common mistake to think that a reduction of the radius of the curvature is compellingly linked with a decrease in yarn tension. One has to distinguish clearly between average yarn tension and real yarn tension peaks. Through the changing effects of false twist and twist contraction, the formation of wrapper fibres changes separately from the changing oscillation characteristics of the yarn on the shell. A navel with a smaller radius of curvature and reduced mean values in yarn tension, but higher tension peaks, clearly generates lower yarn elongation values. However, a combination of different design parameters can compensate for this effect. The groove depth substantially affects the retention time of the yarn in the groove and, with it, spin stability. The groove can be used as a vibration agitator to reduce the average frictional force between yarn and spin element, and to improve the propagation of false twist into the rotor groove. There is a fibre-specific optimal average yarn tension, where end breaks are at a minimum.[58]

The grooves cause the yarn to bounce off the surface of the navel for very brief periods of time. Yarn tensions measured inside the rotor are very close to the theoretical figures, if there are not too many pulses due to yarn riding over the grooves. Unpublished work at NC State University showed that the number of pulses rose with the number of grooves until four grooves were cut. Increasing the number from four to eight gave only four pulses and this was interpreted to mean that the yarn jumped over alternate grooves.[59]

8.7 Control of end breakage rate and twist loss in rotor spinning

In rotor spinning, the end breakages are broadly classified into three main groups[60]:

1. Tension yarn breaks. These are found in spun yarn, normally between the take-off nozzle and take-up rollers. The yarn ends on the yarn package have a blunt appearance, and short broken yarn ends are in general found in the rotor groove.
2. Spinning yarn breaks. These occur in the yarn peel-off zone in the rotor groove when continuous fibre spin-in is interrupted.
3. Yarn breaks that result from sliver breaks, or breaks due to similar interference factors outside the spinning box, which can be easily controlled by proper maintenance and good practice.

Das and Ishtiaque[61] suggested a new seven-fold method of classifying broken ends in rotor spinning, which is performed by collecting broken ends from the yarn package and by observing the rotor groove after end breakage.

8.7.1 Factors affecting end breakage rate

End breakage in rotor spinning is much lower as compared to ring spinning due to lower yarn tension. Rotor and opening roller speed are the two major factors influencing end breakage in rotor spinning.

The effect of rotor speed

Yarn breakage rate increases as the rotor speed increases. A higher rotor speed causes powerful centrifugal forces on the fibres in the rotor groove,[62] resulting in poor spinning stability and an increase in yarn tension, which in turn causes high yarn breakage. Das and Ishtiaque[61] observed no clear trend for breakages with a foreign matter embedded end, with an unopened fibrous end, with a seed coat embedded end or with a tapered end, with increase in rotor speed. It was also observed that the proportion of breaks with a tapered end is very high, showing that a large number of yarn breaks occur due to disturbance in the fibrous strand inside the rotor groove. Breaks with a trash particle embedded end and the breaks with a blunt end both consistently increase with rotor speed. At lower speeds, small trash particles are constantly taken away by the yarn, due to the comparatively low centrifugal force acting on these trash particles. They then become entrapped within the yarn structure, and so do not cause breakage. At higher rotor speeds, these small trash particles accumulate inside the rotor groove due to higher centrifugal forces, and after saturation point all the accumulated particles try to come out along with the fibrous strand, causing yarn breakage. Because the spinning tension increases with rotor speed, a rise in blunt ended breakages occurs due to the increase in centrifugal force on the fibre stand and yarn winding speed.

The effect of opening roller speed

As the opening roller speed increases, the carrying factor increases, which in turn increases the opening efficiency of the opening roller.[63] The carrying factor is the effective number of wire points per unit of time. Owing to the better opening of fibres, it can be expected that the fibre tufts of smaller size and uniform dimensions are fed into the transport tube and thus into the rotor groove. However, too high an opening roller speed results in higher rotor deposition,[64,65] and fibre orientation inside the transport tube also deteriorates drastically,[66] causing higher end breakage.

Das and Ishtiaque[61] observed that the proportions of foreign matter embedded in broken ends and blunt broken ends do not show any specific trend with the increase of opening roller speed. The proportion of breaks with unopened fibrous ends increases with the opening roller speed, due to higher rotor deposition and deterioration of fibre orientation (clustering) inside the transport tube. The occurrence of seed coat embedded broken ends shows a decreasing trend with the increase in opening roller speed, because better separation of seed coats from the fibre reduces the deposition of such seed coat particles on the rotor groove. Breaks caused by trash particle embedded ends show an increasing trend with the increase in opening roller speed. This is mainly due to better separation of trash particles, and thus a higher deposition of such particles on the rotor groove. The maximum proportion of broken ends belongs to the tapered end category, and shows a decreasing trend with the increase in opening roller speed. However, the miscellaneous type of break shows a marginally decreasing trend as the opening roller speed increases.

8.7.2 The mechanisms of twist loss in rotor spinning

The machine selected twist in rotor spinning does not accurately correspond to twist measured in the yarn. One reason for this is that surface layer fibres in open-end yarn have a different level of twist from those forming the body of the yarn. A stage will be reached where some fibres are not yet fully twisted while others are already twisting in the opposite direction. These simultaneous opposing twisting effects will cause the measured twist to be lower than the theoretical twist.[67]

According to Kleinhansl,[68] twist loss in the spinning chamber results from torsional stresses in the twisted yarn acting backward past the separation point, and so imparting twist to the strand of fibres in the rotor groove before its full count has been reached. Fibres incorporated into the yarn after that point are not fully twisted and so the total twist is reduced. London and Jordan[69] concluded that the difference between measured and machine twist is a function of doffing tube design. According to Lord and Grady,[70] the difference between measured and machine twist values, expressed as a twist gradient, is a function of machine design, operating functions and fibre characteristics.

Salhotra[71] has put forward a hypothesis to explain twist loss through the incidence of sheath fibres. According to Salhotra the transmission of twisting torque cannot be one hundred percent efficient; there is always a possibility of leakage through fibre ends at the yarn forming point. In the extreme case of a single fibre in the yarn core at yarn formation point, one can reasonably expect a loss of twisting torque in the yarn core.

8.7.3 The factors affecting twist loss

Control of twist loss is extremely important considering its influence on yarn quality and spinning performance. Major factors that influence twist loss are amount of twist and opening roller speed.

The effect of twist factor

The twist properties of rotor yarn have been the subject of several studies. It was mentioned by Audivert[72] that the difference between adjusted machine twist and measured twist depends on the value of machine twist and yarn linear density, where higher machine twist and higher yarn linear density provides a higher percentage of twist deviation (PTD). Salhotra[71] observed that, as the twist factor is increased from 38.3 to 57.4, there is an almost continuous decrease in twist efficiency and an increase in the percentage of sheath fibres. This is obviously due to the well-known increase in the length of the peripheral twist extent (PTE). The higher the machine twist, the higher the twist loss, and so the higher the PTE.

The effect of opening roller speed

The main purpose of the opening roller is to open the fibre sliver to individually aligned single fibres. Well straightened, aligned and parallelised fibres from the opened sliver should make efficiently twisted rotor yarn bodies. Tyagi[73] reported a marked increase in twist efficiency with an increase in opening roller speed. The increase in twist efficiency is the result of a decrease in the percentage of sheath fibres. Greater fibre separation at higher opening roller speeds reduces the possibility of fibres becoming wrapped simply because they are entangled with fibres undergoing belt formation. Ülkü[74] employed a similar approach, but concluded that there is only a small degree of association between opening roller speed and twist. The slightly increased number of twists for high opening roller speeds can be attributed to the high degree of fibre separation.[75] Salhotra[71] also studied this effect, and observed that the twist efficiency goes up as the opening roller speed increases.

8.8 Future trends

The main goal of the modern yarn manufacturing industry is to achieve competitive yarn quality and price. Current developments in spinning technology are aimed at maintaining high productivity with effective quality control, by selecting equipment and spinning conditions that suit the raw materials.

The trend towards automation of spinning systems and yarn manufacturing machinery is likely to continue, and automated processes will be increasingly integrated into manufacturing units. Spinning machines that combine carding and drawing functions, and ring frames linked with winders, have already been developed. Research into directly spinning yarns in card and magnetic ring spinning is currently in progress. Production rates are predicted to rise as machines become available with even more spindles, and robot-controlled equipment will become standard.

Improvement in the performance of control strategies will result in more consistent production, facilitating process optimisation, and reducing the need for reprocessing and waste production.

Process models underpin most modern control approaches. Different controllers can be synthesised for different models. Even the prevalent Proportional+Integral+Derivative (PID) algorithm can be utilised, though its performance capabilities are limited. More sophisticated strategies for improving process control include adaptive algorithms and predictive controllers. Statistical process control (SPC) techniques are also experiencing a revival due to the modern emphasis on quality. Attempts are being made to integrate traditional SPC practices with engineering feedback control techniques. Recent interest in the development of practicable nonlinear controllers is of particular significance, because they recognise that many real processes are nonlinear and that adaptive systems may not be able to cope with significant nonlinearities. One approach involves designing control strategies based on nonlinear black box models, for example nonlinear time-series or neural networks. A second method relies on an analytical approach, making use of a physical–chemical model of the process. Cheap powerful computers and advances in the field of Artificial Intelligence are also making an impact. Local controls are increasingly being supplemented with monitoring, supervision and optimisation schemes; roles that were traditionally undertaken by plant personnel.

8.9 References

1. Lüdicke, A., Ein Studie über die ringspindel, *Dinglers. Polutech. J.* 1881; **242**, 334.
2. Escher, R., Theorie der ringspindel, *Civil Ing.* 1883; **29**, 448.
3. Gégauf, C., Analyse Complète, Théorique et Pratique, Concernant Les Métiers à Filer à Anneaux, *Ind. Textile* 1901, **17**, 273, 360, 394, 440.
4. Linder, G., Ballonform, fadenspannung und Läfer- stellung am Ringspinnmachinen, *Leipz. Monatschr. Textil Ind.* 1910; **25**(8),123.
5. Baltz, W. E., Ueber die Tätsachliche Ballonform bei der Ringspindel, *Milliand Textilber.* 1933; **14**, 5.
6. Hannah, M., Application of a Theory of the Spinning Balloon, *J Textile Inst.* (a) 1952; **43**, T519; (b) 1955; 46, T1.
7. Mack, C., Theoretical Study of Ring and Cap Spinning Balloon Curves (With and without Air Drag), *J Textile Inst.* 1953; **44**, T483.
8. Mack, C. and Smart, E., Measurements of the Air- Drag of Textile Thread, *J Textile. Inst.* 1954; **45**, T348.
9. Crank, J., A Theoretical Investigation of Cap and Ring Spinning System, *Textile Res J.* 1953; **23**, 266.
10. Grishin, P. F., Balloon Control, *Platt's Bulletin*, **8**, 6/161, 8/240, 11/333 (1954).
11. Bracewell, G. M. and Greenhalgh. K., Controlled Balloon Spinning, *J Textile Inst.* 1954; **45**, T730.
12. De Barr, A. E., A Descriptive Accounts of Yarn Tensions and Balloon Shapes in Ring Spinning, *J. Textile Inst.* 1958; **49**, T58.
13. De Barr, A. E., The Role of Air Drag in Ring Spinning, *J Textile Inst.* 1961; **52**, T126.
14. Mack, C., Theory of the Spinning Balloon. *Quart. J. Mech. Appl. Math.* 1958; **XI**, 196–207.
15. De Barr, A. E. and Catling. H., *The Principles and Theory of Ring Spinning*, The Textile Institute and Butterworths, Manchester and London, 1965.
16. Nerli, G., Mechanical behaviour of yarn in Certain Textile Manufacture, *Textile Tech.* (a) 56(Dec. 1984), (b) 36(Sept. 1985).
17. Batra, S. K., Ghosh, T. K. and Zeidman, M. I., An integrated approach to dynamic analysis of the ring spinning process – Part I: without air drag and coriolis acceleration, *Textile Res J* 1989; **59**, 309–17.
18. Batra, S. K., Ghosh, T. K. and Zeidman, M. I., An integrated approach to dynamic analysis of the ring spinning process – Part II: with air drag, *Textile Res J* 1989; **59**, 416–24.
19. Batra, S. K, Ghosh, T. K. and Zeidman, M. I., An integrated approach to dynamic analysis of the ring spinning process – Part IV: inherent instability of free balloon, *Textile Res J* 1995; **65**, 417–23.
20. Lisini, G. G., Toni. P., Quilghini, D and Di GiogiCampedelli VL., A comparison of stationary and non-stationary mathematical models for the ring-spinning process, *J Textile Inst* 1992; **83**, 550–9.

21. Fraser, W. B. On the theory of ring spinning, *Proc R Soc Lond A* 1993; **342**, 439–68.
22. Fraser, W. B., Air drag and friction in the two-for-one twister: result from the theory, *J Textile Inst* 1993; **84**, 364–75.
23. Stump, D.M and Fraser, W.B., Transient solutions of the ring spinning balloon equations, *J ApplMech* 1996; **63**, 523–8.
24. Zhu, F., Hall, K. and Rahn, C.D., Steady state response and stability of ballooning strings in Air, *Int J Non-linear Mech* 1998; **33**, 33–46.
25. Zhu, F and Rahn, C.D., Limit cycle prediction for ballooning strings, *Int J Non-linear Mech* 2000; **35**, 373–83.
26. Fan, R., Singh, S. K. and Rahn, C. D., Modal analysis of ballooning strings with small curvature, *J ApplMech* 2001; **68**, 332–8.
27. Tang, Z. X., Fraser, W. B. and Wang, X., Modelling yarn balloon motion in ring spinning, *Appl Math Model* 2007; **31**, 1397–410.
28. Tang, Z. X., Wang, X. and Fraser, W. B., Simulation and experimental validation of a ring spinning process, *Simul Model* 2006; **14**, 809–16.
29. Tang, Z. X., Wang, X. and Fraser, W. B., An investigation of yarn snarling and balloon flutter instabilities in ring spinning, *J Textile Inst* 2006; **97**: 441–8.
30. Rong, Yin. and Hong-bo, Gu., Numerical simulation of quasi-stationary ring spinning process linear elastic yarn, *Textile Res. J* 2011; **81**(i), 22–7.
31. Fraser, W. B., The effect of yarn elasticity on an unwinding balloon, *J Textile Inst* 1993; **83**, 603–13.
32. De Barr, A. E., The role of air drag in ring spinning, *J Textile Inst* 1961; **52**, T126.
33. Anbarsan, M., Optimizing spindle speed, *The Indian text. J* 1996; **CVI**(12): 24–30.
34. Catling, H., Traveller-Ring Friction in Spinning, *Journal Mechanical Engineering Science*, 1962; **4**(1), 94.
35. Zheng-Xue Tang, W., Barrie, Fraser, Lijing, Wang and Xungai, Wang, Examining the effects of balloon control ring on ring spinning, *Fibers and Polymers*, 2008; **9**(5): 625–32.
36. Hori, Y., A Statistical Analysis of Yarn Breakage Rates on a Spinning Frame. *J. Text. Mach. Soc. Japan (English Edition)*, 1955; **1**(2), 54.
37. De Barr, A. E., End – breakages in ring spinning, *J Textile Inst* 1959; **50**, T284.
38. Ghosh, A., Ishtiaque, S., Rengasamy, S. and Patnaik, A., The Mechanism of end Breakage in ring Spinning: A Statistical model to Predict the End Break in Ring Spinning, *AUTEX Research Journal*, **4**(1), March 2004, 19–24.
39. Martindale, J. G., A New Method of Measuring the Irregularity of Yarns with Some Observations on the Origin of Irregularities in Worsted Slivers and Yarns, *J Textile Inst.*, 1945; **36**, T35–T47.
40. Lawrence, C. A., *Fundamentals of Spun Yarn Technology*, CRC Press, 2002, pp. 272–3.
41. DeBarr, A. E. and Catling, H., *The Principals and Theory of Ring Spinning*, pp. 136–7, The Textile Institute, Butterworths, 1965.
42. Klein, W., Spinning Geometry and its significance, *Int Text Bull, Yarn Fabric Forming*, 1993; **3**, 22–6.
43. Oxtoby, E., *Spun Yarn Technology*, Butterworth & Co. (Publishers) Ltd., 1987, 111.
44. DeBarr, A. E. and Catling, H., *The Principals and Theory of Ring Spinning*, The Textile Institute and Butterworth & Co. (publishers) Ltd., 1965, p. 129.

45. Axson, M., Observation of the Traveller during Spinning, Letters to Editor, *J Textile Inst* 1954; **45**, T368.
46. Kanai, K., Kanai, H., Sakata, T. and Kamogawa, H., Twist in Ring Spinning and Doubling Part 1: Snail Wire, *J. Textile Mach. Soc. Japan* 1960; **13**(2), 19–23.
47. Gessner, W., *Fadenbruchursachen in der Kammgarnspinnerei*, Textil-Praxis, Vol-19, January, 1964, pp. 58–62.
48. Wegener, W. and Hanslandwehrkemp., *Die Drehungsverleilung im Fadenbeim Spinnprozess*, Textil-Praxis (Ger), Vol-17, October, 1962, pp. 995–1000.
49. Subramanian, T. A., Salhotra, K. R. and Bhaduri, S. N., Twist Flow in Ring Spinning, *Textile Res J* 1995, 1967; **37**, 195.
50. Mack, C. Quart, Theory of winding in centrifuge spinning, *J. Mech. Appl. Math.* 1956; **9**, 75.
51. Stalder, H., Ph.D. Thesis, Eidgenössische Technische Hochschule, Zürich, 1968.
52. Arinc, K. and Grosberg, P., 'Open-end Spinning' in International Textile Seminar, New Delhi. 1973.
53. Krause, H. W. and Soliman, H. A., The Significance of Yam Tension in Open-end Spinning, *J. Text. Inst.* 1976; **67**(5), 166.
54. Chao, S. N and Oxenham, W., *Proceedings of International Symposium on the Theory and Practice of New Spinning Techniques*, InstytutWlokiennictwa, lodz, Poland, 1978, p. 129.
55. Grosberg, P., The Tension Inside the Rotor, *J Text Inst* 1984; **75**(5), T332.
56. Cheng, Y. S. J. and Cheng, K. P. S., Selecting process Parameters that Influence the Strength of Rotor Spun Yarn Formed on the SDL Quickspin System, *Textile Res. J*, September, 2004; **74**(9) 792–6.
57. Derichs, J., Theoretical and Practical Limits of Rotor Spinning in the Production of Fine Yarns, Schlafhorst documentation, No. 16, a141e-3.86. 1–9, 1986.
58. Sonntag, E and Bolze, J., *Design of a navel and its influence on yarn structure*; 2003, Emil BRÖLL GmbH & Co, Austria; www.broell.com
59. Lord, R., *Handbook of yarn production*, Woodhead Publishing Ltd., Cambridge, England, 2003, p. 195.
60. Trommer, G., '*Rotor Spinning*', *Deutscher Fachverlang*, Frankfurt/Main, Germany, 1995.
61. Das, A. and Ishtiaque, S. M., End Breakage in Rotor Spinning: Effect of Different Variables on Cotton Yarn End Breakage *AUTEX Research Journal*, 2004; **4**(2), June © AUTEX.
62. Ishtiaqe, S. M. and Saxena, A. K., Rotor machine variables and yarn structure: part iv – influence of machine variables on configuration of neps and thick places of polyester rotor yarn, *Indian J Fibre Text Res*, 1998; **23**, 209.
63. Ishtiaque, S. M. and Saxena, A. K., Rotor machine variables and yarn structure: part iii – fibre distribution in yarn in relation to rotor machine variables, *Indian J Fibre Text Res*, 1998; **23**, 141.
64. Lawrence, C. A. and Tahiroghi, S., Rotor machine variables and yarn structure: part iii – fibre distribution in yarn in relation to rotor machine variables, *Indian J Fibre Text Res* 1990; **15**, 108.
65. Simpson, J. and Murray, M. F., Effects of Combing Roller Wire Design and Rotor Speed on open-end Spinning and cotton properties, *Textile Res, J,* 1979; **49**, 506.

66. Harlock, S. C., An investigation into feed mechanisms for open-end spinning machines, Ph.D thesis, The University of Leeds, 1977.
67. Hunter, L., The Production and Properties of Staple-Fibre Yarns Made by Recently Developed Techniques, *Textile Progress* 1978; **10**, 1/2.
68. Kleinhansl, E., Beitrag zur Drehungsprufung an OE-Garnen, *Melliand Textilberichte International (Ger)*, 1972; **53**(1), 16–22.
69. London, J. F. and Jordan, G. B, Open-End Yarns and Their Characteristics, *Knitting Times*, 1974; **43**(15), 44.
70. Lord, P. R. and Grady, P. L., Twist Structure of Open End Yarns, *Textile Res. J* 1976; **46**, 123.
71. Salhotra, K. R., The Role of Sheath Fibers in Influencing Twist Loss in Rotor Spinning, *Textile Res. J* 1981; **11**, 710–14.
72. Audivert, R., The Deviation of Measured Twist from Machine Twist in Rotor-spun Yarns, *J Textile Inst,* 1988; **2**, 333.
73. Tyagi, G. K., Effect of fibre profile, spin finish and opening roller speed on properties of polyester OE rotor-spun yarns, *Indian J Fibre Text Res* 2003; **28**, 411, December.
74. Ülkü, Ş., Acar, M., King, T. G. and Özipek B., Fibre Alignment and Straightening in Opening for Open end spinning, *Textile Res J* 1993; **63**(6), 309.
75. Ülkü, Ş., Özipek, B. and Acar, M., Effects of Opening Roller Speed on the Fibre and Yarn Properties in Open-End Friction Spinning, *Textile Res J.* 1995; **65**(10), 557.

9
Maintenance of yarn spinning machines

A. BASU, Central Silk Board, India

DOI: 10.1533/9780857095633.2.225

Abstract: Machines play a major role in determining the productivity of an industry, the quality of its products, power consumption and working atmosphere. All these factors are dependent on the mechanical condition of the machines. In spinning units, it is also very important that the condition of the major machines is maintained by systematic and well-planned maintenance practice. This chapter deals with such important aspects of maintenance management as maintenance scheduling, maintenance costs, the efficiency of maintenance functions, assessment of quality of maintenance, etc. Modern practices such as total productive maintenance (TPM), 5S and computerised maintenance management have also been covered. Proper maintenance management helps companies to improve profitability, product quality and customer relationships.

Key words: ABC analysis, breakdown maintenance, condition monitoring, maintenance cost, predictive maintenance, preventive maintenance, total productive maintenance.

9.1 Introduction

Maintenance of machines is one of the most important functions of any production unit. In the past the major purpose of machine maintenance was to prevent breakdowns. With time the role of maintenance in product quality has been understood. The advent of high-speed machines has made maintenance management an integral part of plant management. The production loss of a machine is proportional to stoppage time as well as to the speed of the machine. The stoppage of high production, high cost machinery costs more than for a low cost machine. In the spinning industry, most machines have become high production and high cost. Hence fewer machines are required in a mill than was the case for earlier machines, and this makes each machine much more important. Long stoppage of any one machine can influence the productivity of the whole unit. Maintenance is an investment that buys/gives increased production time. With the complexity, sophistication and automation of machines, a very serious burden now falls on maintenance engineers with respect to the quality of maintenance, maintenance aids, documentation, etc.

9.1.1 Classification of maintenance system

The maintenance system can be broadly divided into two groups:

1. unplanned maintenance and
2. planned maintenance.

Unplanned maintenance includes:

- breakdown maintenance or emergency repair,
- corrective maintenance,
- opportunistic maintenance.

In a breakdown maintenance system, repair is undertaken only after the failure of the machine. The machine is allowed to run undisturbed till it fails. Of course, lubrication and minor adjustment are done during this period. Only when the machine fails to perform designed functions, or comes to a halt, is any maintenance/repair work done.

Corrective maintenance, as the name implies, refers to maintenance actions for correcting or restoring a failed unit (or a unit about to fail). Its scope is very vast and may include different activities, from minor adjustments and repairs to redesign of machinery. Corrective maintenance actions can be further classified according to priority. For corrective maintenance, each task, once taken up, is generally completed fully. Each corrective job may differ from the others.

In a multi-component system with several failing components, it is often advantageous also to undertake opportunistic maintenance. When a machine is taken down for maintenance or to replace one or a few defective parts, the opportunity can be taken to maintain/change other components even though they have not yet failed. This would probably be more economical in the long run than having to shut down again when those components fail. Normally the cost of replacing several parts jointly is much less than the sum of costs of several separate replacements. However, the cost of loss of the residual life of those parts has also to be taken into consideration in each calculation. For instance, where a machine is supposed to be overhauled once in a year and defects in many parts are observed after 10 months or so the overhauling is brought forward to avoid extra stoppages for replacing the damaged parts only.

A planned maintenance system includes:

- routine maintenance,
- preventive maintenance,
- predictive maintenance,
- design out maintenance.

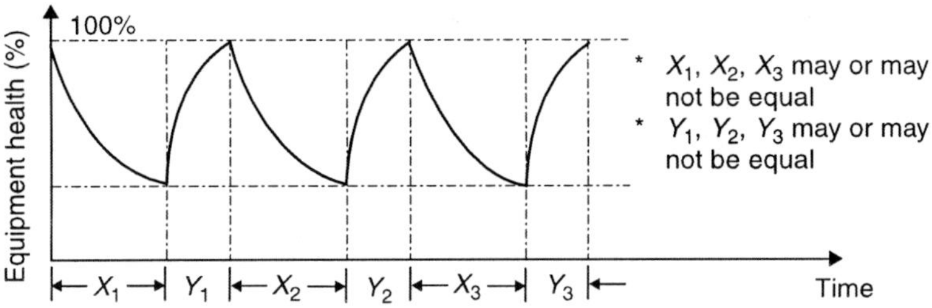

9.1 Frequency of preventive maintenance.

Routine maintenance is the simplest form of planned maintenance, and at the same time very important. In this method minor work is carried out at regular intervals. It involves jobs such as cleaning, lubrication, inspection and small adjustments of pressure, settings, tightening of loose parts, changing of traveller, etc. It also includes inspection of bearings, belts, couplings, etc. The small and critical defects observed during inspection are rectified immediately. Bigger rectification is scheduled for the next shut down. Such maintenance is essential for reduction in machine breakdown. Frequency of this maintenance function may be once in a shift or once in a day.

Preventive maintenance is one of the oldest maintenance systems being practised in industry. Nowadays, corrective maintenance and condition-based maintenance are also added to this concept. Preventive maintenance is the planned maintenance of plant and equipment in order to prevent or minimise breakdown. This can be of two types: (1) fixed time maintenance and (2) condition-based maintenance. Figure 9.1 shows a typical performance cycle of a machine. In this figure the periods X_1, X_2 and X_3 indicate the actual operating periods during which the machine condition deteriorates, and the periods Y_1, Y_2 and Y_3 are the periods of capital repairs or overhauls during which deteriorated equipment is restored to its original condition. In practice, due to difference in working conditions, the X_1, X_2 and X_3 may not be equal, and similarly Y_1, Y_2 and Y_3 may not be equal. The actual duration of the preventive maintenance schedule is derived from the manufacturer's recommendation, the experience of the maintenance manager/engineer on similar machines, and by failure analysis and/or trend monitoring of the condition.

Predictive maintenance implies predicting the failure before it happens, by identifying the root causes for failure and eliminating those causes before they result in extensive damage to the machines. The objective of predictive maintenance is to run the machines in good condition for a long time, so that the machine life and time between overhauls can be extended.

The main differences between preventive and predictive maintenance are as follows. In preventive maintenance, overhauling of machines is done after a fixed period; the frequencies of all activities are decided beforehand. In predictive maintenance, work is undertaken on the basis of condition monitoring, and only before the machine is expected to fail or deteriorate. In preventive maintenance overhaul and repair schedules are made generally on the basis of manufacturer's recommendation, plus the experience and judgement of maintenance manager, and hence sophisticated condition-monitoring instruments are less essential. In predictive maintenance sophisticated condition-monitoring instruments are necessary and continuous monitoring is required.

Predictive maintenance comprises three stages, that is, detection, analysis and correction. Generally the cost of predictive maintenance, including the cost of monitoring equipment, is justified by extending the time interval between shutdowns or overhauls, reducing the total stoppage time overall, and by fewer breakdowns.

The objectives of condition-based maintenance are to intervene before the failure occurs, to do maintenance only when needed, to reduce the number of failures as well as the number of shut downs, and to reduce maintenance costs and cost due to production loss. The condition of the parts of the machines is monitored or the quality of the products is monitored. Whenever there is an indication that any of these are deteriorating the machine is scheduled for maintenance. Recently this method has become more popular because of development of good and reasonably priced condition-monitoring equipment, such as online indicators and portable instruments that detect fault generation very early, development of superior monitoring techniques, and the increased cost of machine downtime. This method is ideal for the spinning plant because most of the vital components earmarked for planned replacement are those that fail gradually and progressively, for example, card wires, ring, traveller, etc.

Design out maintenance is a design-oriented creative measure aimed at rectifying design defect or defects originating from improper installation, poor material choice, etc. For its success it requires a strong maintenance–design interface so that the maintenance engineer works in close cooperation with design engineer. It is more suitable for items/machines of high maintenance cost. The choice to be made is between the cost of redesign and the cost of recurring maintenance.

9.1.2 Maintenance: material and cost control

In maintenance, the materials represent 40–60% of the total maintenance cost. Hence it is very important for any industry to control the material costs. There are certain steps, which can optimise the cost of material for the maintenance:

- analysis of causes of breakdowns,
- standardisation of spares and machineries,
- keeping on top of obsolescence,
- selection of proper suppliers,
- designing proper and simple systems for record keeping, ordering and follow-up.

It is very important to decide what should be the inventory level for a unit. Overstocking will block capital, and in some cases those spares may not be used during the concerned machine's lifetime. On the other hand, too low a stock of spares can seriously increase the machine down time, thereby causing production loss. A balance has to be made between these two situations, arrived at on the basis of the experience of the manager, identification of a supplier within a reasonable distance, etc.

ABC analysis is a tool that can help a manager to concentrate on the 'significant few' rather than the 'insignificant many'. The items are classified as A, B and C depending on their importance, and controlled according to these categories. The items, which are of high or moderate monetary value or high usage, are categorised as 'A' group. Generally this group comprises 15–20% of total items but represents 75–80% of total cost. The items of intermediate value and usage constitute the 'B' group. Generally 25–30% of the items fall in this category, with 15% of total cost. Items of low value and usage are categorised as 'C' group. Though 40–50% of the material comes under this category their value lies within 5–10% of the total cost of spares.

The items that fall under 'A' category need tighter control, regular reviews, close follow-up and very accurate records. Examples of 'A' category items are top comb holders, nipper supports, differential gears, etc. 'B' category items require normal control, good records and regular attention. 'C' category items need simple control, larger inventories and minimal records amongst these three categories. Examples of 'C' category items are anti-balloon control rings, lappet hooks, separators, spindle tapes, spindle locks, separator holders, suction tubes, roving guides, top synthetic clearers, gears, travellers, etc.

In many cases the total maintenance costs for a company include building and other amenities, as well as of machinery. For better control of the maintenance function the concerned manager should keep track of machinery maintenance cost and, within that, the costs of breakdown maintenance and preventive maintenance. The performance of a maintenance manager can be assessed by the ratio of breakdown cost to total maintenance cost and preventive maintenance cost to total maintenance cost. A high percentage of breakdown costs indicates that the maintenance manager is doing mostly firefighting. For a competent maintenance-in-charge, preventive and predictive maintenance cost should be around 80% of the total maintenance costs. However, one should be careful not to overdo maintenance. The cost of providing maintenance within a company is determined by the level of service provided, the usage and cost of maintenance parts and materials, the extent and cost of outside service (e.g. contractors), the nature of maintenance facilities (workshops) and the size of the maintenance labour force.

9.1.3 Maintenance productivity and the people factor

In maintenance of machinery, human factors play the most important role. The performance of the machine depends on how accurately the settings have been made, whether the screws have been tightened, filling of correct lubricating oil in the proper amount, etc. The accuracy of these operations depends on the skill level of the personnel engaged in maintenance activities. The performance of a maintenance team needs to be assessed to manage the maintenance function. One method is to measure the productivity of maintenance workers as well as assessing the quality of work. This can be determined by a combination of three contributing factors: performance, utilisation and work methods. Performance is a measure of the speed with which people work. It is influenced by the level of their innate and acquired skills, the conditions surrounding their work and the effort put in by them. Utilisation is a measure of how much of the available time is spent in working. Starting time, early finishing and extended breaks influence it. Work method is determined by the types of tools, equipment and work sequences and is influenced by the workplace setting and organisation.

Direct performance (%) can be determined when job time standards and time reporting are used. It is calculated as:

$$\frac{\text{Standard time for jobs completed}}{\text{Time taken to complete the job}} \times 100$$

It provides a measure of the effectiveness of the group during the periods when it is actually working. However, to obtain the effectiveness of a total maintenance section, including its supervisors and planners overall efficiency can be calculated.

$$\text{Overall efficiency}\,(\%) = \left(\frac{\text{Standard time for jobs completed}}{\text{Total available time}}\right) \times 100.$$

To keep track of labour cost trends, an output cost-per-unit index should be used. A maintenance department does not produce goods, so its output has to be measured in standard hours, that is, time standards. The cost-control indicator – cost per standard hour – should be produced on a weekly or monthly basis.

Increasing global competition has forced textile units towards cost reduction for each and every activity. One strategy is the implementation of a 'common gang' concept. The term common gang refers to the deployment of common members in cross-functional areas of the maintenance department. The conventional system of working consists of deployment of manpower section-wise such as carding, ring frame, comber, etc. In the common gang concept the preparatory divisions can be grouped as one section and ring frame and rotor spinning division as one. The next step may be to have only one team for the entire mill. Such groupings require considerable multi-skill from maintenance personnel. The advantages of this concept include efficient utilisation of manpower through rationalisation, avoidance of monotony through job rotation, reduced dependence on a few individuals, enhancement of team spirit amongst the maintenance personnel, and reduced production losses due to less downtime of machines.

9.2 Maintenance of spinning preparatory machines

The maintenance of all spinning machines needs proper planning and scheduling. The job planner should have knowledge about machine, job, available techniques and facilities, etc. He/she prepares a step-by-step procedure that would accomplish the work with the most economical use of time, manpower and material. It should include sketches, line diagrams, networks, etc. Most of the maintenance functions can be obtained from instruction manuals, drawings, maintenance manuals, etc. supplied by the manufacturers. In addition many manufacturers provide training facilities for maintenance of their machines. Based on defect lists and condition-monitoring results, a list of all spares and components needing to be changed should be made by the maintenance manager. Availability of all these materials must be ensured before actual start of the job. Planning of maintenance jobs normally falls into two categories: short-term plans and long term plans. Jobs coming under short-term plan include lubrication plans and schedules, small defect rectification, vibration monitoring of critical equipment and small preventive and predictive maintenance jobs. The long term maintenance plan includes major repairs, capital repairs, annual and statutory overhauls, renovation and revamping, modernisation, strategic maintenance planning, etc.

The maintenance department maintains the records and documents pertaining to installations and subsequent maintenance done. These records/data may be in different forms and are stored in the system in such a manner that they can be retrieved when necessary. These records are used effectively to maintain control of maintenance cost, reliability and availability. Some documents are permanent records, such as instruction manuals and drawings, etc., and some are to be regularly updated, such as history cards, etc. Any omission in that would create difficulty in deciding the course of future maintenance.

The spinning division is generally classified into two major sections – spinning preparatory, and spinning. The preparatory division consists of the blowroom, card, draw frame, comber and speed frame machines. These machines influence the properties of yarn significantly; hence it is important that they are properly maintained.

9.2.1 Maintenance of blowroom machines

The main functions of the blowroom are opening and cleaning of fibres. The fibres fed to this machine are in the form of big/medium size, tightly packed lumps. These are opened by the beaters or blades and then cleaned. The maintenance functions of blowroom machines are regularly undertaken as per cleaning and preventive maintenance schedules. The frequency of the functions depends on the importance of that function towards the quality of the products, quantity of waste, etc. The functions include cleaning and greasing of all machines of the blowroom line, grid bar polishing, bypass valve checking, saw tooth beater wire, disc beater, pin roller, beater replacement, etc. The frequency of wire changes or grinding is undertaken after a fixed production level instead of a fixed time. The frequency depends also on the material processed. If polyester or similar fibre is processed, the gap between replacements comes down as compared to cotton processing machines due to the nature of the fibre. Beside the fixed frequencies, condition monitoring also decides the frequencies. For example, the quality of fibres is regularly assessed before and after each beater. If any of the parameters, such as nep formation or short fibre generation, increases significantly as compared to the standard, or if the cleaning efficiency of the line falls, maintenance is undertaken for that particular beater on priority basis. The settings play an important role in the quality of the product, and hence need to be checked periodically. In the latest cleaning machines by Rieter, Trutzschler and others, the intensity of cleaning and the amount of waste can be programmed and adjusted over a wide range without interrupting the operation of the machine. The desired settings can be entered directly on the machine panel or remotely from a central touch screen control panel. With increased usage of electronic control in textile machines, the mechanical complications have been reduced, but the presence of one or two staff with knowledge of electronic controls in a maintenance team has become essential.

9.2.2 Maintenance of cards

The major function of cards is to clean and individualise the fibres. For the perfect working of cards, the settings between different parts, such as cylinder-flat, cylinder-doffer and licker-in-cylinder play very important roles. Hence, for the carding department, as well as lubrication, cleaning and usual maintenance, regular checking of the settings is essential. Commonly the settings are termed as half setting and full setting, based on the extent of checking and correcting. Half setting is mostly associated with general cleaning, primarily involving removal of fluff, fly and accumulated dirt from various parts of the cards such as gears, cylinder and flexible bends, licker-in hood, under-casings and side covers. Half setting involves the dismantling and cleaning of the licker-in, feed plate and pneumafil ducts, and fitting them back and setting of flats, and cylinder, cylinder and doffer, and cylinder and cylinder under-casings. In the case of full setting, in addition to the functions covered under half setting, dismantling and cleaning of cylinder under-casings, back and front plates, cleaning of take-up units and complete resetting of the cards are undertaken. If tinted man-made fibres are processed by the cards, licker-in and cylinder under-casings need to be cleaned during full setting. The frequency of half settings is more than of full settings.

Metallic wires are mostly used in cards, and are also known as card clothing. The sharpness of the card clothing is very important for the proper processing of staple fibres. After long exposure to fibres, these wire points become blunt or damaged. These blunt-edged wires cause damage to the fibres. To improve the sharpness of the card wires, they are ground. The grinding process varies for different parts of cards, such as cylinder wire, dofferwire, etc. Licker-in wire does not need grinding, as it is used as long it maintains its condition, after which these wires are removed and new wires are mounted. For cylinder, flats and doffer the condition of the wires is checked regularly. The frequency of grinding of cylinder and doffer wires is best decided on the condition of the wire (by visual checking) and the quality of the fibre after processing. During each grinding process the tip of the wire, the land area, is increased and the depth reduces. When the depth reduces beyond a certain limit, the space for processing the fibres reduces considerably. Then the clothing must be replaced with new wires. By the time the teeth are ground down by 100 μm, generally the service life of the wire is considered to be over. Besides visual checking of the wire condition if card sliver neps and yarn imperfections start increasing drastically, the wires are replaced immediately with new clothing. In some places the wires are replaced after a fixed amount of production, for example 400 000 to 500 000 kg of sliver. Flat stripping comb, flat chain, cylinder undercasing, etc. are replaced after a fixed working time (e.g. 2 years, 3 years). This frequency is decided based on the recommendation of the machinery manufacturers, experience of the maintenance personnel, and experience of that particular unit. Presently online monitoring of nep level in the card web is available. This has led to the development of the 'Integrated grinding system'. This system automatically grinds the cylinder to maintain optimum sharpness of wire over its entire life span (e.g. Rieter C50 card). All this takes place when the card is running and without operator intervention. Associated software determines the need for grinding and then activates the grinding process.

9.2.3 Maintenance of draw frame, comber and speed frame

The major function of the draw frame is to parallelise the fibres in sliver. The drafting and doubling processes involve this machine. Its normal maintenance function includes greasing and cleaning of moving parts, setting of rollers, cots buffing, stop motion checking, oil changes, etc. The frequency of cots buffing depends on the material being processed, such as cotton, polyester or other fibres.

The slivers and yarns are tested regularly for their mass variation (CV%) by utilising offline and online measurement tools. The instruments used for this measurement generally provide spectrograph in addition to CV%. This spectrograph helps to identify the source of periodic variation if there is any. In addition, it provides some information on drafting waves. If the source of these faults is the draw frame, the identified parts are replaced or repaired.

The lap-forming machine for the comber is a prerequisite for combing. The regular preventive maintenance of this machine includes cleaning, cots buffing, proximity switch setting checking, full setting, etc. After an amount of buffing the cots are changed.

The function of the comber is to remove short fibres and neps. This, being a complicated machine, needs more thorough maintenance work. The checking of settings or gauge is a major part of the maintenance function. It includes brush gauge checking, unicomb gauge checking, proximity switch checking, etc. The settings of these parts play a very important role in the quality and cost of the products. Cleaning and greasing of various parts are also considered regular maintenance work. Various parts of the machine are replaced after a certain period, depending on the condition of the parts, such as nipper-pin change, draw-box cots change, brush change, top comb and unicomb change. The quality of the combed sliver is checked for qualities such as nep removal efficiency, short fibre content regularly. If the results are below expectation the settings are checked and corrected as necessary. The comber noil is also regularly checked for fibre length distribution. If the noil contains good long fibres the settings are checked and corrections are made. After a predetermined period, the machine is overhauled.

In the process sequence, as the process comes nearer to the final product, the influence on the quality and productivity of final product increases. For example if the speed frame does not perform properly, the quality of roving will deteriorate and as a result the quality of the yarn as well as the productivity of the ring frame will be affected. Different maintenance/cleaning functions need to be performed with different frequencies. Cleaning of important parts, such as flyer and bottom apron, and general cleaning should be conducted most frequently for example once in a month. However, certain other maintenance activities, such as spindle gauge checking, bottom roller gauge checking, arbor greasing, top and bottom apron washing, lifter shaft bearing greasing, pressure checking, top roll buffing, etc. can be done once in 6 months. Long term functions include cot change, false twister change, etc. which can be conducted once every 2 years depending on the fibres being processed.

9.3 Maintenance of ring and rotor spinning machines

The spinning machine is the final machine for yarn production. These yarns are woven or knitted to get fabric. The quality of yarn can influence the fabric quality and not too much can be done in fabric formation process to hide the imperfections of the yarns. Secondly, these spinning machines have the highest speed in the whole spinning line. The maintenance of these machines plays a major role in the industry's profitability. Both preventive maintenance and condition monitoring are followed. The yarn quality and utilisation loss of the machinery due to ends down, idle spindle, etc. are checked regularly. The commonly checked yarn qualities are yarn unevenness, imperfections, hairiness and tensile properties. Sudden change in any of these properties or fault in spectrograph analysis helps in identifying the source that is machine parts. Based on the analysis, preferential maintenance is undertaken. Online monitoring appliances such as Ring-data, Rotor-data help the maintenance personnel in identifying 'rogue spindles or rotors'.

9.3.1 Maintenance of ring spinning machines

The major maintenance function of ring frames includes spindle and lappet gauging, top arm pressure setting, jockey-pulley cleaning and greasing, traveller clearer setting, nose bar setting, draft zone cleaning, builder motion setting, pneumatic suction checking, etc. (Nijahawan 2006). The replacement of various parts is undertaken after a particular period of working. This period is determined based on the damages caused during the process, effect on the quality of the product, recommendation of the machinery manufacturers and experience of the maintenance manager. Generally parts such as cots, aprons, bobbin holders, spindle tapes and spindle oil are replaced at a predetermined frequency. The machine is overhauled around every 4 years. Condition monitoring, such as visual inspection of rings, study of yarn hairiness and end breakages, is generally used to determine the time of ring replacement. Similarly, study of yarn imperfections and evenness, and visual inspection, help in determining the time of top and bottom apron replacement.

9.3.2 Maintenance of rotor spinning machines

The rotor spinning is most popular amongst the open end spinning systems. The rotor speed generally ranges from 80 000 to 100 000 rpm and involves many automatic functions such as auto-piecing, auto-doffing, auto-cleaning, and start-stop mechanism. Hence, proper monitoring of this machine and proper maintenance is very important. Beside general cleaning, the maintenance of rotor spinning includes belt tension setting for the rotor drive, combing roller drive, main drive and pneumafil drive, spin box servicing, suction units checking, take-up roller cots buffing, etc. Parts such as the rotor, opening roller, etc. are taken for maintenance after a predetermined period, or if there is any damage. Overhauling is undertaken generally every 3 years.

9.4 Future trends

The importance of maintenance has been accepted by most spinning mills. With high investment in machinery and increased labour cost, various concepts are being developed to optimise the maintenance cost and loss of production due to machine stoppage. Total productive maintenance (TPM), 5-Zero concepts, application of 5S and reliability based maintenance (RBM)/reliability centred maintenance (RCM) are a few. More and more units are applying these concepts, and in future most will opt for one of these concepts.

9.4.1 Total productive maintenance (TPM)

TPM is productive maintenance carried out by all employees through small group activities. It involves keeping the current plant and equipment at its highest productive level through cooperation of all areas of the organisation. In fact, TPM is an extension of the total quality management (TQM) philosophy to the maintenance function. Despite increased automation and unmanned production, one cannot do without human labour (Ratnam and Chellamani 2004). Maintenance still depends heavily on human output. The first goal of TPM is preventive maintenance, performed to stop failures before they have a chance to happen. This is to machinery what preventive medicine is to human beings. The first important step is to ensure that the machine is not stricken with disease, through the performance of daily preventive activities, specifically cleaning, lubricating, tightening and inspection are carried out correctly. The overall goals of TPM are: maintaining and improving equipment capacity; maintaining equipment for life; using support from all areas of operation; encouraging input from all employees and; using teams for continuous improvement.

In addition to preventive maintenance, TPM also stresses the need of corrective maintenance, performed for the purpose of increasing overall reliability and maintainability. In addition, it calls for taking steps to ensure that new machinery will be maintenance free. A TPM activity board is generally posted in the work place so that everyone will be able to comprehend immediately the state of the TPM programme; tags are put on the machines at areas where problems occur; amounts and times of lubrication are posted at the lubrication area on machines; the direction of rotation is pasted on valves and rotating members of the machines; pipes are colour-coded for easy distinction; and other similar steps are taken to satisfy the need for knowledge based on 'here-and-now-ism' in TPM.

The major characteristics of TPM are known as 'autonomous maintenance by operators'. Autonomous maintenance denotes that all operators will maintain their own equipment, performing the preventive maintenance practices described above to ensure their machines do not suffer serious breakdown. The idea behind this approach is that the person actually using a machine is the one best qualified to perform the daily preventive measures – cleaning, lubricating, tightening and inspecting – needed to keep it running properly. Generally the first task is to break the traditional barriers between maintenance and production personnel so they are working together. There is one more advantage of TPM, in addition to its major contributions, which is the elimination of failures, defects and accidents – the machines and work environment in shops where there is a thriving TPM programme tend to be almost unbelievably clean, making them more conducive to human endeavour. TPM requires vigorous training at every level, from top management personnel to the machine operator.

The main objectives of TPM are zero breakdown and zero defect. When breakdowns and defects are eliminated, equipment operation efficiency improves, costs are reduced, inventory can be minimised and, as a consequence, labour productivity increases. One firm reduced the number of breakdowns to a fiftieth of the original number. Some industries showed 17–26% increase in equipment operation rates, while some showed a 90%

reduction in process defects. Labour productivity can increase by 40–50% as a result of successful implementation of TPM. One textile unit in India claims the following increments in machine availability per annum due to implementation of TPM: Carding – 340 h; Comber–150 h; Fly-frame – 148.5 h; Ring frames – 181.7 h; Ring-doubling – 111 h; Autoconer – 76 h.

9.4.2 Application of '5S' concept

The '5S 'concept has become popular in many industries as a quality management tool (Basterfield *et al.* 2008). Similarly, this concept can greatly improve the efficiency of maintenance function. '5S' has come into existence in different parts of the world, though it was conceived and first applied in Japan. The name derives from five Japanese words; Seiri, Seiton, Seiso, Shitsuke and Seiketsu. *Seiri* means organisation, that is, putting things in order – organising them – in accordance with specific rules or principles. In 5S terms it means to distinguish between the necessary and unnecessary, and to implement stratification management to get rid of the unnecessary. *Seiton* means neatness. In general usage as well as 5S usage, this means having things in the right place or right layout, so that they can be used immediately. It is a way of eliminating searches. Once everything has a right place, so that it is functionally placed for quality and safety, one can have a neat workplace. *Seiso* means cleaning. In 5S terms, it means getting rid of waste, grime and foreign matter and making things clean. Cleaning can have a tremendous impact on downtime, quality, safety, morale and every facet of the operation. *Seiketsu* means standardisation. It means continually and repeatedly maintaining the organisation, neatness and cleaning. Innovation and total visual management are used to attain and maintain standardised conditions so that one can act quickly. *Shitsuke* means discipline. It means instilling the ability to do things the way they are supposed to be done. Whether it is emergency procedures, standard operating procedures, it is crucial that every effort be made to get staff to do each and every step each and every time.

It can be seen from the above that the 5S concept involves cleanliness, discipline and standardisation, and all these are very important parts of maintenance management. If the spare parts are kept in orderly manner, and different places are marked for different parts, that is, pulleys of different sizes, belts of different size and types, bearings, etc. are kept in marked places separately, it saves a lot of time and energy for the maintenance personnel. Similarly staff engaged in maintenance activities should be equipped with right tools and a trolley should contain all the necessary tools, placed in orderly manner. This will improve the efficiency of the maintenance team considerably. Hence, many units are using 5S tools for improvement in efficiency of maintenance team.

9.4.3 Computer managed maintenance system

This is not any new system of maintenance but is making use of computers for quickly and efficiently planning and organising various jobs for undertaking systematic plant maintenance. For large and complex organisations, manual collection and analysis of the following information and records for ideal maintenance systems become extremely difficult and time consuming:

- Correlating causes and effects for defect and jobs done earlier.
- Enough statistics for taking a proper decision.
- Proper and undistorted communications of machine health, defect list and other information.
- Accurate and timely generation of information such as jobs done, resources used, special problems faced, jobs not done, etc.
- Planning and scheduling of various jobs/actions well in advance.
- Cost incurred and budget aspects.
- Safety, environment and statutory requirements.

Some of the main difficulties in achieving these requirements are: high volume of data, chances of human error during recording of same/similar data manually, difficulties in communication of information involving different departments, checking and analysis of statistical information, and the requirement of more manpower to handle huge volumes of data (Hartmann 1987).

A Computerised Managed Maintenance System generally covers the total maintenance management system, consisting of equipment classification/information, maintenance planning, material planning, captive engineering shop planning, inventory control, down time information, work order management, maintenance cost, maintenance audit and performance measurement, etc. The payback on the expense of computerisation of the maintenance management system comes in different ways. One of the most significant benefits is the time saving on historical data retrieval. Planning and scheduling become considerably easier. All identified work is immediately accessible, allowing maintenance supervisors to change their schedules easily if an outage occurs.

9.5 Sources for further information and advice

All standard machinery manufacturers provide information on preventive maintenance and lubrication requirements, along with the machine manuals. Those manuals should be consulted before overall planning of maintenance functions. In addition, many manufacturers conduct regular training programmes for maintenance personnel at their premises. Maintenance personnel should be sent to attend similar training programmes in rotation. This will update their knowledge and they will be motivated.

9.6 References

Hartmann, E. (1987) *Maintenance Management*, U.S.A., Industrial Engineering and Management Press.

Basterfield, D.H., Basterfield-Michna, C., Basterfield, G.H. and Basterfield-Sacre, M. (2008), *Total Quality Management*, Delhi, Dorling Kindersley (India) Pvt. Ltd.

Nijahawan, N. (2006) *Comprehensive Handbook on Spinning Maintenance; Part I: Maintenance Management*, India, The Textile Association.

Ratnam, T.V. and Chellamani, K.P. (2004) *Maintenance Management in Spinning*, India, The South India Textile Research Association.

Part III

Process control in fabric manufacture, coloration and finishing

10
Process control in knitting

S. C. RAY, University of Calcutta, India

DOI: 10.1533/9780857095633.3.243

Abstract: Acceptability of a knitted product to the consumer largely depends on its quality. The manufacturer has a moral and ethical duty to ensure that the customer receives material of quality proportionate with the price he or she is paying for it. For this reason, inbuilt process control by the manufacturer is essential. This can only be achieved by regular inspection and testing of the raw material, the knitting process, the knitting machine and the finished product. Control at different stages of manufacture is also important. To be effective, the process control scheme for knitted fabric must begin with checking incoming yarn and extend through and beyond the knitting process. From the consumer's point of view, the important parameters of a knitted fabric that need to be maintained for a specific quality and yarn count are the weight per square metre (GSM), width and expected shrinkage during future wet treatment. The difference in loop length within a fabric can produce visible defects, the most common being the occurrence of widthwise bars or streaks. In fact, the most important criteria for producing quality knitted fabrics are aimed at controlling loop length. All the checks and process controls applied throughout the knitting process are highlighted in this chapter.

Key words: checking, defect, loop length, process control, quality, spirality.

10.1 Introduction

The acceptability of a knitted product largely depends on its quality. Consumers have become more quality conscious. The manufacturer has a duty to ensure that the customer gets material of a quality commensurate with the price they are paying. The manufacturer also needs to maintain a uniform standard of quality, as well as to meet the required specification. Effective process and quality control are essential. This can be achieved by regular checking and testing of the raw material, knitting processes and the finished product. Control at all these different stages of manufacture is important.

To be effective, the quality control system for knitted fabric must begin with checking the incoming yarn and continue throughout the knitting process. The dimensions of knitted fabrics are determined by the number and size of stitches per unit area; this in turn is determined by loop length. A difference in loop length within a fabric can produce appearance defects, the most common being the occurrence of widthwise bars or streaks. In fact, the most important criteria of quality control in knitting are aimed at controlling the loop length.

The structure and culture of the knitting industry, particularly in India, is quite different from other textile industries involved with spinning, weaving, etc. With a few exceptions, most knitting houses are decentralised and small scale. Although the knitting houses have been achieving their own quality targets through necessary checking and testing for the past few years, common standards and norms are not well established and are not available for specific knitting processes and knitted products. However, in order to satisfy the quality demands of consumers the manufacturers have to control parameters such as GSM and the width of the fabrics. The knitting industry is split into three sub-sectors – flat bed (both hand and power driven) weft knitting, circular weft knitting and warp knitting. These three methods vary widely in their features and requirements. Based on the information available,[1–9] the scope of process control in all the sub-sectors of knitting is discussed in detail in this chapter.

10.2 Key control points in knitting

The quality of the fabrics/garments produced in knitting largely depends on the quality of the yarn, knitting machine condition, process parameters of knitting and ultimately the process control scheme adopted in the industry. So attempt should be made to identify the key points in knitting and control those as per norms.

10.2.1 Checking and testing of yarn

Checking and testing of yarn involves parameters such as:

- count,
- unevenness (U%),
- strength,
- twist,
- elongation,
- appearance,
- coefficient of friction,
- bending rigidity.

With the exception of the coefficient of friction and bending rigidity, all properties listed need to be tested and verified. These parameters are common and easy to measure. Standards and normal measurements to draw comparisons with are available for spinning. The values of these parameters are usually specified by the yarn supplier. To assess a yarn's suitability for the

Table 10.1 Parameters for 40^s Ne cotton yarn suitable for circular weft knitting

Strength	16 g/tex
Elongation	7%
Unevenness	10.5%
Twist factor	30.23 (3.16 in an English cotton system)
Hairiness of a typical cotton yarn of 40^s Ne (14.8 Tex)	6.5

purpose of knitting, these parameters need to be measured. Some change in yarn count and/or count mixing will certainly cause variation in GSM and impact on the quality of the end fabric. Higher values of unevenness can result in various defects in the fabric including yarn breakage and ultimately loss in machine efficiency. Twist can affect the softness and absorption properties of the yarn, air permeability of the knitted fabric, and the comfort of the wearer. Generally twist multiplier of the knitting yarn is maintained at a lower level than the weaving yarn of a similar count. Therefore the count, tensile properties and twist of the yarn need to be rigorously checked before use in knitting processes, to confirm that the requirements are met. The desired parameters can be seen in Table 10.1.

The coefficient of friction and bending rigidity are very important for understanding the mechanics of the knitting process, as well as the performance of a particular yarn during knitting. The techniques used to determine these two parameters are likely to be mandatory in the near future in the engineering of all knitted technical fabrics.

Relationship between yarn properties and knitted fabric qualities

There are a number of links between yarn properties and the resulting quality of the knitted fabric. The appearance of a knitted fabric is therefore strongly affected by any defects in the yarn used. Some of these relationships have been summarised in Table 10.2.

10.2.2 Checking of machinery

Checking of machinery involves inspecting the following:

- stop motion,
- machine speed,
- feeding arrangement,
- lubrication system,
- take-down mechanism.

Table 10.2 Effect of yarn defects on fabric appearance

	Cause (yarn defect)	**Effect** (fabric appearance)
a.	Very uneven yarn	Cloudy fabric
b.	Yarn with very low strength and/or too many thin places	Holes or cracks
c.	Yarn with long term unevenness	Fabric with stripes
d.	Bad dyeing or blending	Horizontal stripes
e.	Insufficiently paraffinated or rough yarn	Drop stitches and holes
f.	Too large or weak knots or bad splicing	Yarn end separation or breakage during knitting
g.	Uncleared thick places	Irreparable faults in the fabric
h.	Too much yarn hairiness	Diffused stitch appearance and fluff build-up
i.	Very high twisted yarn	Spirality or distorted stitches

There are three or more types of stop motion used in knitting machines. In the case of yarn break stop motion, as soon as a yarn breaks the machine stops automatically. The operator is not constantly required to monitor whether any yarn is breaking or not; after the machine stops he simply repairs the yarn breakage and restarts the machine. On the other hand, if the yarn breakage is not sensed and the machine is not stopped automatically by the stop motion the machine will continue to run with broken yarn and fabric will drop down. Restarting the machine after repairing the defect may take a few hours, resulting in a very high loss of machine efficiency. Furthermore, the process of repairing the problem will give rise to a major defect in the fabric.

For various reasons, such as slackening of the belt, jamming of machine parts, etc., the machines are sometimes run at comparatively lower speeds than the target, which slows down production. Machine speeds must be checked at regular intervals and rectified if required.

Proper setting, alignment and cleaning of the yarn feeding arrangement ensures smooth supply of yarn for even running of the machine. Deposits of fluff and dirt in the feeder or misalignment of the feed plate can cause obstruction of yarn flow and can even cause the yarn to break. The take-down mechanism not only brings down the fabric but also controls the courses per inch in the fabric. Improper setting of the take-down mechanism may cause defects such as skew, horizontal lines and dropped stitches in the fabric. So the take-down mechanism must be checked and regulated to achieve the desired courses per inch in the fabric with minimum other defects related to take-down. Lubrication of the machine, as per the schedule provided by the machine manufacturer, is important for smooth running of the machine with minimum wear and tear, vibration and heat generation. As the oil used for lubrication could stain the fabric, it needs to be tested beforehand to check that the oil washes out.

Setting up processes and machine parameters

The most important parameters and settings for knitting machines with regard to quality control are:

- yarn input tension,
- stitch cam setting,
- take-down load,
- rate of yarn feeding,
- number of feeders,
- yarn patterning in feeders,
- needle gating,
- tucking and floating arrangements.

Yarn input tension plays a vital role in influencing the mechanics of knitting and the robbing back phenomenon which determines the ultimate loop length.[2,3] Very low and very high input tensions are detrimental for knitting. In general, loop length decreases with an increase in input tension. Input tension can easily be adjusted with the help of the yarn tensioner and can be measured using a yarn tension-meter. A new concept in this field is the incorporation of a built-in tension-meter on every yarn feeder for continuous measurement and control.

The stitch cam setting is the most important parameter for varying loop length in the knitted fabric. Loop length always increases with increase in the stitch cam setting in both single jersey and double jersey machines, with the exception of a few cases under delayed timing (3-Needle or more delay) in double jersey machines. Circular weft knitting machines are provided with a large number of stitch cams; the cam setting must be uniform across the circumference otherwise an uneven loop length will occur and fabric will be defective. Positive yarn feeding is important to help achieve a uniform setting for all stitch cams. Arrangement of a centralised cam setting from a particular point can solve this problem.

The checking and control of the yarn feeding rate and take-down load are also important, as both these parameters influence the yarn tension during the knitting process and the resultant loop length in the knitted fabric.

In order to increase production more feeders can be added to a circular weft knitting machine. However, due to mechanical defects or shortage of material, one or more feeders are sometimes kept idle. In this situation knitting continues, but the fabrics in the non-feeding or idle stitch cam zones become prone to defect.

The order of needle gating depends on the design of the knitted fabrics. If the order changes anywhere in the needle bed the design will vary and the fabric will be defective. For this reason it is essential to inspect the order of needle arrangement in the needle bed and rectify any changes that take place.

Sometimes tuck or float loops are created in a regular pattern to make various derivatives of regular designs to enhance the aesthetic value of the knitted fabric. Therefore it is also essential to inspect the mechanism of tuck and float loop formation at regular intervals, otherwise defects may occur.

10.3 Quality control of knitted fabrics

Some important factors to check with regards to quality control of knitted fabrics are:

- loop length,
- GSM,
- courses per inch,
- fabric width,
- wales per inch,
- fabric defects,
- stitch density,
- fabric tightness factor,
- yarn count,
- fabric construction (design),
- yarn type.

Loop length is the most important specification for fabric quality in knitting. Most scientific studies regarding quality control of knitting and knitted fabrics are based on loop length. The manufacturers of knitted products are also gradually becoming more aware of the value of consistent loop length. The consumers of knitted fabrics are very much concerned about loop length, as are traders of knitted fabrics, making it financially one of the most important quality control factors. For this reason loop length is discussed in more detail later in the chapter.

The physical properties of the yarn, and ultimately the properties of the knitted fabric, are influenced by the yarn count, fibre composition and the settings of the spinning machinery used to make it.

The GSM of knitted fabric can be adjusted very easily by varying yarn count, without altering the stitch density or loop length. Today, most trading of circular weft knitted fabric is based on GSM and width. If count is not specified, manufacturers of knitted fabric can achieve the target GSM using coarser yarns with lower stitch density. The consumer is subsequently deprived of the desired quality of fabric. Furthermore, count variation between the feeders will not only produce uneven fabric but could also result in defective fabric.

Courses per inch (cpi), wales per inch (wpi) and stitch density are basic parameters of knitted fabric; variation of these specifications within the fabric will result in uneven thickness. The courses per inch and wales per inch are interrelated[4] and mainly depend on loop length governed by machine gauge, stitch cam setting, yarn properties and yarn input tension. These three parameters should be strictly controlled to produce high quality, even fabric.

The fabric tightness factor is the measure of compactness of the knitted fabric. It is calculated using yarn count and loop length and is directly proportional to the coarseness of the yarn, but inversely proportional to the loop length. Therefore, the tightness factor can be increased by using coarser yarn or smaller loop length, and vice versa.

Generally, the variety of structures available using circular weft knitting is limited, but there is scope to produce complex designs using Jacquard and multiple cam tracks on a knitting machine. One of the unique features of flat bed weft knitting and warp knitting is the variety of ornamented structures that can be created. It is important to check that the desired design has been produced correctly by analysing the design after processing.

10.3.1 Controlling GSM in knitted fabrics

As mentioned above, GSM is broadly dependent on stitch density (cpi × wpi), loop length and yarn count. In general, if the stitch density is high, if the yarn diameter is large, or if the yarn is heavy, the GSM will increase proportionally. However, if the loop length is high then the GSM will decrease, as stitch density decreases at a higher rate to the increase in loop length. For double jersey fabric, the additional factors of knitting timing and the gap between the two beds can also cause variation.[3,5]

GSM is affected by variation in a large number knitting parameters. Therefore, in order to control GSM, these parameters need to be controlled. Depending upon the requirements of the consumer, GSM is mainly controlled by changing the stitch cam setting, the yarn input tension and the yarn count.

10.3.2 Testing the quality of knitted fabric

Some of the specifications which need to be considered with regards to testing the quality of knitted fabrics include:

- fabric yield,
- fabric extension,
- fabric appearance,
- air permeability,

- fabric pilling,
- fabric bow and skew,
- dimensional changes,
- abrasion resistance,
- angle of spirality,
- bursting strength.

Knitted fabrics are used to create various end products, all of which require a high quality fabric to work from. The fabric will have to undergo further processing before it reaches the consumer, therefore it is necessary to assess the fabric throughout to maintain quality and consistency.

Fabric yield is expressed as a percentage of the ratio of weight of fabric produced (output of the knitting machine) compared with the weight of yarn used (input of the knitting machine). As some wastage of yarn is unavoidable in continuous processing, the yield will almost always be less than 100%. However, due to moisture gain (higher moisture content in the ultimate fabric than the moisture content in the yarn used for knitting), the loss in yield due to wastage of yarn is mostly compensated for. Steps must be taken during the knitting process to ensure that yarn wastage is kept to a minimum.

Knitted fabric properties such as fabric extension under a predetermined load, pilling, abrasion resistance and bursting strength are tested and compared with standard values. These values are set by the manufacturer, based on assessment of performance in the end use of the fabric. For example, pilling is a very common defect in the end application. It is therefore recommended to produce fabric on a laboratory scale to test pilling tendency for different yarns, before moving on to large scale production.

In order to identify the presence of defects such as unevenness (thick and thin places), bow and skew, and spirality, the fabric is passed over an inspection table. Arrangements are then made with regards to the removal/rectification of the defects and any further action which needs to be taken.

10.3.3 Important quality aspects of fabrics from consumers' point of view

The dimensions of some relaxed, knitted fabrics can change when they are subjected to wet treatments, such as scouring, bleaching, dyeing and compacting. This change in dimensions, has the ability to affect many of the knitted fabric parameters, including courses per inch, wales per inch, stitch density, GSM and fabric width/circumference. With the exception of loop length, the parameters of knitted fabrics will continue to change throughout post knitting operations. With this in mind, the parameters of knitted fabrics are only measured in their finished state (usually after compacting).

Shrinkage during subsequent wet treatments, that is, performed by the consumer, also needs to be taken into account. The tolerances for a change in GSM (plus or minus) are only 4–5% of the ordered GSM value. For example, against the ordered value of 150 GSM, the manufacturer is required to produce a fabric with a GSM in the range of 144–156, otherwise orders may be cancelled.

Consumers of knitted fabrics are also concerned with the final width of the fabric (in a two-fold state for circular fabrics and in an unfolded state for flat fabrics). The tolerances here (plus or minus) are again only 4–5% for wide width fabrics, which are cut and opened, before being transformed into the end products. However, for body size fabrics for innerwear (e.g. vests), the reasonable tolerances (plus or minus) are only 0.5 in, irrespective of the diameter. So if the consumer requires fabric for making a vest 36 inches in circumference, the manufacturer must supply knitted fabric with a circumference in the range of 35.5–36.5 in. In addition, the finished fabric must not shrink more than 3% during subsequent wet treatments and throughout the life of the end product. In recent years the leading consumers, particularly from Europe and the USA, have also specified requisite yarn parameters, such as CV% of yarn count, number and magnitude of yarn faults, etc. in addition to the yarn count, unevenness percentage, tensile properties, and the acceptable limit of spirality in the finished fabric.

It has become apparent that more norms and standards for knitted fabrics are to be established in the near future, keeping in mind the satisfaction of consumers on the one hand and the performance of the end product on the other.

10.4 Control of knitted loop length

The basic unit of a knitted structure is the loop and the length of yarn contained in a knitted loop is called loop length. The length of yarn going in to a loop during knitting depends on many factors but it is controllable. The properties of knitted fabric mostly depend on the loop length.

10.4.1 The importance of maintaining loop length

It has been observed by many researchers in the field of knitting that the dimensional properties of knitted fabrics are mainly governed by two parameters, namely the length of yarn in a loop and the shape of the loop. Although the shape of the loop is finalised upon relaxation treatment of the fabric, the length of loop is usually defined by the knitting machine during loop formation. This relationship between loop length and dimensional properties is applicable to any knitted fabric, irrespective of type and count of yarn, or type and gauge of machine. The three basic laws which govern the behaviour of knitted structures as laid out by Munden[4] are

1. Loop length is the fundamental unit of weft knitted structure.
2. Loop shape determines the dimensions of the fabric and this shape depends upon the yarn used and the treatment which the fabric has received.
3. The relationship between loop shape and loop length may be expressed in the form of simple equations.

10.4.2 Measurement of loop length

The simplest way of checking loop length in a knitted fabric is by measuring the uncrimped length of yarn unravelled from a knitted fabric with a known number of stitches (wales). Unravelled yarn from a knitted fabric is usually crimped, therefore the yarn has to be straightened, but not stretched, by applying a load (say 0.2 g/tex) in order to measure the exact length. The decrimping load can be calculated using a standard tensile tester. The measurement of the straightened length of yarn can be carried out using any simple apparatus; the HATRA Course Length Tester may be used for the purpose. Of course, this method of measurement is destructive and has to be carried out after the knitting process. This is popularly known as 'Off-Machine' measurement.

For measurement of loop length during knitting rather than at inspection, a variety of yarn speed meters and yarn length counters are available. These are used to measure yarn consumption rates in relation to the speed or number of machine revolutions in circular knitting; the loop length is then worked out using a simple calculation. Measurement of loop length during the knitting process is known as 'On-Machine' measurement.

More recently, techniques have been developed using image analysis to determine the loop length in any fabric with the help of computer. The fabric sample is scanned or viewed using a magnifying lens; the photographs of the components of yarn which make up the loop are then processed by the computer, either in 2-D or in 3-D form, in order to calculate loop length and further analyse the loop structure. This image analysis technique is non-destructive, but still produces accurate results.

In warp knitting it is more difficult to gain a direct measurement of loop length. The yarn consumption of each guide bar is measured for 480 knitting cycles; this is known as the run-in per rack. This measurement is then used to calculate the loop length. The yarn quantity can be measured using special measuring equipment or simply by marking one of the yarns close to the warp beam and then measuring its position after 480 cycles (1 rack). The run-in per rack may vary from guide bar to guide bar, and the relative amount of yarn fed from each beam can also have an influence. This relationship is known as the run-in ratio and it affects different fabric constructions and qualities in different ways. Therefore, the length of warp wound onto the beams is always dependent on both the run-in per rack and the run-in ratio.

10.4.3 Control of loop length

The size of the loops and the dimensions of the knitted fabric are influenced by the amplitude of the kinking movements used in loop formation. For this reason, the knitting machines are provided with adjustable cams that enable this amount to be varied. The control exerted by such fine adjustments is not in itself sufficient to dictate the amount of yarn drawn from the supply package.

Yarn tensioner and the fabric take-down setting, which can be adjusted by the knitter, exert a significant effect on loop length. There are also many other factors, including temperature, humidity, yarn extension, package hardness and build, which are largely outside the control of the knitter, that can produce variations in yarn friction and tension, thus affecting loop length.

Attempts have been made to produce constant tension devices that could eliminate the effect of the variations mentioned above, as well as allow the needles to draw an amount of yarn determined solely by the cam setting on the knitting machine.

These trials were only partially successful. The positive (measured) feed devices subsequently developed to perform this task have resulted in a significant reduction in course length variation and dimensional variation. Fabric made using a positive feed device has a more even appearance and less barré. Adjustment of the positive feed device is quick and simple. The required course length is obtained by changing the roller diameter and/or gearing. The contact drive between the friction surface and the yarn is maintained on a nip or capstan roller principle.

For flat bed knitting machines or circular knitting machines with Jacquard, where positive feed devices cannot be attached for loop length control, the storage feed assembly can make a significant contribution to knitting efficiency by improving yarn control and allowing the economic use of effect yarns that were formerly regarded as difficult to knit because of draw-off problems.

10.5 Common faults in knitted fabrics

Although the basic aim is to produce fault free fabric, occurrence of some faults is unavoidable in the fabrics during knitting. Such faults or defects are caused mainly due to defective raw materials, improper settings of the knitting machine and lack of supervision.

10.5.1 Sources of faults

Some sources of faults in fabrics are:

- raw material (yarn),
- yarn feeding and feed regulator,
- machine setting including pattern,
- climatic condition,
- machine maintenance,
- miscellaneous.

As fabric is constructed of yarn, any fault or defect in the yarn could produce faults in the knitted fabric (holes, horizontal lines, bars and streaks, press-off, etc.). For this reason, storage, procurement of appropriate yarn, and quality checks of the yarn are essential in order to eliminate faults in the knitted fabric.

Improper feeding of yarn can cause tension variation, leading to defects in the final fabric, yarn breakage and loss of efficiency. Monitoring and regulation of the feeding system are also important from quality point of view.

Machine settings, such as the stitch cam setting, tensioner settings, the take-down mechanism and pattern/design settings, play a critical role with regards to the quality of the fabric and efficiency of the machine. Therefore control of these settings is of prime importance for producing quality fabric without faults.

Climatic conditions, such as extreme hot, cold or humidity in the room where the knitting processes are taking place can cause the knitting machinery to deteriorate, which could cause faults in the fabric produced. An air conditioned room with regulated humidity is always advisable for the smooth running of the knitting machines.

Proper maintenance of knitting machines is also extremely important for the smooth running of processing without causing faults. Maintenance includes proper lubrication, adequate repair of major and minor defects in the machine, replacement of broken and defective parts, etc. For example, a bent needle or trick can cause obstruction in the movement of the needle inside the trick during loop formation, raising the yarn tension during loop formation to a very high level. This can result in variation of loop length and even yarn or needle breakage. Maintenance work needs to be carried out on regular basis by qualified and skilled personnel using proper tools, proper machine parts or elements, and lubricants.

10.5.2 Types of faults in knitted fabrics

Common faults that can occur in knitted fabrics, along with probable causes, are shown in Table 10.3. Once the cause of the defect is ascertained the necessary remedy can be provided. The defects in Table 10.3 mainly occur during the knitting process and are observed in grey knitted fabrics.

After wet processing (scouring, bleaching, dyeing, printing, finishing), such defects generally become more prominent, and new defects may also appear on the knitted fabric. Defects which are commonly found in wet processed fabric are:

1. metamerism,
2. tiny holes,
3. stains,
4. dried up marks,
5. white patches,
6. dyeing patches and lines,
7. stitching puncture,
8. pin holes.

10.5.3 Spirality in knitted structures

Spirality or skew is a common defect that is generally found in single jersey structures. Ideally courses and wales should be at right angles to each other. Skew occurs when wales are displaced from their vertical position; this is known as wale skew. Skew can also occur when courses are displaced from their horizontal position (course skew). Usually, wale skew is caused by yarn and course skew is caused by feeders. It is important to note that skew caused by yarn and skew caused by the number of feeders in the machine can combine to create more skew, or alternatively it can offset the skew. This distortion of either the wale line or the course line is popularly known as spirality. Spirality is a dimensional distortion in the circular knitted fabric. It is undesirable as it can lead to displacement of seams, mismatched patterns and other sewing difficulties.

The main source of spirality is the twist liveliness of the yarn used. Loop formation involves both twisting and bending, resulting in twist redistribution in the arms of the loop. If the yarn is twist lively, so that it tends to snarl upon itself, then the loop shape is affected as the yarn in the fabric is prevented from snarling by its contact with adjacent loops. The net result is that all the loops in the fabric take up an inclined position, giving the fabric a skewed or spiral appearance, and the wale lines are no longer at right angles to the courses.

Table 10.3 Types of faults in knitted fabrics

S. No.	Name of the fault	Probable causes of the fault
1	Vertical lines	• Defective needles and tricks • Needles loose or tight in the trick • Mixing of needles/sinkers of different types
2	Horizontal lines	• Uneven yarn • Uneven yarn input tension • Uneven take-down load • Mixed yarn • Loose stitch cam • Uneven twist in yarn • Poor unwinding of yarn
3	Holes and cuttings	• Weak yarn • Yarn with knots, slubs, etc. • Lint in yarn path • Very high speed of the machine • Rough or defective sinker • Needles too tight in the tricks • Higher yarn/fabric tension • Unsuitable or very small loop length
4	Dropped stitches	• Low yarn tension • Low take-down load • Stiff needle • Wrong stitch cam setting
5	Distorted stitches (spirality)	• High twist (torque) in yarn • Incorrect positive feed setting • Uneven yarn input tension • Bent trick wall • Wrong stitch cam setting • Bad or bent needles
6	Press-off	• Faulty stop motion • Poor yarn quality • Plugged yarn guide • Machine running fast • Weak knots and slubs
7	Bursting	• Too tight fabric (very small loop length) • Stiff needle latch • Uneven yarn input tension
8	Barré (bars or streaks in course direction)	• Uneven yarn input tension • Defective yarn • Improper needle action
9	Off pattern	• Defects in design elements • Incorrect feeding arrangement
10	Bias/Skew	• Uneven take-down load • Mixed yarn
11	Soiling	• Bad handling • Dirty working environment
12	Oil stains	• Oil drippage from machine parts • Bad lubrication technique

S. No.	Name of the fault	Probable causes of the fault
13	Big knots, slubs, etc. on the fabric surface	• Bad yarn quality
14	Motes and foreign matter	• Poor yarn quality, motes and foreign matter present in the yarn • Dirty working environment
15	Snagging	• Mechanical strain during knitting in case of continuous filament yarns
16	Colour fly	• Single fibres, bunches of fibres or yarn pieces in varying colours stuck on the yarn or knitted into the fabric
17	Winder line (lines at the two edges of a tubular knit fabric)	• Too high take-down load • Long storage of fabric in roll form after knitting • Dense fabric structure

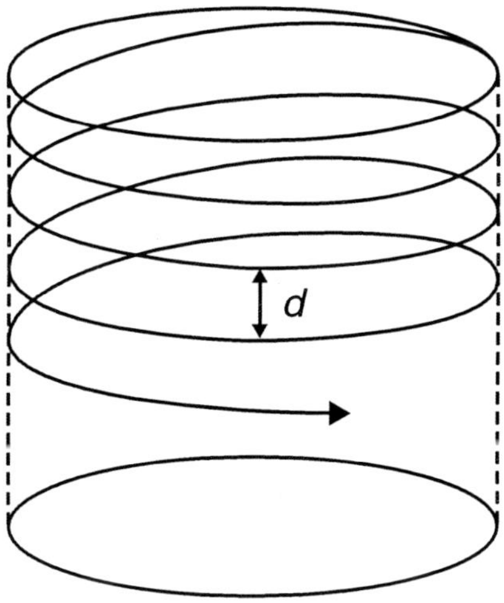

10.1 Spirality in course direction due to multiple feeders.

On a single jersey knitting machine, for each feed of yarn one revolution of the machine will make one course of fabric. The more yarn feeders that are used, the more courses will be created in one revolution of the machine, which means that the courses are stacked one on top of each other after each revolution. This creates a spiral line (Fig. 10.1). The distance between the spiral lines represents the production of courses for one revolution of the cylinder. For example, if one revolution of the cylinder produces 1.5 inches of linear metres of cloth, then there will be 1.5 linear inches of skew in the course that is generated. Machines with a large number of feeders can create substantial skew or spirality in the fabric.

Spirality of more than 5° is clearly visible in knitted fabrics, and this is considered to be a fault. The spirality value generally increases after washing due to relaxation of the residual torque of the yarn. Yarn with higher twist multiplier always produces higher spirality than a yarn with low twist multiplier.

The probable factors which affect spirality in knitted fabrics are:

- twist liveliness (spirality increases with increase in twist multiplier),
- feeder density in the machine (spirality increases with increase in number of feeders),
- tightness factor of the fabric (spirality decreases with increase in tightness factor),
- machine gauge,
- yarn linear density,
- variation in knitting tension and yarn frictional properties,
- spinning technology (Friction > Ring > Rotor >Air jet),
- residual torque of the yarn,
- combination of Z-twisted and S-twisted yarns in alternative order reduces spirality,
- twist setting of yarn in autoclave,
- plied yarns produces lesser spirality than single yarn.

Spirality in the wale direction can be measured in terms of degree of spirality [$\theta = \tan^{-1} (D_w/L_w)$] where θ = spirality angle, D_w (AA') = displacement of the wale from a normal line to the course of a fabric measured at a distance L_w (OA) (Fig. 10.2). Similarly, the degree of spirality in the course

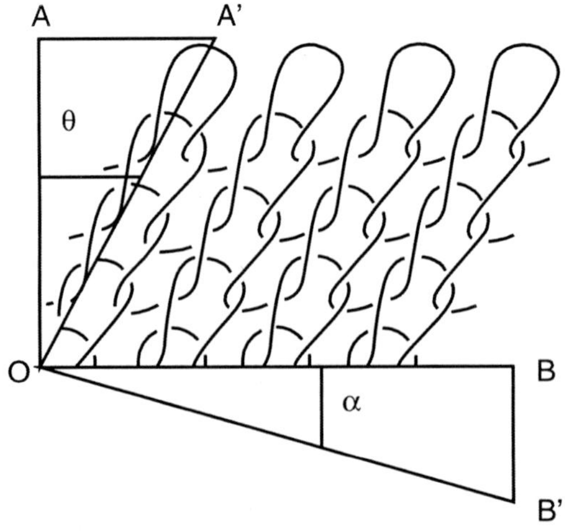

10.2 Spirality angle in single jersey structure.

direction will be [$\alpha = \tan^{-1} (D_c/L_c)$] where α = spirality angle, D_c (BB') = displacement of the course from a normal line to the wale, of a fabric measured at a distance L_c (OB) from the identified wale line.

Spirality becomes more prominent when widthwise striping is produced using two or more coloured yarns in a machine with a very high number of feeders. In that case the value of '*d*' will be *F*/cpi where *F* is the number of feeders.

Practical problems arising from spirality encountered in garment production are displacement or shifting of seams, mismatching of patterns, sewing difficulties, etc. Spirality has an obvious effect on both the aesthetic and functional performance of knitted structures and garments produced from them.

Spirality can be eliminated by setting the twist in the yarn or by using balanced two-fold yarns where possible. However, with single yarns of natural fibres, the set is usually not permanent and can be reversed by washing. The effect of the direction of machine rotation in relation to the direction of twist in the yarn has some effect on spirality, but it becomes negligible after the washing of the fabrics. The easiest and most popular technique for minimising spirality in knitted structures is to use 'S' twisted and 'Z' twisted yarns in alternative feeders during knitting. The neighbouring yarns, which are twisted in opposite directions, act in an opposing manner and neutralise the spiral formation. Plating is another effective way to produce spirality free fabric.

10.6 Other process control factors in knitting

In addition to the detection and minimization of the common faults through process control, attempts are being made to identify the factors which may result either some undesirable properties in the end product or loss in efficiency of the knitting process.

10.6.1 Main factors affecting the dimensional properties of knitted fabrics and garments

The main factors affecting the dimensional properties of knitted fabrics are:

- Fabric structure – different structures relax in different ways.
- Type of fibre(s) – fabric or garments made from different fibres relax differently.
- Stitch length – the length of yarn in a knitted loop is the determining factor for all structures.
- Relaxation/finishing route – the fabric dimensions vary according to relaxation/finishing sequence.
- Yarn linear density – yarn diameter affects the dimensions slightly but affects the fabric tightness, area density and other physical properties.

Table 10.4 Prerequisites for faultless production in knitting

(a)	The machine must be installed on a true horizontal floor or surface, as far as possible, without any vibration.
(b)	Yarn package holders must be mounted in such a way that the yarn should not rub against the sides of the packaging when it is expelled.
(c)	Yarn should be guided from the package up to the knitting zone without unnecessary deviations in order to avoid additional increase in tension and entanglement with the neighbouring yarn.
(d)	Yarn packages must be well built with the proper package (cone) angle. The quality of the yarn must fulfil the requirements for knitting.
(e)	If basic knitted structures are to be produced on a large scale, the machinery should be equipped with yarn feeding units that generate a constant yarn tension and deliver uniform yarn length.
(f)	Yarn guides must be flawless, eyelets made of porcelain or sintered ceramic must have a smooth surface without any furrows.
(g)	The needles must also be flawless. Their shape must be adapted according to the machine gauge and yarn count – this is especially important for hooks.
(h)	The condition of needle bed(s) should be well maintained without wear and tear, particularly in the case of tricks.
(i)	The drive to the needle bed(s) should be smooth and free of play in between dial and cylinder.
(j)	The needle beds must be exactly centred towards one another.
(k)	The fabric take-down and wind-on mechanism must be capable of being set individually in order to maintain the desired tension.
(l)	The operator must be thorough in regards to checking and maintaining the machinery and should be quality conscious.
(m)	The machine must be equipped with and well maintained highly sensitive stop motions.
(n)	The machine must be oiled and lubricated regularly. Automatically operating lubrication systems are absolutely necessary to get a high machine performance and a good operating reliability.
(o)	The machine must be cleaned, i.e. the deposition of fluffs and dirt over the machine as well as deposition of residual paraffin on the tension discs and yarn guides should be removed regularly.
(p)	The knitting plant must have clean working environment with proper temperature and humidity, and preferably it should be air conditioned so that yarn does not dry up during knitting.

Note: With regards to the state of the machines, the above mentioned points must be kept in mind to achieve faultless production of knitted fabric.

Table 10.4 describes the process control points to be considered in order to produce a faultless knitted product.

10.6.2 The role of the supervisor in quality control

The supervisor has a major role in ensuring that the required product quality is being achieved. The supervisor will be unable to play that role if, firstly, they are not aware of precisely what is required and the means of monitoring it, and secondly, they are not convinced of the necessity for it. Even if the above conditions are satisfied, their efforts will be inhibited if they are required to devote the greater part of their time to other tasks which should be undertaken by service personnel. In short, a supervisor must be encouraged and permitted to supervise. As a result of systematic development the supervisor should, amongst other facets of supervision, be:

- thoroughly familiar with the specification of the product,
- thoroughly familiar with, and skilled in, techniques of monitoring the consistency of the product,
- understanding of the importance of achieving a certain level of quality.

They can then communicate the importance of these factors and pass on their subject knowledge to their operatives, building the foundation for an overall improvement in workroom standards.

10.6.3 'Snap study' of the knitting process

Many reputable knitting houses have been gradually implementing the concept of 'snap study' to determine the overall efficiency in knitting and process control.

To undertake a 'snap study' or snap reading one must walk down machine alleys making a tally of which machines are currently stopped against a list of the causes of machine stoppage. As the word 'snap' indicates, the recording of the incidence should be spontaneous; the person who takes snap rounds should not anticipate an incipient stoppage, nor should they wait for it to occur as they go round. 'Snap study' is a technique that aids analysis of the various causes of loss in efficiency, and helps workers to estimate the percentage loss due to each cause by accurately calculating production.

In addition to aiding efficiency, a 'snap study' can help to identify breakages, including their cause and location, as well as idle times of the machinery and the causes of this. It can also track the involvement of operators with the machinery, along with many other factors. In this way a 'snap study' can help to improve the quality and quantity of production, with minimum defects and wastage of material.

It is also possible to evaluate the performance of one particular machine using this technique. The total number of breaks is generally converted into breaks per 1000 revolutions; this is then compared with the individual machines set norms. The norms are likely to vary depending on the type of machine, yarn quality and count, number of feeders in the machine, etc. As, for example, for a particular knitting house the breaks per 1000 revolutions ≤ 1.0 is very good, > 1.0 but < 1.5 is good, > 1.5 but < 2.0 is average, > 3.0 but < 4.0 is poor and > 4.0 is very poor.

The material handling systems adopted and the prevailing work culture also influence the production of knitted fabric. 'Snap study' can be extended to evaluate these two factors in order to help monitor and improve the quality of production.

10.7 Future trends: online quality control

The modern trend in quality control during knitting processes is the use of online systems and computers. Online systems employ many sensors and/or transducers to detect faults and deviation from set values. This detection, or signal, goes to the central computer, which processes and interprets the results. The computer then sends the information to the controllers, which are usually electronically controlled devices. These controllers either control the process directly or stop the process temporarily. This online control system can be attached to a group of machines and can produce cumulative quality results covering any particular period; these can then be stored for future applications. Some of the sensors and controllers found in modern knitting machine are:

- yarn breakage and speed monitoring sensor,
- yarn tension monitoring sensor,
- monitoring of group needle selection during shaping,
- fabric quality monitoring sensor,
- machine speed control servomotors,
- production monitoring and display unit,
- automatic lubrication monitoring and controlling unit,
- needle monitoring and electronic needle selection device.

This modern type of online control system used in knitting is equivalent to the Ring-data or Loom data system, widely popular in other areas of the textile industry.

For a few of the online control systems, both experimental studies and commercial models are discussed below.

Process control for total quality in circular knitting has been proposed by Araujo *et al.*[6] The proposed approach is based on yarn input tension analysis, which is a reflection of the whole knitting process for a given yarn feeder. By using this method it is possible to observe the whole process of loop formation, thus enabling the detection of abnormalities, along with their position in the appropriate knitting element and possible cause diagnosis. The presence of a defect can be detected, identified and located with high accuracy using this technique, which constitutes a major step in reducing repair time. The results obtained through the application of statistical tools in one case, and direct comparison in another, suggest that it is possible to detect and distinguish the defects by means other than by observation. The use of control charts can be a valuable tool for evaluating the general working condition of a knitting machine.

The Knit+Integrated system,[7] developed in Universidade do Minho, is a low cost online knitting process control system that can be easily installed on circular or flat knitting machines. It can be used for fault detection and for classification during the knitting process. At the same time, the system can provide relevant production information, such as an instant value of the yarn input tension. It can work independently or connected to a central unit that collects the information from all systems.

Online control of knitted fabric quality (loop length control) was also carried out by Semnani and Sheikhzadeh[8] using a circular knitting machine. Variation of yarn tension between different knitting tools can cause differing loop length of stitches during the knitting process. The authors applied a new intelligent method to control loop length of stitches in various tools, based on the ideal shape of stitches and the real angle of stitch direction. To measure deviation of stitch direction against variation in yarn tension, image processing techniques were applied to pictures of different fabrics with constant front light. Afterwards, the rate of deformation was translated to determine which required compensation of stitch cam angle to rectify stitch deformation. A fuzzy control algorithm was applied to the loop length modification in the knitting tools. The presented method was tested on different knitted fabrics of various structures and yarns. The results obtained showed that the developed technique could be utilised for loop length variation control when using different knitting tools based on stitch deformation, as well as for various knitted fabrics with different fabric structures, densities and yarn types.

Real time monitoring, planning and quality control for knitting mill: KnitMaster,[9] the BarcoVision's leading Manufacturing Execution System (MES) for the knitting industry, monitors and synchronises all logistical activities within the knitting mill, from yarn purchasing and inventory up to the shipment of the finished fabric. Powerful analysis tools enable quick identification of weak points and bottlenecks, resulting in an optimal usage of production capacities. With today's requirements for low quantity orders and rapid delivery times, scheduling has become a critical function for the textile mill. KnitMaster offers the planning department an interactive tool allowing them to optimise machine loading based on real time information. Optional modules are available for online monitoring of the yarn feed rate, automation of oiling and doffing functions, connection of weighing scales, integration of the fabric inspection department and interface with the company's ERP system.

Most weft knitting (circular and flat) and warp knitting machines can be connected to the KnitMaster by means of DU8P Data Units. These Data Units are available for wired and wireless networking. The large display allows the users to view all production and planning information. Nine predefined 'quick entry' keys facilitate fast entry of the most frequent reasons for stoppages. All other data entry is conducted through software driven function keys. Besides count inputs and digital inputs for automatic stop detection (needle breakage, scanner stop, yarn breakage, etc.), the DU8P is also equipped with two serial interfaces which can be used to connect weighing scales, printers, barcode scanners, etc. Through bi-directional communication, the DU8P can stop the machine and activate a light or any other machine function, such as oiling and doffing, by using one of the five digital relay outputs.

10.8 References

1. Sadhan Chandra Ray (2011), *Fundamentals and Advances in Knitting Technology*, Woodhead Publishing India, New Delhi.
2. Knapton J. J. and Munden D. L., A study of the mechanism of the loop formation on weft knitting machinery, Part I: The effect of input tension and cam setting on loop formation, *Textile Research Journal*, 1966, **36**, 1072–80.
3. Ray S. C. and Banerjee P. K., Some preliminary investigations into the mechanics of 1x1 rib loop formation on a dial and cylinder machine, *Indian Journal of Fibre and Textile Research*, 2000, **25**, 97–107.
4. Munden D. L., The geometry and dimensional properties of plain knit fabrics, *Journal of Textile Institute*, 1959, **50**, T448–71.
5. Little T. J., Factors influencing the exact replication of double knit fabrics, *Textile Research Journal*, 1978, **48**, 361–5.
6. Mário de ARAÚJO, André CATARINO and Hu HONG, Process Control for Total Quality in Circular Knitting, *AUTEX Research Journal*, 1999, **1**, 21–9.
7. André Catarino, 2C2T-Centre for Textile Science and Technology, Universidade do Minho.
8. Dariush Semnani and Mohammed Sheikhzadeh, Online control of knitted fabric quality: Loop length control, 2011, http://www.fibre2fashion.com
9. KnitMaster, Barco Vision in Pursuit of productivity, 2011, http://www.Visionbms.com

11
Process control in weaving

V. K. KOTHARI, Indian Institute of Technology Delhi, India

DOI: 10.1533/9780857095633.3.265

Abstract: A weaving unit must have a satisfactory system of process control that focuses on loom production, quality monitoring and the cost of fabric production in order to achieve quality and production targets at the lowest possible cost. This chapter discusses factors that affect loom productivity and efficiency and what needs to be done to optimise these factors to produce fabric of required quality at the lowest cost. Online quality and process control in weaving are discussed and factors affecting the weaving cost are analysed.

Key words: weaving process control, productivity and weaving efficiency, fabric defects, online process monitoring and control, weaving cost.

11.1 Introduction

Woven fabric consists of two sets of yarns called 'warp' and 'weft' as shown in Fig. 11.1. The fundamental processes involved in weaving are:

- Shedding, that is, dividing the longitudinal 'warp' yarns into two sheets to create a space called a 'shed'.
- Picking, that is, insertion of the transverse 'weft' or 'filling' yarn into the space created by the division of the warp yarns.
- Beating, that is, pushing the inserted weft yarns.

These basic processes have remained unchanged over centuries, whether the technology is a handloom, power loom, automatic loom or shuttleless loom. Shuttleless looms have been developed to overcome the inherent problems created by the dynamics of the picking mechanism on conventional shuttle looms and make use of entirely different method of weft insertion. Air jet, water jet, rapier and projectile looms are the various types of shuttleless weaving machines, named after the method employed for weft insertion. Shuttleless looms have a very high productivity; in the case of a single phase jet loom, weft insertion rates can exceed 2000 m/min. Since shuttleless machines have a much higher production rate and are more costly, process control in weaving using these machines is even more important, as any loss

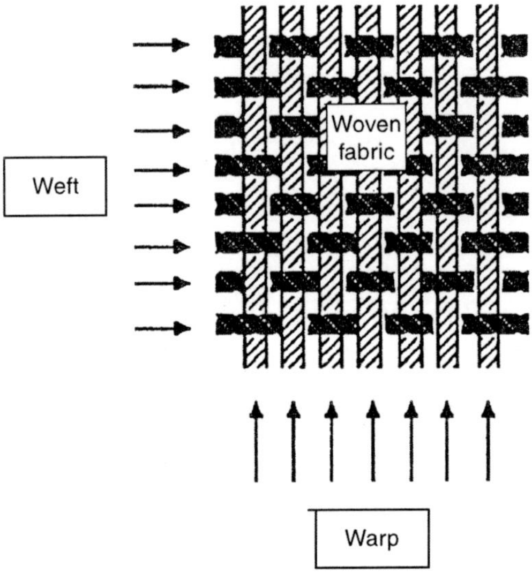

11.1 Weaving as a system for producing fabrics.

in production or quality would have a much higher influence on the economic viability of the production process.

Fabric production is a very competitive industry, and the principal objective of any weaving unit is to produce fabric of suitable quality at an acceptable cost. For this to be achieved, a thorough understanding of processes, machines, materials and staff is required. The mill's success also depends on the choice of product mix and the marketing skills of the management. To achieve quality and production targets at the lowest possible cost, a weaving unit must have a satisfactory system of process control, namely one which focuses on loom production, quality monitoring and the cost of fabric production (Fig. 11.2).

The practical parameters for any given woven fabric are the warp and weft threads per unit length and the crimp of the yarn used. Each of these parameters, together with the type of fibre, will have an influence on the quality of the fabric. If the loom is running under normal conditions, the weft density of the grey fabric should be consistent. However, the process can be affected by various disturbances, leading to variations in weft density. Disturbances during weaving can usually be traced back to the take-up of fabric and the let-off of warp ends. During weaving, the warp and weft yarns enter the system and grey fabric comes out; for quality fabrics, this process should be continuous and dynamically balanced. This balance is primarily determined by the fabric specifications, and dictates the set-up of the loom. Any interference with the balance could lead to an alteration of the fabric's characteristics, which, in turn, could lead to a fault in the fabric. It is therefore crucial that

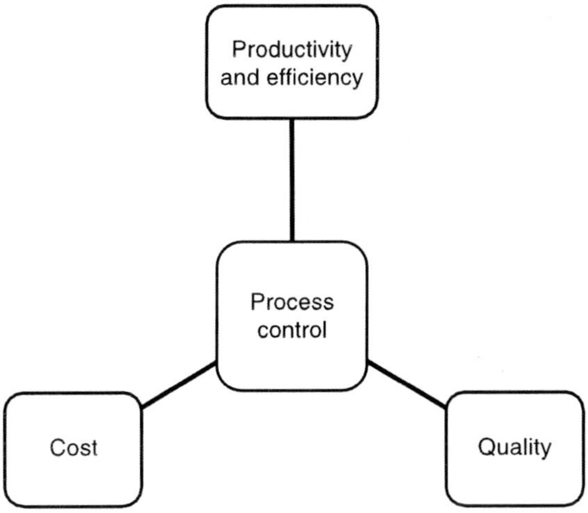

11.2 Main elements of process control in weaving.

the appropriate conditions are maintained during weaving; in particular, that beat-up distance and the level of warp tension remain constant.[1]

11.2 Controlling loom productivity, efficiency and fabric quality

Control of loom productivity, efficiency and fabric quality are essential for producing fabrics at acceptable cost with selected yarns and to succeed in the competitive environment.

11.2.1 Loom productivity and efficiency

The productivity of a loom is reliant on loom speed, loom width, and pick density.[2] The theoretical production levels that these factors would facilitate, however, are rarely achieved, due to such issues as yarn breakages, machine failures or necessary changes of the warp beam, all of which may cause loom stoppages. The weaving efficiency, therefore, influences the actual loom production. Efficiency, which can be quantified as the ratio of actual production to potential production, is affected by the following factors:

1. loom type, speed and width,
2. yarn type, quality and yarn preparation,
3. fabric structure, density and style,
4. weave room conditions,
5. weaver's skill and workload.

Warp breakage rate is an extremely important factor in loom efficiency. According to a number of studies, warp breakage rate is largely governed by yarn quality, yarn clearing and sizing.[3] Looking more closely at the main causes of weaving machine stoppages in cotton yarn weaving, a rate of 3–5 overall stops per 100 000 picks is achievable on modern shuttleless machines. Of these breaks, roughly 20% are related to problems in weft while the remaining 80% are due to deficiencies in the warp. Generally speaking, most warp stoppages are the result of either yarn defects (20–30%) or warp and weaving preparation (30–40%); the remaining 30–40% are related to weaving process itself.[4] A detailed understanding of reasons for warp stoppages can be obtained only by a proper stoppage cause analysis at the loom itself, but it is clear the overall process can be improved by optimising yarn quality, yarn preparation along with loom parameters, and weave room conditions.

A major task in minimising the stoppage rate is to optimise yarn stresses so that they are kept as low as is necessary. During the weaving process, warp is subjected to both static tension and frictional stresses, along with cyclic elongations that create tension peaks in the yarn. Reliable information about the forces acting on the warp ends can be obtained only through a warp tension trace, for which specialist warp-tension sensing devices that create online tension recordings are now available. Though such warp tension measurements are useful in controlling warp let-off, more precise control of the warp tension can be achieved through the use of positively controlled warp tension devices. These are particularly useful for the following processes:

1. improving warp separation,
2. reducing warp tension,
3. minimising tension peaks for yarns thus improving the performance of weak yarns,
4. increasing weft density,
5. controlling the position of the cloth fell,
6. controlling fabric structure and its appearance.

During pick insertion, the peaks in weft tension are extremely brief, but become more frequent with higher machine speed or weft insertion rate on any given machine. Optimisation of peak tension during weft insertion could result in fewer weft breaks. By utilising the appropriate process control systems to monitor weft tension online during the weaving process, a new generation of high-performance weaving machines is now being developed, providing better conditions for minimising yarn breaks.

Weaving efficiency also depends on the number of looms per weaver; this can vary from 4 to 150, depending on the number of loom stops per hour and the repair time of each stop. A lower loom allocation increases idle labour time and thus the overall labour cost, while higher loom allocation increases idle machine time and thus overall machine cost. Loom allocation is thus dictated by the appropriate balance between these two costs, which, in turn, is dependent on conditions such as temperature and humidity.

11.2.2 Control of fabric quality

The quality of any given fabric is usually measured by both its defects and its general properties. These include fabric specifications such as fabric width, ends and picks per unit length, weight per unit area, as well as the required functional properties of the fabric. The process of yarn preparation should minimise yarn faults, which could otherwise result in unacceptable fabric appearance or defects. Defects incurred during the weaving process itself must also be kept to a minimum, as the cost of fabric defects can be very high, with potentially substantial reductions in the value of the product. The tolerance limit of non-repairable faults per 100 m of fabric has been considerably reduced (from 15 to 5) in recent years, and is forecast to reach as low as 3 in the future.[4]

11.3 Online process control, quality control and monitoring in weaving

Manual process and quality control systems in fabric production are costly, inefficient and not suitable in a competitive environment. In the context of high speed machines and technological developments, online monitoring and controls are increasingly used during the fabric production.

11.3.1 Online process control

To maximise both the quality and weavability of the fabric, yarn stresses must be kept to a minimum. With careful assessment of the weaving process, the appropriate technological development of the loom can be identified and adopted. By electronically synchronising the let-off with the take-up system, for example, it becomes possible to ensure a dynamic cloth fell correction after each machine stop, and thus avoid leaving starting marks in the fabric. Electronically controlled warp let-off and cloth take-up units ensure a high degree of fabric regularity, while electronic monitoring systems have made it easier to identify any problems, allowing swift, corrective action to be taken. Settings for a fabric can be stored and loaded back on the control unit of the weaving system when required.

The possibilities offered by online process control mechanisms can only be exploited if the machines are able to convert electrical signals into the corresponding technological functions of the weaving itself. This is usually achieved with the use of elements such as servo-, stepper- and linear motors, or alternatively lifting and stepping magnets. Advanced solutions, such as the use of highly accurate linear motors, have been found for controlling thread clamps of weft yarns to be presented to the rapier.

Many weaving machine manufacturers offer a quick style change (QSC) system. The basic principle of this system is to prepare a module outside the weave room, so as to replace the empty module inside the weaving machine as quickly as possible. With the QSC system, German company Lindauer Dornier GmbH are known to have operated changes from a fine worsted fabric to a cashmere fabric in less than 30 min. Almost all major manufacturers offer their own version of the system. Dornier also offers a Fast Dobby Change (FDC) system, which allows a mill to exchange a positive cam motion for a dobby so as to increase versatility and increase shedding machine speeds. The exchange time in an FDC system is usually no more than 1.5 h.

In rapier machines, electronically controlled weft tension devices can reduce yarn tension during insertion, while the opening and closing time can be selected – usually at yarn pick up – according to the material used. In the event of a weft break between the package and the weft feeder, automatic package switching devices prevent the machine from stopping mid-process. Another important development is the emergence of new, pneumatic tuck in motions. This involves the use of air to first hold the weft end in place, before forcing it to be tucked in the next shed. The elimination of the tuck in needle with the use of a pneumatic tuck in motion enables the loom to run much faster in comparison with purely mechanical devices. These numerous technological developments have, in short, allowed mills to increase machine productivity, improve system controls, reduce breakages and ultimately facilitate a rise in quality. It is safe to conclude that, with the rise of new generation machines, process control is becoming considerably easier.

11.3.2 Quality control and monitoring

Visual inspection is a necessary but costly process for textile manufacturers. It is undertaken at a number of stages during the production of a textile, with different commercial justifications at each stage. The final inspection is carried out for quality control purposes, whereas inspection at other stages is generally used to identify defective material, so that it can be either mended or removed from the production process before unnecessary costs are incurred. Figure 11.3 shows an example of a visual inspection unit; such

11.3 Fabric inspection unit.

systems may vary from simple inspection tables to semi-automatic inspection tables with automated fabric movement and fault marking.

With regard to articles, colours, pattern and structure, manual cloth inspection remains the ideal method for judging fabric, as an inspector can utilise his or her own experience and expertise to subjectively judge the severity of a given fault. Manual visual inspection of textiles does have a number of widely acknowledged limitations, however. The subjective nature of the process gives rise to inconsistencies from one inspector to the next, while basic human error is, from time to time, inevitable, meaning that all faults may not be detected.

These limitations can be overcome by automatic fabric inspection systems, which provide consistently objective and reproducible assessment of the fabrics. The inspection data is provided directly in electronic form and can therefore be processed quickly and the analysis can be immediately transmitted to the people concerned.

Fabric inspection, though, has proven to be one of the most difficult of all textile processes to automate. It has taken decades for computer and scanning technology to develop to the extent that practical, consistent and reasonably user-friendly systems can be produced. Automatic inspection systems are designed to increase the accuracy, consistency and speed of the detection of defects in the manufacturing process of fabrics.

In the last 20 years progress in the design of the textile machinery has been remarkable, most notably in the improvement of productivity, automation and efficiency. On modern, high speed machines, constant monitoring of the machine, the weaving process and the fabric quality is highly desirable, so as to avoid faults and defects, achieve a high level of efficiency, and reduce the production of inferior quality fabric to a minimum.

To ensure quality, fabric inspection should ideally be both fast and exact. For this reason, high speed automatic inspection systems have increasingly become the focus in terms of development. Fabricscan (Zellweger Uster), I-TEX (Elbit Vision Systems Ltd.) and Cyclops (Barco) are all examples of systems based on machine vision, and in some cases, fabric inspection stations can be integrated with loom monitoring system allowing for the automatic recall of weaving data. I-TEX, Cyclops, Zellweger Uster, among other companies, now offer on-loom inspection systems for all types of weaving machines.[5] This rise in the commercial availability of such automatic fabric inspection systems is largely due to the following reasons:

- Image acquisition systems have vastly improved and are now available at a reasonable cost.
- High-performance parallel processors are now available.
- Image processing systems have become more advanced and more versatile.
- Large data storage capacity is now easily available at very low cost.

The inspection system is operated from an operating terminal, where article-specific inspection criteria and the necessary individual data are entered or read in via a bar code reader. The reports can also be called up via the operating terminal. Detected faults can, at any time, be called up on the screen and analysed quickly and easily. This is particularly useful in assessing whether a problem is recurring, frequent or anomalous. Ultimately, the visual display makes the assessment and the implementation of corrective measures considerably easier.

Automatic cloth inspection, with the advantage of quick, objective and reproducible measurements of the fabric characteristics, provides us with an effective instrument of modern quality management in the following ways:

- The continuous process optimisation; a system which allows the user to quickly identify and adjust any weaving machines that produce an increasing number of disturbing faults.
- The process control function is provided with a complementary alarm system, ensuring that deviations from the specified limit values are directly indicated and recorded.
- The determination of fabric quality, through common quality parameters such as the number of faults per piece or unit of length, or the so-called demerit point system. The systems allocate point values according to the size of a fault (with the Uster Fabriclass system, the faults can be classified based on the length and intensity of deviations in the direction of the warp or weft).

11.4 Uster Fabriscan inspection system.

The automatic fabric inspection system from Zellweger Uster shown in Fig. 11.4 uses a neural network to learn the characteristics of the flawless fabric.[6] The system then detects marks, records and classifies the faults automatically. The fabric passes over a two-component, illumination module, which allows for an inspection in reflected or transmitted light. The selection of the illumination type depends on the fabric density, the special fault types or the textile process stage at which the inspection is carried out. Depending on the inspection width, there are between 2 and 8 special CCD high-resolution line cameras installed above the light source. With these, it is possible to inspect fabrics which fall between the standard widths of 110 and 440 cm. The cameras continuously scan the fabric for deviations. During the normal inspection process, the system, with speeds of up to 120 m/min, looks for local deviations from the normal fabric appearance; the characteristics of these local deviations are then analysed. Depending on the conclusions of the analysis, the system initiates a marking of the fabric, records the event or carries out a fault classification.

The entire learning phase of the neural network, and its subsequent transmission to the evaluation unit, takes approximately one minute, and has to be carried out once for each article. All following pieces of that particular article will then be inspected according to the same standard.

The company's current system, Fabricscan, can inspect fabric at speeds up to 120 m/min while offline (the inspection speed of an online system is approximately 30 m/min), and can detect defects down to a resolution of 0.3 mm. The Uster Fabriscan is available for both transmitted and reflective light, allowing it to recognise a wide range of faults (oil spots, e.g. can only be seen in reflective light whereas start marks can only be seen in transmitted light).

I-TEX from Elbit Vision Systems Ltd. (EVS) is a well-established automatic fabric inspection system. The system consists of an image acquisition unit, computers and a post-inspection evaluation work station for review and analysis of defect report, along with a video album of all defects. The system can detect defects as small as 0.5 mm on fabric widths of up to 330 cm, and at speeds of up to 100 m/min. On any unicolour fabric, I-TEX detects diverse spinning, weaving, dyeing, finishing and coating defects. Virtually any visible defect, from yarn and weaving faults (holes, missing threads, starting marks, broken yarns), to water and dyestuff stains, can be detected by the system. For grey and technical fabric inspection, I-TEX can be configured for extra-high resolution and extra-wide fabrics, while for unicolour, dyed finished fabric inspection it uses both transmitted and reflected light, as well as two sets of cameras.

EVS also offer an in-line system called Shade Variation Analyser (SVA), which utilises a calibrated, travelling spectrophotometer to measure shade consistency in textiles during fabric flow. The spectrophotometer readings are compared to a reading at the beginning of the roll to detect side-to-side and beginning-to-end shade variation. The SVA can be integrated with an EVS I-TEX system or can operate as an independent, standalone unit. For printed designs, EVS has developed the in-line fabric monitoring and defect detection system PRIN-TEX, which can be mounted on a rotary screen printing machine. This enables online detection of recurring printing defects. Upon detection, an alarm is raised and the fabric fault and its location are displayed in real-time on a video monitor, enabling swift corrective action and an overall improvement in quality. Figure 11.5 shows the on loom Inspection system of Elbit Vision System.[7]

The Cyclops automatic loom inspection system from Barco detects warp and weft defects by means of one or two mobile scanning heads. These heads consist of a CMOS camera and infrared LED illumination unit, installed either on the off-loom take-up or above the cloth roll. At a number of positions, an image of the fabric is taken and transferred to the image processing unit. The software analyses the texture of the fabric and detects deviation from the normal texture. Any detected defect is signalled to the loom. All defect information is also sent to a fabric quality data base through the Barco Sycotex loom monitoring systems (Fig. 11.6), allowing the production of defect maps and various types of quality reports. Based on the defect analysis, the grey inspection may or may not be bypassed (a fabric judged to be first grade, for example, can be sent directly for further processing).

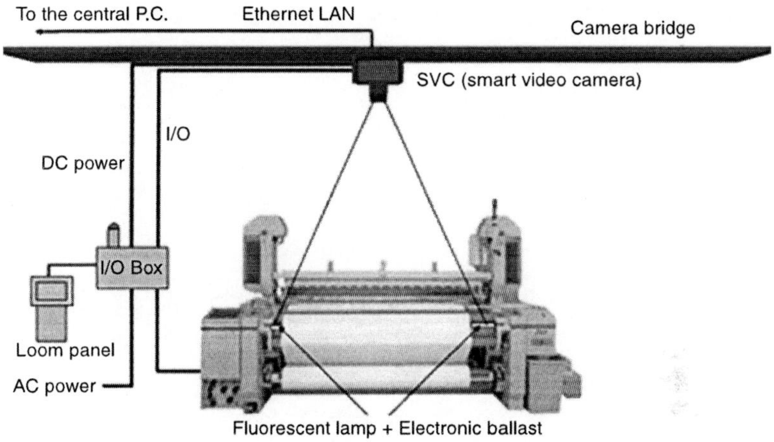

11.5 On-loom inspection system of Elbit Vision System.

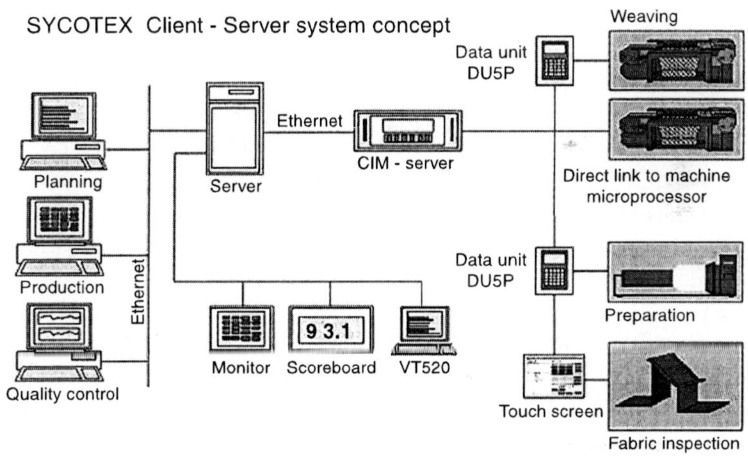

11.6 Barco automatic loom inspection and monitoring systems.

Setting up the Cyclops system is very simple. The scanning range adjusts itself to the fabric position and width through automatic detection of the fabric boundaries. Illumination and camera settings are optimised by the calibration software module in accordance with the optical characteristics of the fabric. Furthermore, the structure of the fabric is automatically identified, so as to calculate the algorithm parameters for optimal defect detection.[8]

The latest version of the Cyclops system (Fig. 11.7) has now also been optimised for the inspection of carbon, Kevlar and glass fabrics. Barco Vision has also launched a brand new sensor for air jet weaving looms, entitled the

11.7 New Cyclops automatic on loom fabric inspection system.

Kinky Filling Detector (KFD). This detector, a combination of laser and camera technology, continuously checks the fabric for the presence of kinky weft yarns (loops) and can stop the loom in case of excessive occurrence of these defects within a certain length of the fabric.[8]

Online detection not only reduces the amount of defective produce, it also reduces the amount of required handling. If a roll being removed from the machine is of high quality, the roll might be immediately prepared for shipment, and thus undergo minimal handling. If the roll has one section that is defective, moreover, the operator will, with the help of online detection, know precisely where to find it, thus minimising the time it takes for removal, resulting in a significant reduction in cost. As machine production increases every year, so does the importance of online quality monitoring.

Effective fabric inspection can reduce the fabric defects in the final fabric and is thus highly advantageous for process control in weaving.

11.4 Cost control in weaving

The ultimate aim of any mill is to make profit, a goal dependent both on sales income and production expenditure. While sale price depends on product positioning, supply and demand and other factors, a product is rarely able to dictate price in a competitive market. It is, therefore, necessary to optimise the production costs, the key factors of which are the respective costs of raw materials and weaving itself. More than two-thirds of all production cost derives from the cost of raw materials, which can vary considerably depending on market conditions, and are thus somewhat difficult to control. It is therefore essential that a mill optimises weaving cost in order to remain competitive in the market. Weaving cost per metre of the fabric produced may be divided into the following three main elements: (a) labour cost, (b) fixed cost, and (c) other costs.

Labour cost is the sum of cost of all personnel directly involved in operations from preparation to inspection of the final fabric. The labour cost depends on the type of work, skill and efficiency of the operator, the extent of the automation and condition of the machines, and finally the work allocation. In high wage countries, labour cost can be a significant component of conversion cost while in low wage countries, the proportion of the total conversion cost dedicated to labour cost will be smaller. A higher number of looms per worker reduces the labour cost, but will also decrease loom efficiency and thus reduce production and increase the fixed cost. There has to be a compromise, then, to achieve the lowest overall cost.

Fixed costs are those that do not depend on the production and include interest, depreciation, overhead, maintenance, space, air-conditioning and management costs. With more expensive machinery and other installations, it becomes necessary to spread the fixed cost over a higher production. Any increase in fixed costs should be accompanied by increased productivity and machine efficiency otherwise the weaving cost will rise. To operate looms at higher speed while also ensuring a low end breakage rate, production can be increased; to achieve this, though, it is necessary to spend money on good preparation and better quality yarn. Once again, a compromise is necessary to achieve the lowest possible cost.

There may be costs associated with the inventory which will add to the production cost. To minimise such costs, it is necessary to reduce the inventory to the lowest level practicable. Continuous monitoring of loom data, proper planning of yarn inventory and other preparatory processes can provide more accurate predictions of such factors as the time when a new beam will be required on a given loom, which will in turn reduce the down time otherwise caused by the beam change, and ultimately increase the efficiency.

Control of hard waste is another essential step in minimising the conversion cost. All areas of a weaving mill will create certain amount of waste during the processing of material. This waste can be divided into two types (i) process waste, and (ii) incidental waste. Process waste is unavoidable waste linked to the type of equipment and the nature of the process; it cannot be reduced below a certain level unless equipment changes and major process changes are adopted. All other waste is incidental and completely avoidable. Waste resulting from poor package quality, work practices and material handling must be eliminated or reduced to bare minimum. A good example would be the need for extra ends in a beam to take care of missing ends. Better control of lappers at sizing will reduce missing ends and thus the necessity for large numbers of replacements. Proper setting of the loom and even tension in warp ends can further help in reducing the end breaks. In the case of formation of cloth with fringe selvedge, optimisation of the weft insertion system can also help to reduce waste. A higher number of fabric defects not only affects the fabric quality and leads to a drop in value, but can also lead to a higher quantity of hard waste. Savings through waste reduction can be substantial and can reduce the overall cost of fabric production.

11.5 References

1. Chen, X., Characteristics of Cloth Formation in Weaving and Their Influence on Fabric Parameters, *Textile Research Journal*, April 2005, 281–7.
2. Lord, P.R. and M.H. Mohamed, *Weaving: Conversion of Yarns to Fabrics, Chapter 17: Weave Room Management*, Woodhead Publishing Co., Cambridge, England, 2001.
3. Paliwal, M.C. and P.D. Kimothi, *Process Control in Weaving*, ATIRA, Ahmedabad, 1983.
4. Legler, F., Is process control in weaving possible or just a dream?, *Textile World*, February 2003, 26–9.
5. Dockery, A. 'Automated Fabric Inspection: Assessing the Current State of the Art', 2001. Available from: http://www.techexchange.com/library/Automated%20Fabric%20Inspection%20-%20Assessing%20The%20Current%20State%20of%20the%20Art.pdf
6. Jose Fuster, S.A., 'Uster Fabriscan Product Catalogue', 2011. Available from: www.fuster.com
7. Elbit Vision Systems Ltd. (EVS), 2012. Available from: http://www.evs.co.il/index.php?
8. Belgium Monitoring Systems (BMS) Vision, 'Cyclops Automatic on-loom fabric inspection', 2012. Available from: http://www.visionbms.com/textiles/en/products/product.asp?element=1315

12
Process control in nonwovens production

D. MOYO, A. PATANAIK and R. D. ANANDJIWALA,
CSIR Materials Science and Manufacturing, South Africa, and Department of Textile Science, Nelson Mandela Metropolitan University, South Africa

DOI: 10.1533/9780857095633.3.279

Abstract: This chapter deals with process control in the field of nonwoven manufacturing. A brief background about different nonwoven manufacturing processes is provided. Different manufacturing processes such as needle punching, hydroentanglement, melt blowing and spunbonding are discussed. Roles of different processing parameters and their influence on the final product are covered. Various ways to control the processing parameters are discussed. Some future directions are also highlighted.

Key words: hydroentanglement, melt blowing, needle punching, nonwoven, process control, spunbonding.

12.1 Introduction

This chapter deals with process control in different nonwoven manufacturing processes, such as needle punching, hydroentanglement, melt blowing and spunbonding. The performance of needle punched nonwovens is dependent upon material and manufacturing process parameters. The needle punching parameters can be broadly divided into three categories: material parameters (fibre type), needle parameters (needle type, shape, etc.) and process parameters (depth of needle penetration, stroke frequency, punch density, etc.). Various ways to control the needle punching process parameter are discussed. Their influence on the properties of the final product is discussed. In hydroentanglement, the influence of processing parameters, such as feeding speed of the web, web area density, condition of nozzle and jet strip, manifold pressure, condition of perforated belt and distance between the nozzle and perforated belt, on the final product and ways to control some of these parameters are discussed. In the melt blowing process the role of processing parameters, such as air temperature, die-to-collector distance, airflow, collector speed, die temperature, die hole size, air gap, air angle on the final product, and ways to control these parameters are discussed. In the spunbonding process, the role of processing parameters, such as primary air temperature, quench air rate, air suction speed, collection speed, bonding temperature and pressure, on the final product and ways to control these parameters are discussed. Some future trends are also addressed.

12.2 Needle punching: process variables and process control

Needle punching can be defined as a physical method of mechanically interlocking fibre webs by using barbed needles to reposition some of the fibres from a horizontal to a vertical orientation forming a three-dimensional intermingled structure (Rawal and Anandjiwala, 2006; Russell, 2007; Smith, 2000). Although needle punching is now a mature technology, a great deal of work has been done in relating the fibre parameters, structural factors and needling parameters with the mechanical properties of nonwovens (Hearle *et al.*, 1968; Patanaik and Anandjiwala, 2008, 2009; Watanabe *et al.*, 2004). The results obtained by these researchers add to the understanding of the needle punching process.

The needle punching operation integrates the fibre lay-down and needling processes in continuously to consolidate the structure. The fibre web formed during fibre lay-down is transported on a conveyor belt to the needle punching machine, where fibres are mechanically entangled by the penetration of the needles to form a coherent nonwoven fabric. The barbs on the needles pick up fibres on their downward stroke and carry them into the structure, thereby compacting the web and providing high frictional resistance to fibre withdrawal (Hearle and Sultan, 1968b; Patanaik and Anandjiwala, 2009; Patanaik *et al.*, 2007). The rollers pull the batt through the needle loom, which consists of two plates – a stripper plate on top and a bed plate at the bottom as shown in Fig. 12.1 (Smith, 2000). During processing, the needle beam connected to the needle board mounted with needles moves up and down with the needles carrying the fibres during their downward motion. The fibres are carried downwards by the barbed needles and reoriented from a predominantly horizontal direction to a vertical position to create fibre interlocking. The fibres are released by the barbs during the upward needle stroke. The unhooking and releasing of fibres take place when a part

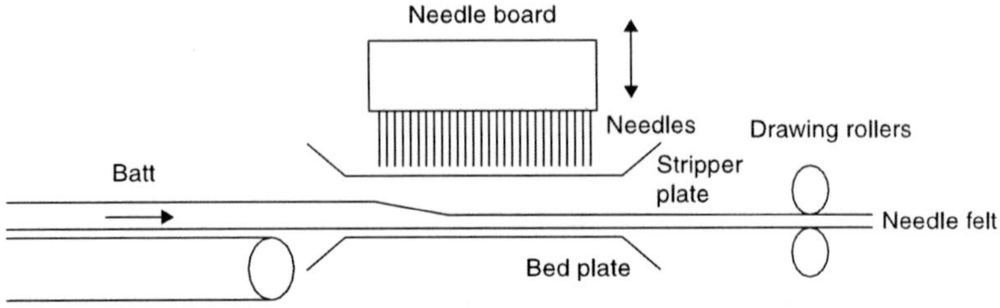

12.1 Diagram of a needle loom. (*Source*: Smith, 2000. With permission from Woodhead Publishing Limited, UK.)

of the tension in the fibre, produced by the downward motion of the needle, is resisted by the equal and opposite frictional force built up between fibres within the structure of the web (Cannity, 1962).

The performance characteristics of needle punched nonwovens are dependent upon the fibre and structural mechanics of the fabric developed during the manufacturing process. The needle punching parameters can be divided into three categories; material parameters (fibre and web types), needle parameters (needle type, shape, arrangement and number of barbs, etc.) and the machine parameters (depth of needle penetration, stroke frequency, punch density, etc.). These parameters are interrelated and approach limits dictated by process economics and fibre properties (Hearle and Sultan, 1968a, 1968b; Hearle *et al.*, 1968; Patanaik and Anandjiwala, 2009; Ramkumar and Roedel, 2003; Ramkumar *et al.*, 2004; Roedel and Ramkumar, 2003; Russell, 2007; Smith, 2000).

12.2.1 Web parameters

The mechanical properties of needle punched nonwovens are dependent upon laying techniques used for the production of web structures or the initial web structure (Russell, 2007; Thirlwell and Treloar, 1965). A needle punching machine can process a wide range of fibrous webs from different systems, which influences the fibre arrangement within the web. Depending on the feed rate employed, the web density can be varied, and with higher feed rate thicker and denser nonwovens are formed. The resultant properties of needle punched nonwovens directly relate to the web density; for instance, lower feed rates increases the permeability characteristics of the nonwovens. It is due to the presence of fewer fibres per unit volume. Relatively fewer fibres in the structure results in the formation of larger pores, especially at lower stroke frequencies. However, increases in both feed rate and stroke frequency reduce the permeability of the needle punched nonwovens, as the pore size decreases with the increase in the number of fibres. The isotropy or anisotropy of the subsequent nonwoven is dependent upon the web laying technique (Kiekens and Zamfir, 2002). Control of the web is vital in the needle punching process because the fabric's structural properties, such as thickness, basis weight, bulk density and air permeability, are directly related to the web features, such as fibre orientation, web density, web thickness and web homogeneity (Russell, 2007; Smith, 2000).

12.2.2 Depth of needle penetration

The depth of needle penetration in needle punching refers to the longest distance that the first barb reaches below the lower surface of the web or bed plate (Hearle *et al.*, 1968). The depth is set by lowering the bed plate to increase the needle penetration or raising it to decrease the depth of penetration. During the needling process the barbs on the needles carry fibres through the web thickness from the surface to the base. An increase in the depth of needle penetration causes an increase in the relative frequency of the fibres oriented in the machine direction for cross-lapped webs, due to the fact that the fibres have to take a longer path because of a higher depth of needle penetration from surface to the thickness direction. In the process, some of the fibres would be released and recover from the stress and strain. On recovery, the fibres, preferentially oriented in the cross-machine during web laying reorient in the machine direction. The depth of needle penetration in the needle punching process is probably the most significant processing variable influencing the mechanical properties, dimensional stability and fabric density. Generally, the mechanical properties, such as tensile strength, improve with the increase in the depth of penetration due to the greater extent of fibre reorientation and contact points. The tensile properties steadily improve with the increase in depth of needle penetration, but reach a limit beyond which it begins to decrease. This decrease with excessive needle penetration is explained by the fibre damage that occurs at those levels resulting in the weakening of the structure. The fabric thickness, on the other hand, decreases with the increase in needle penetration because of the improved consolidation of the structure with greater depth of needle penetration (Patanaik and Anandjiwala, 2009; Rawal and Anandjiwala, 2006).

Needle penetration determines the number of fibres carried on the down stroke; the deeper the needle penetration, the higher is the number of fibres carried by the needle. The needle configuration is also another factor influencing the number of fibres carried on the down stroke. At a higher depth of penetration the lower barbs on the needle carry the fibres through the entire web thickness, which remain protruding beyond the bottom surface, whereas with a low depth of penetration the punched loops do not protrude through the web thickness. If the penetration is too great, tufts are punched right through the previous loop to give a pseudo-knitted entanglement of fibres. The linkage of such punched fibre tufts results in an increase in the tensile modulus and tenacity, but a reduction in the breaking extension of the fabric. The change in mechanical properties is largely due to the intensity of fibre entanglement (Hearle *et al.*, 1968; Miao, 2004).

12.2.3 Stroke frequency

Stroke frequency refers to the rate at which the needle board moves per second forcing the needles through the bed plate and penetrating the web. An increase in the stroke frequency results in a higher number of fibres being reoriented from the horizontal to the vertical direction, which significantly reduces the larger pores of the structure. This process parameter also determines fabric density and tensile strength since, at higher stroke frequencies, there is improved consolidation of the web. The frequency of the strokes and linear speed of the web must be balanced to achieve the desired degree of consolidation. The key parameter representing the entanglement of fibres is known as penetration per square inch (PPSI). PPSI is directly proportional to the density of the needles on the needle board and to the frequency of strokes, but inversely proportional to the speed of the web (Wizeman, 2000).

12.2.4 Amount of needling (punching density)

Punching density is the amount of needling that the web receives from the barbed needles when passing through the needle punching machine. The amount of needling received by the fibrous web determines the degree of entanglement of the fibres, thus influencing such properties as tensile strength, strain and fabric density (Rawal and Anandjiwala, 2006). With the increasing punching density, fabric breaking strength increases to a maximum and then decreases. The punching density also directly affects the packing factor. An increase in the punching density takes place, resulting in a corresponding increase in the packing factor as more fibres are entangled and a higher level of structural consolidation. This results in an increase in the compressive modulus, and the fabric also becomes stiffer. A stiff nonwoven has a low compressive strain, which means that the fabric has a low deformation under compression stresses. When the punching density is beyond its optimum, the strength of the nonwoven falls due to fibre damage inflicted by the excessive needling. The use of the optimum punching density parameters, at which the nonwoven has maximum tensile strength, is important to obtain the best possible nonwoven structure. The effect of punching was shown to change the mechanical properties, and the greater the punching density, the higher will be the tensile strength and modulus, until a maximum is reached. Excessive punching density was shown to result in damage to the fibres, which lowers the mechanical properties, such as fabric tenacity and initial modulus of the fabric. These results were corroborated by Maitre (1989), who reported that mechanical factors, such as stroke frequency, needle density, material advance and depth of needle penetration, are interrelated and approach limits.

The amount of needling is a parameter that affects the thickness, weight and the strength of the fabric. The needling process results in fibre entanglement and reorientation of the fibres in the thickness direction, thereby binding different fibre layers into a coherent self-locking structure (Debnath *et al.*, 1994). Needling the fibrous web reduces the web thickness due to a collapse of the spaces between fibrous layers and entanglement of the fibres by the action of the needles. Fibres are pulled from the top layer or surface to the base of the web during the down stroke of the needles. This consolidates the fabric, and its weight decreases with the increase in the needling density due to the drafting and spread of fibres during punching. The web is drafted by the nip rollers when it is drawn through the plates in the needling zone, thereby increasing the length of the structure and reducing its weight. The fabric also experiences recovery of the fibres previously drawn to the base of the structure when needles are withdrawn from the web causing some fibres to be pulled up and recover from compression. Thus, fibre recovery leads to the spreading out of the fibres in the structure and loss in area weight. The amount of needling is varied by altering the rate of web advance, changing the stroke frequency of the needle board, or increasing the number of passes of the nonwoven through the needle punching machine. The stress/strain behaviour of the nonwoven produced by needle punching is also influenced by the amount of needling used to produce the structure. Tensile strength and modulus of the fabric increase with the increase in needling, up to a limit, after which further needling tends to decrease the modulus and tenacity. This decrease beyond the maximum is caused by the damage to the constituent fibres at high needling intensities (Hearle *et al.*, 1968).

12.2.5 Needle type

Much technical consideration should be taken with regard to needles, as they ultimately influence several properties of needle punched nonwovens. The correct type of needles should be selected to produce any specific type of fabric. For instance, when barbs are close, more fibres are carried across the fibre web, and the C and F types of barb spacing therefore form very compact products. However, the surface of such products is not smooth and does not have a good appearance, because of the penetration holes showing on the surface caused by excessive needling actions. The best surface is obtained with regular and medium barb needles, as they cause relatively small holes on the surface. The closer the barbs are the more fibres they transport during the downward stroke (Falk, 2000). Needles are utility components in the needle punching machine and prolonging their lifespan is important for reducing operating costs. The main factors that influence needle damage are non-alignment of needles, fibre quality (waste, regenerated or unopened fibre tufts), needle punching condition (e.g. if the speed of the web per stroke is too high it results in sideways deflection of needles by the fabric), needles penetrating a thick web too aggressively and encountering high friction between needles and fibres, deep barbs that increase the load on the needle during penetration, machine vibrations during operation, and improperly designed hole sizes in the bed and stripper plates (Foster, 1985).

12.2 Hydroentanglement manifold system. (Source: CSIR, 2011.)

12.3 Hydroentanglement: process variables and process control

The hydroentanglement technique is an established mechanical method of bonding fibres together to produce nonwoven fabrics using collimated high pressure water jets issued from a series of parallel jet heads (manifolds). The high pressure jets entangle loose fibres within fibre webs carried on a perforated belt, and eventually produce a fibrous structure of high integrity (Patanaik and Anandjiwala, 2010; Patanaik *et al.*, 2009; White, 1990; Xiang, 2007; Xiang *et al.*, 2007; Medeiros, 1996).

Much research has gone into achieving efficient optimisation of the process, which involves energy transfer, a costly resource, and utilising it efficiently. The focus on the geometry of the nozzles led to the proof of the superiority of the cone-down nozzle geometry over other geometric forms because it forms water jets that have a longer intact length. The breakup of the water jet is undesirable and leads to a loss of energy, and as such it is vital that the design of the nozzle geometry ensures that the water jet remains long and collimated (Begenir *et al.*, 2004; CSIR, 2011; Patanaik and Anandjiwala, 2010; Patanaik *et al.*, 2009; Pourmohammadi *et al.*, 2003; Tafresh and Pourdeyhimi, 2004). Figure 12.2 below shows the water jets issued by the manifold of an Aquajet hydroentanglement machine (CSIR, 2011).

The hydraulic energy of the water jet is utilised in the compression of the web, permanent deformation of the web, frictional work resulting from fibre displacement and orientation, fluid drag resistance of the fibres, and being absorbed by the standing water in the web. The energy to entangle the fibres depends on fibre properties, such as fibre type, modulus, linear density, amount of crimp, bending rigidity, length and inter-fibre friction, and on web characteristics such as area density, thickness and fibre arrangement. Also the energy that the water jet possesses depends on the uniformity of flow from the nozzles, frictional losses and coherence of the flow (Begenir *et al.*, 2004; Ghassemieh *et al.*, 2001, 2003; Mao and Russell, 2005; Patanaik and Anandjiwala, 2010; Patanaik *et al.*, 2009; Tafresh and Pourdeyhimi, 2004).

The extent of the entanglement of the fibres during hydroentanglement is the most significant factor influencing both the physical and the mechanical properties of hydroentangled fabrics. The entanglement of fibres can be influenced by several different processing parameters. The main processing parameters, such as the water jet pressure, feed rate of the fibres and conveyor belt speed, can be manipulated to control the structural features or degree of fibre entanglement in the fabric produced (Ghassemieh *et al.*, 2001, 2003; Mao and Russell, 2005; Patanaik and Anandjiwala, 2010; Patanaik *et al.*, 2009; Tafresh and Pourdeyhimi, 2004).

12.3.1 Water jet pressure

The hydroentanglement process is basically an energy transfer method where energy is given and stored by the water during pump pressurisation. The energy is converted into kinetic energy when the water jets are formed and is used to transform the loose fibres into an entangled mass having structural strength. The pressure intensity of the jets is responsible for determining the degree of fibre entanglement and fabric strength (Patanaik and Anandjiwala, 2010). Increases in pressure result in a corresponding increase in fibre entanglement and subsequently also in fabric strength. The fabric extensibility is reduced by the higher degree of fibre entanglement. This reduction can be explained by the continued consolidation of the fabric as the pressure increases, which reduces the capability of the structure to stretch (Russell, 2007).

The rate of increase in fabric strength, due to pressure, increases to a point after which a further increase in pressure produces only a marginal increase. This point depends on the fibre type and web weight. It is important to determine this critical pressure, because beyond it increased water pressure has no significant effect on fabric properties. Generally, water jet pressure affects the strength, absorbency, thickness and permeability of nonwovens. The nonwoven fabric tensile properties increase directly with increasing jet pressure till they level off beyond critical pressure level. The fabric extensibility decreases linearly with increasing water jet pressure (Ghassemieh *et al.*, 2001; Pourmohammadi *et al.*, 2003).

12.3.2 Water jet pressure profile

Profiling the water jets means that the pressure of the water on each injector manifold gradually increases from the first injector head to the last, in systems with multiple jets. In most hydroentanglement machines, nozzle pressure can be increased sequentially; for example, it may have a sequence of 40, 80, 100, 120, 160, 180, 220 bars for the first to the seventh jet, respectively (Pourmohammadi *et al.*, 2003). The effect of the water jet pressure profiles on fabric thickness, strength, bending rigidity and surface quality has been investigated. The selection of an appropriate water jet pressure profile can minimise energy consumption while optimising fabric properties (Patanaik *et al.*, 2009). The effect on the fabric thickness is more apparent, especially for airlaid webs having a low initial density, which collapse easily under the jet pressure (Pourmohammadi *et al.*, 2003). This shows that by using profiled pressure in the jets, the density of the fabric increases with an accompanying reduction in permeability and thickness. Different pressure profiles can be tried, to obtain the most suitable combination for the product without increasing the water pressure of the jet and thereby minimising the consumption of water. The cost of production is thus minimised by using low pressure profiles to obtain the required fabric density that would otherwise need a much higher nozzle pressure. The necessity to increase the water pressure is eliminated by using the same low pressure jets if they are set at progressively higher pressure profiles. The bending rigidity of the products follows a similar trend to the effect the water jet pressure profile has on the density of the fabric. As the fabric density increases, so does the bending rigidity. Pressure profiling of water jets increases the fabric strength when compared to a hydroentangling system that employs even water jet pressures for all the jet heads despite the total pressure for both methods being kept constant. Fabric strength increases when pressure profiling is used due to better consolidation of the fibrous structure. However, the results showed that specific energy alone may not accurately predict fabric strength; other influential factors, such as web structure and fibre characteristics, are also important. Although the total jet pressure may be constant, the use of water jet profiling reduces the total energy expenditure while optimising the fabric properties in the hydroentanglement process (Pourmohammadi *et al.*, 2003).

12.3.3 Feed rate

The feed rate determines the number of fibres that are fed to a carding machine as per the process requirement. A higher feed rate means more fibres are fed to the carding machine and the carded web produced would be much denser and thicker than it would be in the case of low feed rates.

This is explained by the increased number of fibres present in the horizontal direction at higher feed rate. The density of the precursor web ultimately determines the hydroentangled fabric density as well, since it is directly related. Generally, whatever system is used for producing the web, be it spunlaid or airlaid, the hydroentangled fabric density is determined by the density of the web. A heavy dense web produces a relatively denser fabric. Fabric properties are directly related to the fabric density; for example, high maximum strength is obtained if web weight is increased. All generic web types such as spunlaid, meltblown, carded, airlaid and wetlaid can be hydroentangled. The precursor web which is fed to the hydroentanglement machine largely determines the isotropy and quality of the final fabric (Russell, 2007).

12.3.4 Conveyor speed

The speed of the conveyor determines the amount of hydroentanglement energy absorbed per unit area of the fabric. As the conveyor speed on which the loose fibrous web is carried is increased, the lower is the amount of energy absorbed by the fibres, as they are exposed to the water force for less time. As a consequence of reduced energy transferred at higher speeds, the fabric strength decreases as a result of reduced entanglement of the fibres. The increase in conveyor speed reduces the fabric strength unless it is also accompanied by a proportional increase in jet pressure (Pourmohammadi *et al.*, 2003).

12.4 Melt blowing: process variables and process control

Melt blowing is a single-step process in which a thermoplastic resin is heated in an extruder and then extruded through an orifice die tip, whereupon stream is drawn by high velocity air onto a collector screen to form a fine fibrous and self-bonding web (Bhat and Malkan, 2007; Bresee and Ko, 2003; Bresee and Qureshi, 2004; Lewandowski *et al.*, 2007; Yin *et al.*, 2000). The aerodynamic drag of the air jets on the melt polymer provides the attenuation force that draws the polymer streams into fine diameter fibres as shown in Fig. 12.3 (Bhat and Malkan, 2007). This process is used to produce microfibres of diameters 2–4 μm, which is much smaller than those from fibre spinning (Dahiya *et al.*, 2004a).

The melt blowing process variables are interrelated and complex. These variables influence the characteristics of the web structure and they can be classified broadly into three main groups, namely machine/online variables, offline variables and material variables (Bhat and Malkan, 2007; Dahiya

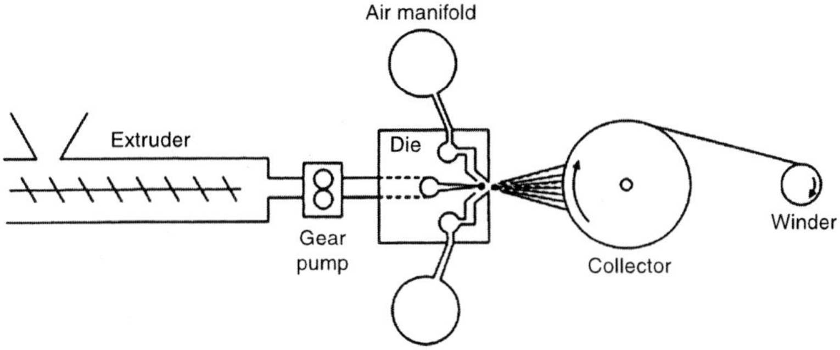

12.3 Melt blowing process. (Source: Bhat and Malkan, 2007. With permission from Woodhead Publishing Limited, UK.)

et al., 2004a; Lewandowski *et al.*, 2007). The adjustment of the parameters is done to control the size of the fibres because it is the fibre size that primarily accounts for the properties of the final nonwoven (Ellison *et al.*, 2007).

12.4.1 Operational/online variables and their control

Variables such as air temperature, die-to-collector-distance, air flow, collector speed, die temperature and hot air velocity are changed when the production line is running or online. The fibre diameter and web uniformity depend upon the proper selection of these parameters.

Air temperature

During the melt blowing process, the temperature is set according to the properties of the polymer being processed, such as the melting point or glass transition temperatures. The heat softens and melts the polymer resin to the required viscosity so that it can be successfully extruded. If the temperature in the extruder is lower than required, the web formed will show some blobs of polymer within its structure, giving it an uneven or a non-uniform structure. If the temperature is too high, the web will be too soft and fluffy and the air flow in the attenuation zone will cause a lot of breakages (Butin *et al.*, 1974).

Die-to-collector distance (DCD)

The distance of the collector from the die is one of the most important process parameters, which can be adjusted to obtain the desired melt blown webs. When the molten polymer is extruded, the fibre entanglement begins within a centimetre from the die before being fixed on the collector surface. The DCD influences the extent of fibre entanglement, fibre orientation and the pore structure. Fibre entanglement is dependent on the DCD at all primary airflow rates, although the effect is more pronounced on the fibre diameter. Increased DCD results in increased percentage of coarse fibre bundles, although generally melt blown webs exhibit a very broad fibre diameter distribution. The fibre diameter generally increases slightly with the increase in DCD. At higher DCD the pore size increases, which means that the cover of the large pores is relatively higher than that at low DCD. The closer the die is to the collector surface the better is the uniformity of web basis weight. The quality of the web largely depends on the DCD, where short distances produce a dense, hard and tightly connected web because of self-bonding of the fibres since they are still hot. If the distance is too great, it gives weak and unconnected fibre webs (Bresee and Qureshi, 2004; 2006; Butin *et al.*, 1974; Yin *et al.*, 2000). Basically the DCD influences largely the openness and basis weight of the web (Dahiya *et al.*, 2004a).

Air flow

High velocity primary air flow produces fine fibres with small diameter, which gives a web structure with increased fibre cover and reduced pore cover. Large air flows increase fibre entanglement, leading to decreased pore size structure, since generally the airstream controls the entire web uniformity and it should be adjusted so that fibres are uniformly distributed. The airflow determines fibre diameter, fibre entanglement, basis weight and the attenuation zone. Low air flow rate results in coarse fibres because of the limited drawing of the polymer stream. Turbulence in the airstream, if not controlled, results in defects called roping – narrow, elongated, thick streaks of fibres in the web. Ropes develop when the air flow rate is out of adjustment or insufficient, such that the attenuated fibres come into contact and are then laid down as collected aggregates. The air flow is adjusted for a fixed polymer flow rate so that they can form continuous fibres. Violent blowing of air results in a defect called fly – fine nanofibres not trapped on the drum and therefore showing up on the surface of the web (Butin *et al.*, 1974, 1976; Bresee and Qureshi, 2006; Bresee *et al.*, 2005; Moore *et al.*, 2004; Sloan *et al.*, 1981).

Collector speed

The collector surface is continuously moving across relative to the fibre being laid on it. Despite the random fibre entanglement occurring towards the surface, there is a small bias in the direction of the machine direction, due to some directionality imparted by the movement, hence the speed selection should be such that little bias is introduced (Yin *et al.*, 2000). For the web to attach to the surface of the collector there is a vacuum that draws the air through the fibre web and forming wire surface. The vacuum pressure level should always be adjusted to draw all the air and lock the fibres (Butin *et al.*, 1974; Yin *et al.*, 2000).

Die temperature

During processing the whole die assembly is constantly heated to maintain a selected operating temperature range. It is important to maintain the die temperature, which ranges between 215°C and 340°C, closely in order to produce a uniform web. The temperature setting is dependent on the polymer type being processed and it should be maintained to prevent any variations in the properties of the web. The viscosity of the melt polymer is adjusted by varying the nozzle tip temperature and, as the temperature is raised, the viscosity decreases. The flow rate of the hot polymer melt is dependent upon the nozzle design and speed of the extruder (Sloan *et al.*, 1981).

Hot air

The high velocity hot air (also called primary air), with a temperature range of 230–360°C, is blown through the slots in the die. The temperature and velocity are maintained at a level suitable for the polymer being processed. The heat of the air and its drag are responsible for polymer attenuation close to the die exit hole. Further away from the die the secondary air drawn from the surrounding cools and solidifies the fibre. The effect of the temperature setting is observed in the web uniformity. If it is incorrectly set, the web exhibits defects in the form of slots, fly and roping (Dahiya *et al.*, 2004a).

12.4.2 Off-line processing variables and its control

Offline processing variables, such as the selection of the die hole size, die setback, air gap, angle of air supply, web collection type and polymer/air distribution, are changed when the production line is idle and not running. The fibre size is mainly influenced by the die hole size, die hole design parameters and die setback. The air gap controls the air exit pressure, which affects fibre attenuation and breakage, depending on the polymer type being processed. Poor die design results in a non-uniform web.

Die hole size

The size of the die holes determines the size of the polymer stream ejected through the hole before it is further drawn and reduced in diameter by the high speed hot air. The initial size ultimately controls the possible fibre diameter obtained, and it is achieved by the die hole size. A small size produces relatively small diameter fibres, giving the web properties that derive from such fibres (Dahiya *et al.*, 2004a).

Air gap

High velocity hot air is blown and exits through the slots in the die at a pressure that draws the polymer streams. Air pressure needs to be controlled at a certain level, as it affects the degree of fibre breakage. The pressure of exiting air is controlled by the air gap (Dahiya *et al.*, 2004a).

Angle of air supply

Hot air can be supplied or set to exit towards the polymer stream at different angles ranging between 30° and 90°. At 30° the fibres are directed towards the collector lying parallel to each other with minimum entanglement. The web that is formed from these parallel fibres usually shows undesirable characteristics of loose coiled bundles, high breakage of fibres and such defects as roping. When the angle is moved towards the other extreme, 90°, there is a high degree of fibre separation and random laydown. It is vital to control the air angle to produce the desired web structure (Moore *et al.*, 2004).

12.4.3 Material variables

Online and offline processing parameters are set according to the type of polymer being processed. A wide variety of polymers can be processed by the melt blowing method. The basic properties of such polymers are that they are fibre forming with acceptable low-melt viscosity, of narrow molecular weight distribution, and can solidify before landing on the collector surface. The choice of polymer to be processed depends on the end-use of the product. Some common fibres forming polymers are polypropylene, polyethylene, polyamide, polycarbonate and polystyrene (Dahiya *et al.*, 2004a).

Generally, the polymer melt flow rate, air flow rate and apparent viscosity of the molten polymer, have to be controlled and correlated to obtain a good product (Butin *et al.*, 1976). The fibre is continuously deposited on the surface and the product withdrawn at a rate synchronised with collection, so that the web built up and weight are maintained (Sloan *et al.*, 1981).

12.5 Spunbonding: process variables and process control

Spunbonds are nonwovens produced by depositing extruded, spun filaments onto a collecting belt in a uniform random manner, followed by bonding of the fibres (Dahiya *et al.*, 2004b; Russell 2007; Smith, 2000). Spunbonding combines the spun laying and the bonding processes in a short continuous nonwoven fabric production process from polymer to fabric. In the spunbonding process, the filaments are extruded from the molten polymer by forcing them through the spinnerets and then drawing them, to obtain the desired molecular orientation, by using high velocity air streams. The air is

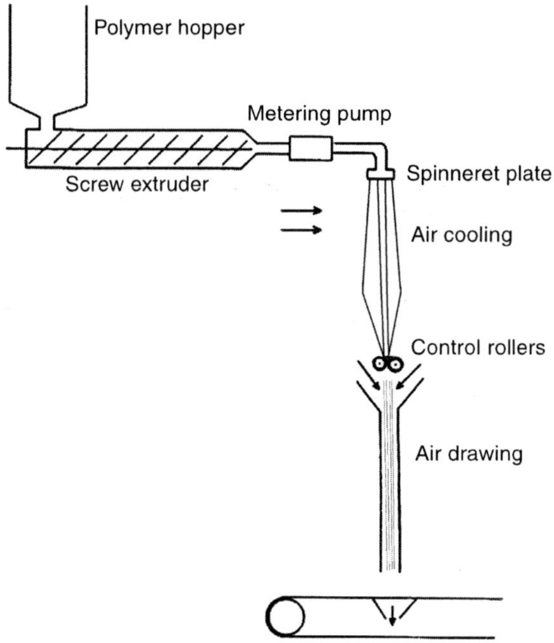

12.4 Diagram of spunlaid process. (Source: Smith, 2000. With permission from Woodhead Publishing Limited, UK.)

then blown at high velocity past the filaments, thereby stretching them and achieving the desired toughness. The drawn filaments are directed onto the conveyor belt and their movement produces a very strong orientation of the fibres in the machine direction, and consequently a high tensile strength in that direction. The laid filaments on the conveyor belt are transported to the bonding machine, which imparts integrity and strength either by needle punching, hydroentanglement, chemical or thermal bonding to form a nonwoven fabric. The ultimate fabric use determines the selection of the bonding method (Russell 2007; Smith, 2000). Figure 12.4 shows a simplified schematic diagram of the spunlaid process followed by the bonding process (Smith, 2000).

The primary process that needs to be controlled in the production of spunbond fabrics is the control of four simultaneous, integrated operations, namely filament extrusion, drawing, lay-down and bonding. These variables are related to either the material or operation and they influence fibre diameter, fibre structure, web lay-down, physical properties or the tactile properties of the web. These variables can be classified broadly into three main groups, namely machine/online variables, off-line variables and material variables (Dahiya *et al.*, 2004b; Malkan, 1995; Russell 2007).

12.5.1 Operational/online variables and their control

Some of the operational/online variables are: primary air temperature, quench air rate, air suction speed and venturi gap, collection speed, throughput, bonding temperature and pressure. Various ways to control these parameters are discussed below.

Primary air temperature

The temperature of the primary air has a significant effect on the diameter of the extruded filaments. Fibre diameter is known to decrease with increase in temperature of the primary air. The drawing force generated determined by the speed of the primary air and higher primary air speeds increases the degree of drawing, thus yielding fine fibres (Bo, 2010; Russell 2007).

Quench air rate

The extruded filaments pass through quench chambers, where air cools the molten polymer, stretching and solidifying it. The drawing effect resulting from the air-drag force leads to the polymer molecular orientation. The quench air velocity controls the attenuation, while its temperature controls the cooling effect, both influencing the development of the microstructure of the filaments (Bo, 2010; Lim, 2010).

Air suction speed and Venturi gap

The speed of the suction air draws the filaments causing fibres to be finely attenuated. The higher the air suction speed, the finer the fibres produced. Air suction plays a vital aerodynamic role of holding the web together on the conveyor belt (Bo, 2010). By increasing the suction speed of the air, the anisotropy of the filament distribution decreases within the structure from enhanced alignment in the machine direction (Salvado *et al.*, 2006). The Venturi gap is a significant variable that plays an important role in the polymer drawing. A larger Venturi gap yields finer fibres because it causes fibres to be better attenuated (Bo, 2010).

Throughput

Throughput refers to the amount of the molten polymer forced through the metering pump per unit time. Throughput is maintained throughout the process because the flow rate of the molten polymer ultimately controls the final fibre diameter of the filament. When throughput is increased, the fibre diameter of the resultant fibres also increases. At lower throughput rates, finer fibres are produced (Bo, 2010; Russell 2007).

Collection speed

The speed at which the conveyor belt move controls the final lay-down of the multi filaments after the initial drawing and electrostatic filament separation. Increasing the speed of the belt relative to the deposition introduces a machine direction orientation bias (Malkan, 1995).

Bonding temperature and pressure

Bonding temperature and pressure have a great influence on the tensile properties of the final fabric. High temperatures increase the bonding of the structure, as more fibre filaments fuse together and form an increasingly stiffer structure. The pressure of the bonding rollers has the same effect, forming a much stiffer product at higher roller pressures. Operation pressures and temperature is dependent on the resin material (Lim, 2010; Russell 2007).

12.5.2 Off-line processing variables and its control

These variables include the selection of the die hole size, die setback, web collection type, etc. The fibre diameter is mainly influenced by the die hole size, die hole design parameters and die setback. The poor die design results in a non-uniform web. Controlling these variables is also important as they play a significant role in the final quality of the produced nonwovens (Lim, 2010; Malkan, 1995).

12.5.3 Material variables

The spunbonding process requires polymers with higher molecular weight and a broad molecular weight distribution to produce uniform webs. The temperature of the polymer melt determines the size of the fibres produced, because the extent to which the melt polymer is drawn is dependent on the flow characteristics. Thus, changes in the initial temperature of the polymer melt cause changes in the rate of fibre attenuation, and this means that higher temperatures yield finer fibres (Bo, 2010; Lim, 2010; Malkan, 1995; Russell 2007). Different polymers possess different characteristics, and the choice of is influenced by the application of the nonwoven. Some commonly used polymers are polypropylene, high density polyethylene, polyamide, and polyester (Dahiya *et al.*, 2004b).

12.6 Future trends

Even with today's sophisticated automation, process control still relies on human skill. Each nonwoven manufacturing process is different, which requires consideration of many variables in the materials and processing stages. So, for a particular product, the selection of parameters and its control always play a role and this will continue into the future. If a database were built for storage of existing knowledge, a better understanding of the process would be readily available for the required product and accordingly the process could be more efficiently controlled.

12.7 Sources of further information

For detailed information on nonwoven manufacturing processes, their control and applications, readers are referred to the published literature (Russell, 2007; Smith, 2000) that are listed in the reference section. The above references are good source of information for the needle punching process and its process control. Similarly for melt blowing, spun bonding published literatures (Lim, 2010; Malkan, 1995) and for hydroentanglement the thesis (Xiang, 2007) are useful resources for further information.

12.8 Acknowledgement

The authors acknowledged Woodhead Publishing Limited for granting permission to reproduce some figures used in this chapter from the following books:

Technical fabric structures – 3: nonwoven fabrics by Smith P A, in *Handbook of technical textiles*, Horrocks A R and Anand S C edited, 2000.

Polymer-laid web formation by Bhat G S and Malkan S R in *Handbook of nonwovens*, Russell S J edited, 2007.

12.9 References

Begenir A, Tafresh H V and Pourdeyhimi B (2004), 'Effects of nozzle geometry on hydro-entangling water jets: experiments observations', *Text Res J*, **74**, 178–84.

Bhat G S and Malkan S R (2007), 'Polymer-laid web formation', in *Handbook of nonwovens*, (Russell S J edited), Woodhead Publishing Limited, Cambridge, UK, 143–195.

Bo Z (2010), 'Mathematical modelling of poly (ethylene terephthalate) air drawing in spunbonding nonwoven process,' *Ind J Fibre Text Res*, **35**, 145–53.

Bresee R R and Ko W C (2003), 'Fibre formation during melt blowing', *Int Nonwovens J*, 12, 21–8.

Bresee R R and Qureshi U A (2004), 'Influence of processing conditions on melt blown web structure: part 1 – DCD', *Int Nonwovens J*, **13**, 49–55.

Bresee R R and Qureshi U A (2006), 'Influence of process conditions on melt blown web structure, part IV – fibre diameter,' *J Engg Fibres Fabrics*, **1**, 32–46.

Bresee R R, Qureshi U A and Pelham M C (2005), 'Influence of process conditions on melt blown web structure: part 2 – primary airflow rate,' *Intl Nonwovens J*, **14**, 11–18.
Butin R R, Keller J P and Harding J W (1974), 'Nonwoven mats by melt blowing', US Patent No. 3849241, November 19.
Butin R R, Keller J P and Harding J W (1976), 'Melt blowing process', US Patent No. 3978185, August 31.
Cannity E G Jr (1962), 'Needle felting – past, present and future', *Canadian Text J*, **79**, 72–3.
CSIR Report, 'Energy transformation in hydroentanglement nonwovens,' Polymers and Composites, Port Elizabeth, South Africa, 2011.
Dahiya A, Kamath M G and Hegde R R (2004a), *Melt blown technology*, Tennessee, Available from: http://www.engr.utk.edu/mse/pages/Textiles/Melt%20Blown%20Technology.htm [Accessed 01 February 2011].
Dahiya A, Kamath M G and Hegde R R (2004b), *Spun bond technology*, Tennessee, Available from: http://www.engr.utk.edu/mse/pages/Textiles/Spunbond%20Technology.htm [Accessed 05 February 2011].
Debnath S, Kane C D, Kadole P V and Patil C A (1994), 'Needlepunched nonwoven blankets from polyester', *Indian Text J*, **105**, 72–80.
Ellison C J, Phatak A, Giles D W, Macosko C W and Bates F S (2007), 'Melt blown nanofibres: fibre diameter distributions and onset of fibre breakup', *Polymer*, **48**, 3306–16.
Falk F (2000), 'Felting and structuring needles', *Nonwoven Report Intl*, **9**, 30–1.
Foster J H (1985), 'Illustrating a point about an increasing technical segmentation,' Foster Needle Co, Manitowoc, USA, Available from: http://www.fosterneedleusa.com/tch_pprs/techseg.html [accessed 04 January 2011].
Ghassemieh E, Acar M and Versteeg H K (2001), 'Improvement of the efficiency of energy transfer in the hydro-entanglement process', *Compos Sci Tech*, **61**, 1681–94.
Ghassemieh E, Versteeg, H K and Acar M (2003), 'Effects of nozzle geometry on the flow characteristics of hydro-entangling jets', *Text Res J*, **73**, 444–50.
Hearle J W S and Sultan M A I (1968a), 'A study of needled fabrics, part III: the influence of fibre type and dimensions', *J Text Inst*, **59**, 137–47.
Hearle J W S and Sultan M A I (1968b), 'A study of needled fabrics, part V: the approach to theoretical understanding', *J Text Inst*, **59**, 183–201.
Hearle J W S, Sultan M A I and Choudhari TN (1968), 'A study of needled fabrics, part II: the effects of needling process', *J Text Inst*, **59**, 103–16.
Kiekens P and Zamfir M (2002), 'Non-wovens from cotton fibres for absorbent products obtained by the needle-punching process', *Autex Res J*, **2**, 166–74.
Lewandowski Z, Ziabicki A and Jarecki L (2007), 'The nonwovens formation in the melt-blown process', *Fibres Text Eastern Europe*, **15**, 77–81.
Lim H (2010), 'A review of spunbond process', *J Text Apparel Tech Mang*, **6**, 1–13.
Maitre K G (1989), 'Frontiers of needle punching', *Nonwovens an Advanced Tutorial*, **38**.
Malkan S R (1995), 'An overview of spunbonding and meltblowing technologies,' *Tappi J*, **78**, 185–90.
Mao N and Russell S J (2005), 'Hydro-entanglement intensity: towards a quantification of the structural consolidation in hydro-entanglement and its influence on

the permeability of fabrics, *EDANA nonwovens research academy proceedings*, Geneva, Switzerland, 121–31.

Medeiros J (1996), 'Spunlace/hydro-entanglement methods and products', in *INDA-TEC'96: international nonwovens conference*, September 11–13, Crystal City, VA, USA, 1–15.

Miao M (2004), 'An experimental study of the needled nonwoven process, part II: fibre transport by barbed needles', *Text Res J*, **74**, 394–8.

Moore E M, Papavassiliou D V and Shambaugh R L (2004), 'Air velocity, air temperature, fibre vibration and fibre diameter measurements on a practical melt blowing die', *Intl Nonwovens J*, **13**, 43–53.

Patanaik A and Anandjiwala R D (2010), 'Hydroentanglement nonwoven filters for air filtration and its performance evaluation', *J Appl Polym Sci,* 117, 1325–31.

Patanaik A and Anandjiwala R D (2009), 'Some studies on water permeability of nonwoven fabrics', *Text Res J*, **79**, 147–53.

Patanaik A and Anandjiwala R D (2008), 'Water flow through the polypropylene-based geotextiles', *J Appl Polym Sci,* **108**, 3876–80.

Patanaik A, Anandjiwala R D and Boguslavsky L (2009), 'Development of high Efficiency particulate absorbing filter materials', *J Appl Polym Sci,* **114**, 275–80.

Patanaik A, Anandjiwala R D, Gonsalves J and Boguslavsky L (2007), 'Modelling pore size distributions in needle punched nonwoven structures using finite element analysis', paper presented in finite element modelling of textiles and textile composites conference, 26–28 September, St. Petersburg, Russia.

Pourmohammadi A, Russel S J and Hottele S (2003), 'Effect of water jet pressure profile and initial web geometry on the physical properties of composite hydro-entangled fabrics', *Text Res J*, **73**, 503–8.

Ramkumar S S, Umrani A S, Shelly D C, Tock R W, Parameswaran S and Smith M L (2004), 'Study of the effect of sliding velocity on the frictional properties of nonwoven substrates', *Wear*, **256**, 221–5.

Ramkumar S S and Roedel C (2003), 'A study of the needle penetration speeds on the frictional properties of nonwoven webs: a new approach', *J Appl Polym Sci*, **89**, 3626–31.

Rawal A and Anandjiwala R D (2006), 'Relationship between process parameters and properties of multifunctional needlepunched geotextiles', *J Ind Text*, **35**, 271–85.

Roedel C and Ramkumar S S (2003), 'Surface and mechanical property measurements of H1 technology needle-punched nonwovens', *Text Res J*, **73**, 381–5.

Russell S J (2007), *Handbook of nonwovens*, Woodhead Publishing Limited, Cambridge, UK.

Salvado R, Silvy J and Drean J -Y (2006), 'Relationship between fibrous Structure and spunbond process, *Text Res J,* **76**, 805–12.

Sloan C H, Wright J A and Morie G P (1981), 'Melt blowing process', US Patent No. 4267002, 12 May.

Smith P A (2000), 'Technical fabric structures – 3: nonwoven fabrics,' in *Handbook of technical textiles* (Horrocks A R and Anand S C edited), Woodhead Publishing Limited, Cambridge, UK, 130–151.

Tafresh H V and Pourdeyhimi B (2004), 'Simulating cavitation and hydraulic flip inside hydro-entangling nozzles', *Text Res J*, **74**, 359–64.

Thirlwell B E and Treloar L R G (1965), 'Non-woven fabrics, part VI: dimensional and mechanical anisotropy', *Text Res J*, **35**, 827–35.

Watanabe A, Miwa M, Yokoi T and Merati A A (2004), 'Predicting the penetrating force and number of fibres caught by a needle barb in needlepunching', *Text Res J*, **74**, 417–25.

White C F (1990), 'Hydroentanglement technology applied to wet-formed and other precursor webs', *Tappi*, **73**, 187–92.

Wizeman G (2000), 'Felting and structuring needles for various application fields', *Melliand Intl*, **6**, 62–3.

Xiang P (2007), 'Numerical modelling and experimental investigation of the hydroentanglement process', PhD thesis, North Carolina State University, Available from: www.lib.ncsu.edu/thesis/available/etd-08082007-172253/unrestricted/etd.pdf [Accessed 15 January 2011].

Xiang P, Kuznetsov A V and Seyam A M (2007), 'Fibres caught in the knuckles of the forming wires: experimental measurements and physical origins of the force of peeling in the hydroentanglement process', *J Engg Fibres Fabrics*, **2**, 1–9.

Yin H, Yan Z, Ko W C and Bresee R R (2000), 'Fundamental description of the melt-blowing process', *Int Nonwovens J*, **9**, 25–8.

13
Process control in dyeing of textiles

S. M. SHANG, The Hong Kong Polytechnic University, China

DOI: 10.1533/9780857095633.3.300

Abstract: This chapter introduces the general technology for the dyeing of textile materials (yarn and fabric of cotton, polyester, nylon, and blends) with respect to dye applications (reactive, disperse, acid, etc.), dyeing methods (batchwise and continuous) and dyeing machines (package, overflow/jet and continuous). The corresponding dyeing processes and the process controls are discussed in detail.

Key words: textile materials, dye applications, dyeing methods, dyeing machines, and dyeing process control.

13.1 Introduction

Dyeing and printing processes are value-added treatments for most textile materials. A dyeing process is the interaction between a dye and a fibre, as well as the movement of dye into the internal part of the fibre. Generally, a dyeing process involves adsorption (transfer of dyes from the aqueous solution onto the fibre surface) and diffusion (dyes diffused into the fibre). In addition to direct absorption, dyeing may also involve the precipitation of dyes inside the fibre (vat dyes), or chemical reaction with the fibre (reactive dyes). From the view of colouration, printing can be considered as partial dyeing with different colours on fabric to form an attractive pattern. Table 13.1 shows some typical dyes applied in the dyeing or printing of various textile materials. A dyeing or printing process is complicated, since it involves fibre kinds, yarn or fabric structures, dyes and chemical auxiliaries, as well as dyeing technology. In order to achieve the required dyeing or printing quality, all factors that may influence the dyeing or printing process must be precisely controlled (Table 13.2).

The water quality for dyeing and printing is very important (Yeung and Shang, 1999), and it must meet the requirements as listed in Table 13.3. Generally, purification of the water is required to avoid such unpleasant dyeing defects as unevenness, dye precipitation, shade dulling, harsh handle or chalking. The preparation and pretreatment of yarns or fabrics also have significant effects on dyeing quality. For example, poor scouring and bleaching can lead to serious unevenness in dyeing, and the non-uniform winding of a yarn package can lead to colour differences in package dyeing. As a

Table 13.1 Classification of dyes according to usage

Dye class	Cellulose (cotton, viscose, rayon)	Protein (wool, silk)	Polyester	Nylons	Acrylics
Direct	**	*	–	*	–
Reactive	**	**	–	*	–
Sulphur	**	–	–	–	–
Vat	**	–	–	–	–
Disperse	–	–	**	*	**
Acid	–	**	–	**	–
Basic	–	–	–	–	**

** The most important dye applied in dyeing and printing.
* Less important dye applied in dyeing and printing.
– No practical application.

Table 13.2 Influence factors in dyeing or printing process

Influence factor	Fabric dyeing			Yarn dyeing (batch wise)		Printing	
	Batchwise						
	Overflow dyeing	Jig dyeing	Continuous (pad dyeing)	Package dyeing	Hank dyeing	Screen printing	Digital printing
Fibre	**	**	**	**	**	**	**
Yarn type	*	*	*	**	**	*	*
Fabric structure	**	**	**	–	–	**	**
Water quality	**	**	**	**	**	**	**
Yarn or fabric preparation	**	**	**	**	**	**	**
Dye selection	**	**	**	**	**	**	**
Temperature	**	**	**	**	**	**	**
Time	**	**	**	**	**	**	**
pH	**	**	**	**	**	**	**
Fabric moving speed	**	**	**	–	–	**	**
Load capacity	**	**	*	**	**	*	*
Liquor ratio	**	*	–	**	**	–	–
Heating and cooling rate	**	**	*	**	**	*	*
Dosing of dyes and auxiliaries	**	**	**	**	**	**	**
Dye solution flow rate and circulation time	**	**	*	**	**	–	–
Pad pressure	–	–	**	–	–	–	–

** The most important influence factor in dyeing or printing.
* Less important influence factor in dyeing or printing.
– No practical application.

Table 13.3 The basic requirements of water quality for dyeing and printing process

Item	Requirement
Colour	Colourless (dilution factor ≤ 10) and without turbid and suspended solid
pH	6.5–7.5
Total hardness (as $CaCO_3$)	< 50 mg/L (for general use) < 17.5 mg/L (for dyestuff dissolution)
Iron	< 0.1 mg/L
Manganese	< 0.1 mg/L
Transparency	> 30 cm

Note: The requirements were recommended by China Association of Dyeing and Finishing Industry (Xi, Chen and Ma, 2006).

result, sufficient and uniform preparation and pretreatment can make the dyeing quality more controllable and predictable. Therefore, process control in dyeing and printing is of significance in achieving high-quality products and increasing dyeing production efficiency. Dyeing technology and dyeing process control will be discussed in detail in this chapter, and printing technology and printing process control will be discussed in next chapter.

13.2 Dyeing of cotton

Cotton is the most important nature textile fibre. Cotton fibres are composed of cellulose, and can be dyed with reactive dyes and direct dyes, as well as vat, sulphur and azoic dyes in proper processes to meet the diversified demands in end-use, such as shades and fastness standards.

13.2.1 Reactive dye and dyeing technology

Reactive dye is the dye that can react with a fibre to form a covalent link, that is forming a permanent attachment in the fibre and could not be removed by repeated treatment with boiling water under neutral conditions. Consequently, the dyes become parts of the fibre, leading to outstanding colour fastness to wash.

Due to the advantages of full colour ranges, brightness, high fastness, low cost, easy application, etc., reactive dye became the predominant dye for cotton dyeing and printing in textile industry since it was invented. Compared with direct dye, reactive dye is applied as easy as direct dye but has very high levels of fastness, especially for wet fastness.

Properties of reactive dye

The characteristic structure of a common reactive dye includes the following components:

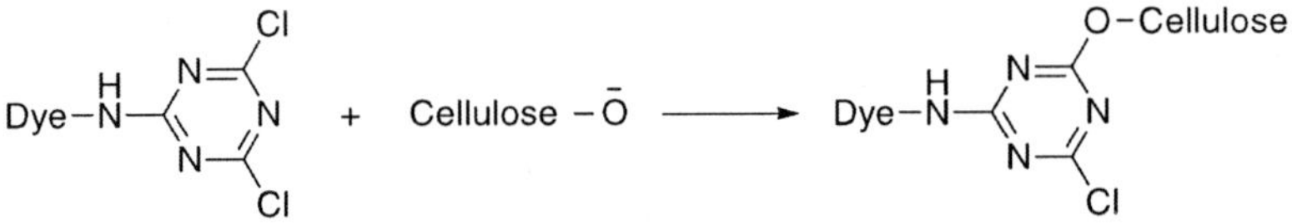

13.1 Substitution reaction between cellulose and reactive dye.

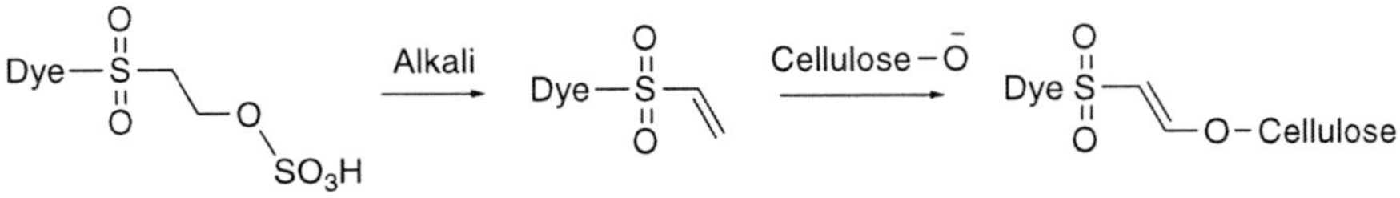

13.2 Addition reaction between cellulose and reactive dye.

1. chromogen, contributing to the colour display;
2. reactive group(s), enabling the chemical reactions between fibre and dye;
3. bridging link, linking the reactive group with chromogen;
4. solubilising group(s), enhancing the solubility of dye.

As the name implies, reactive dye has reactive group(s) to form covalent bonds chemically with cotton and become part of it, rather than as an independent coloured substance within the fibre. Under mild alkaline conditions, the reactive group(s) on the dye molecule can react with the oxygen atom in the cellulose hydroxyl group, either by nucleophilic substitution, or by addition, and sometimes by both mechanisms for a dye with two or more reactive groups (Broadbent, 2001). The representative substitution and addition reactions between dye and cellulose are shown in Fig. 13.1 and 13.2, respectively (Gordon and Hsieh, 2007).

Both the chromophore and reactive group in a reactive dye govern its application characteristics, such as reactivity, substantivity/affinity, diffusion coefficient and solubility. Substantivity, that is the tendency of dye to transfer from dyebath to fibre substrate, is determined by both but the former has a more direct influence. Reactivity and the stability of the dye–fibre bond are determined by reactive group(s). Recently, reactive dyes with two, even three, reactive groups have been widely used for achieving a greater fastness and deeper shade. Based on the varied reactive groups, reactive dyes can typically be classified as dichlorotriazine (DCT), difluorochloropyrimidine (DFCP), vinylsulphone (VS), monochlorotriazine (MCT), dichloroquinoxaline (DCQ), and the mixtures of MCT + VS, and MCT + MCT, etc. Typical structures of reactive dyes with one reactive group are shown in Fig. 13.3, and the major properties of the different types of reactive dyes are listed in Table 13.4.

Table 13.4 The major properties of the different types of reactive dyes

Type of reactive group	Dichlorotriazine (DCT)	Difluorochloropyrimidine (DFCP)	Vinylsulphone (VS)	Bifunctional (MCT + VS)	Monochlorotriazine (MCT)	Bifunctional (MCT+ MCT)	Dichloroquinoxaline (DCQ)
Fixation rate	High	High	Medium	Medium	Medium	High	Medium
Reactivity	Very high	High	Medium	Medium	Low	Medium	High
Wash-off property	Medium	Good	Good	Medium	Medium	Good	Good
Covalent bond stability to acid	Poor	Good	Good	Medium	Medium	Medium	Poor
Covalent bond stability to alkali	Medium	Excellent	Poor	Medium	Good	Medium	Medium
Substantivity	High	High	Medium to high	Medium to high	Very high	High	High
Typical example	Procion® MX	Levafix® EA	Remazol®	Sumifix supra®	Procion® H	Procion® H-E	Levafix® E

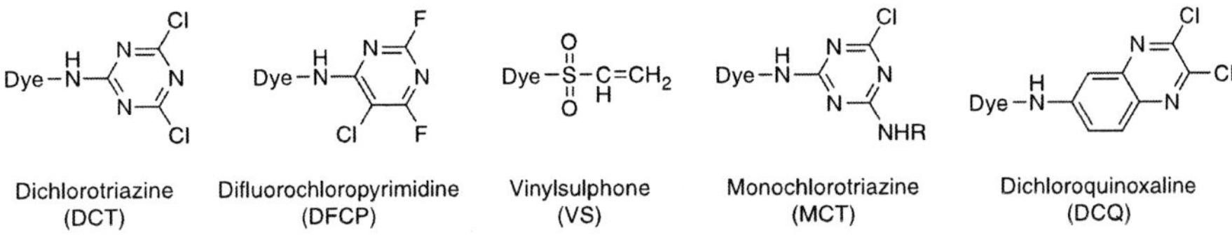

13.3 Typical structures of reactive dyes with single reactive group.

A conventional dyeing process of reactive dyes contains three stages: (1) adsorption and diffusion, (2) fixation, and (3) wash-off. Reactive dyes can be used in both batchwise and continuous dyeing methods. Generally, dyes with relatively high substantivity and slow diffusion are more appropriate for exhaust dyeing in the batchwise method, such as yarn dyeing with a package dyeing machine or hank yarn dyeing machine, as well as fabric dyeing with an overflow dyeing machine or jig dyeing machine. Unlike exhaust dyeing, dyes with low substantivity and quick diffusion are more suitable for the continuous pad dyeing method. Generally, there are two types of the continuous pad dyeing process, the simpler single-pad in which dyes and alkali are padded in one dyebath, and the versatile double-pad in which dyes and alkali are padded separately and sequentially.

Although exhaust dyeing and continuous dyeing are quite different in terms of process control, as listed in Table 13.2, there are common fundamental technology parameters, apart from the machinery variables, which influence the dyeing procedure. These parameters include temperature, time, pH, liquor ratio and auxiliaries.

Temperature influence

In a reactive dyebath, temperature has profound effects on both dye and cotton in aqueous solution. The increased temperature may bring about better dye penetration, more rapid diffusion, better evenness, but may reduce dye substantivity and increase the risk of dye hydrolysis. Also, raised temperature leads to the opening-up of the cellulose structure, which activates the dye–fibre interaction (Ibrahim and Sayed, 1993). Therefore, the dyeing temperature is determined by both the substantivity and reactivity of the dye and the structure of fibre. For high twisted yarn or tight fabric, or for those dyes being not easily level dyed, temperature could be raised to 98°C to promote migration and penetration in the adsorption phase with low reactivity dyes, and cooled to 80°C for fixation by adding alkali. For dyeing loose fabric, such as knitgoods, generally a warm dyebath with temperature between 50°C and 65°C is preferred. Good results can be obtained by carefully controlling the temperature increase rate during heating-up.

The pH influence

The pH primarily influences the concentration of the cellusate anion (cellulose-O^-) on the fibre, as well as the hydroxyl ion (-OH) concentration in the dyebath and on the fibre. Reactive dyes consume some alkali for both dye fixation and dye hydrolysis. The internal alkalinity of fibre also absorbs alkali. These lead to the dyebath pH at the dyeing end being always lower than its initial value. Generally increasing dyeing pH in the fixation stage can accelerate the reaction rate between dye and fibre. For batchwise dyeing method, the dyebath pH is recommended to be controlled ranging from 10.5 to 11. Even with the low reactivity dyes, a pH exceeding 11 is still not appropriate since an unduly high pH will enhance dye hydrolysis and reduce dyeing efficiency in terms of depth and fixation. NaOH, Na_2CO_3, $NaHCO_3$ or a combination of these are the conventionally used alkalis, among which Na_2CO_3 is the most commonly applied. The selection of alkali is usually related to the dyes applied and the dyeing method adopted. The fixation rate of dyes and the dyeing evenness can be controlled by the dosing rate of alkali. Dyes with high reactivity are sensitive to alkali concentration in dyebath, and have an optimal temperature range of between 40°C and 60°C. Typical examples are DCT, DFCP, DCQ and VS.

Electrolyte effect

When cotton fibre is in an aqueous solution, the fibre surface presents negative charge, mainly due to the dissociation of accessible cellulose hydroxyl (Cell-OH) groups and the rearrangement of the charge groups at the interface between the fibre and the water. Reactive dyes, as well as other soluble dyes for cotton, carry negative charges because of sulphate group(s) on their molecules for solubility. Therefore, the electrostatic repulsive force between fibre surface and dyes has to be overcome in order to diffuse the dyes through the fibre–water interface. The most common method to overcome the electronic repulsion in exhaust dyeing is to add large quantity of electrolyte (so-called salt), sodium chloride (NaCl) or sodium sulphate (Na_2SO_4), in the dyebath. The presence of electrolyte in the dyebath reduces the extent of the surface charge, which leads to a reduction in repulsion between ionised dyes and fibre, thereby increasing the substantivity of dyes. Trichloropyrimidine and aminochlorotriazine dyes are typical dyes exerting high exhaustion without alkali, thus it is necessary to control the salt-dosing rate carefully with those dyes to ensure levelling. There are four methods of adding salt in general: (1) portion adding/dosing salt; (2) adding salt at the start of dyeing (especially for dyeing dark shade or nonsensitive colour); (3) adding salt and soda at the start of dyeing; and (4) all-in process. The first two methods are used widely in production.

Liquor ratio

Water is an essential medium for a dyeing process in most dyeing methods. The liquor ratio in an exhaust dyebath is the ratio of the weight of the dry material being dyed to the water weight of the dyebath. Example, a liquor ratio of 1:10 implies 1 kg fibre is dyed in 10 L water. A large liquor ratio benefits dyeing levelness, but has negative impacts on production costs and environment, and results in more carbon emission. Therefore, a low liquor-ratio dyeing machine, such as 1:5 for cotton, is being applied more broadly (Shang, 2002; Shang and Zhuo, 2003).

Auxiliaries

Suitable auxiliaries in the dyebath could enhance dyeing qualities. Both the anionic surfactant and the non-ionic surfactant can increase dyeing evenness. The former can improve dye uptake, and the latter may decrease the exhaustion, but both may slow down the hydrolysis rate of reactive dye. The presence of triethanolamine can improve the swellability of the cellulose structure, and thereby enhance colour strength. Urea could increase the solubility of reactive dye in water and enhance the swellability of cellulosic fibre, which is critical in continuous dyeing (Shamey and Hussein, 2005).

Reactive dye selection and dye compatibility

Dye selection is the first important issue in dyeing, considering both the dyeing behaviour and production cost. Table 13.5 lists the majority of considerations in dye selection.

Nowadays, colour matching with three dyes, such as with the dyes of three primary colours in red, yellow, and blue, is very common in practice. Therefore the compatibility of the selected three dyes should be good, that is the dyes should have the similar dyeing behaviour, including tone-on-tone build-up properties, equal robustness to different dyeing conditions, and no blocking effect between them. Figure 13.4 shows a set of the typical commodity dyes of three primary colours in red, yellow, and blue applied for matching medium and deep shade. There are also sets of commodity dyes of three primary colours with higher colour fastness applied for matching light shade.

Reactive dyeing process control in the exhaust dyeing process

The exhaust dyeing method is mainly applied in the batchwise dyeing machines, such as jig or overflow dyeing machine for fabrics (Shang and Chong, 2002; Shang and Zhuo, 2003), and the package or hank dyeing machines for yarns.

Table 13.5 The key issues concerned for dye selection

Issue	Requirement
Dyeing behaviour	Shade consistency
	Evenness
	Exhaust and fixation rate
	Colour fastness
	Reproducibility
	Compatibility
Production cost	Dye price
	liquor ratio
	Auxiliary agents required
	Dyeing temperature and time
	Washing effectiveness
	Short dyeing process
	Right-first-time
Others	Environmental friendly

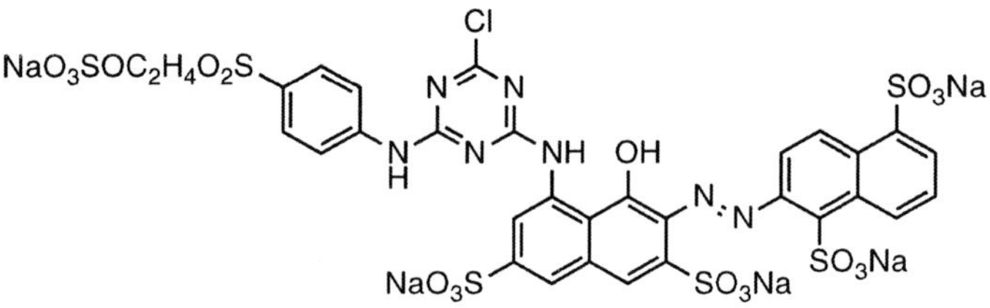

C.I. Reactive Red 239

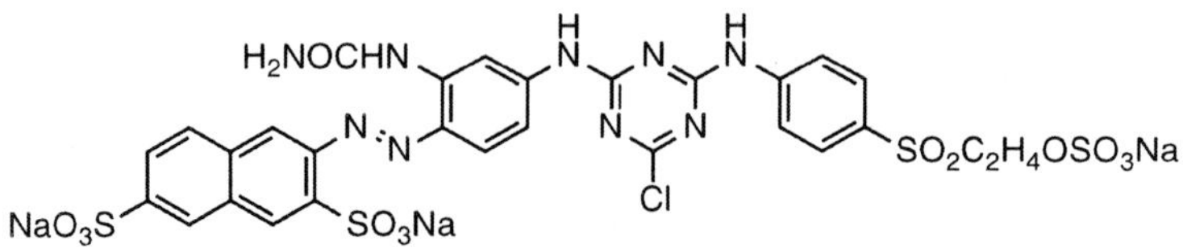

C.I. Reactive Yellow 176

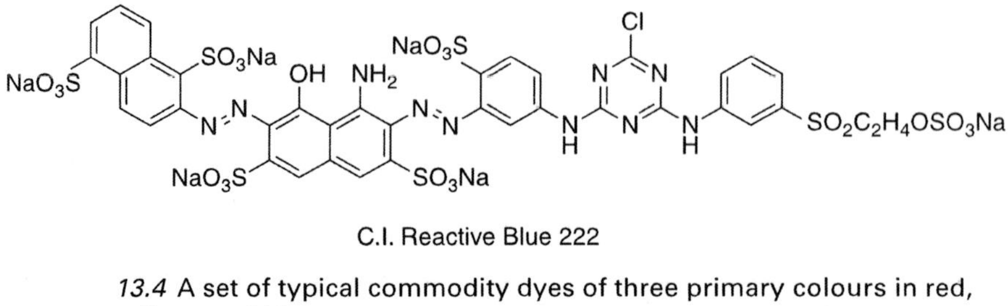

C.I. Reactive Blue 222

13.4 A set of typical commodity dyes of three primary colours in red, yellow, and blue.

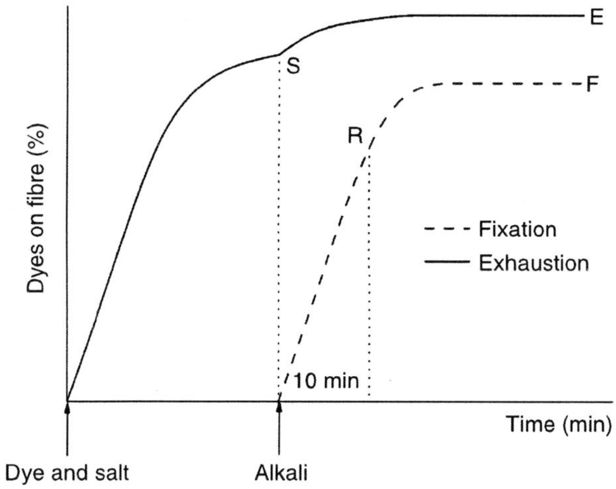

13.5 The SERF value in exhaust dyeing.

In order to identify the tone-on-tone build-up properties of reactive dye in exhaust dyeing, the concept of SERF value is developed. As shown in Fig. 13.5, *S* is the exhaustion value before the addition of alkali, indicating the substantivity; *E* is the final exhaustion value, indicating the exhaustion ability; *R* is the fixation value when fixation lasts 10 min (another explanation is that *R* is the fixation rate when fixation lasts half of the total fixation time when fixation reaches the maximum value), indicating the fixation rate; and *F* is the final fixation value, indicating the reactivity.

Depending on the difference between *E* and *S*, the dyeing curve for exhaust dyeing can be classified into several categories.

For dyes with a large (*E–S*) value, it means the substantivity of these dyes is low. Salt can be added at the beginning of dyeing to increase substantivity, which accelerates the dye adsorption rate. Fixation can be controlled by either the rate of alkali dosing or temperature increase.

The dosing unit is widely equipped with dyeing equipment, and the dosing rate in some advanced dyeing machines can be controlled progressively or retrogressively (Fig. 13.6). The term 'dosing' here not only means simply the addition of chemicals by the dosing system, but also indicates that much more attention has to be paid to the 'dosing' stage, and a versatile dosing programme is necessary in order to obtain the best results.

Figure 13.7 shows the alkali-controllable process, in which alkali is added more carefully than with regular dosing, in a constant-temperature dyebath. Figure 13.8 shows the controlled temperature process in which alkali is regularly dosed at room temperature (assuming 30°C) and the temperature rising is controlled more carefully. The other dyeing profiles are shown in Table 13.6. The symbol of the right-angled triangle '◿' in all dyeing curves

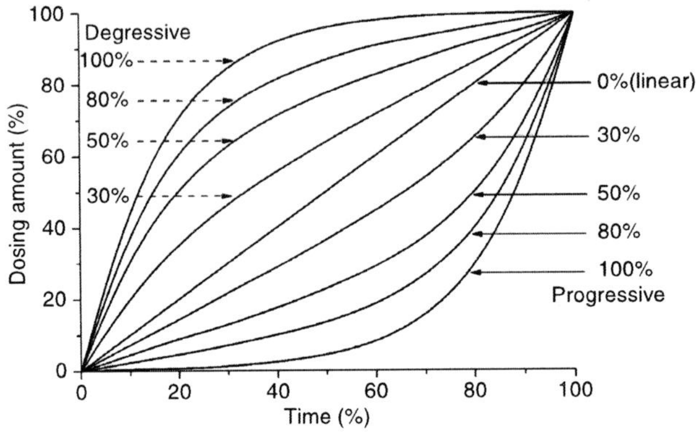

13.6 Progressive and degressive control of dosing amounts.

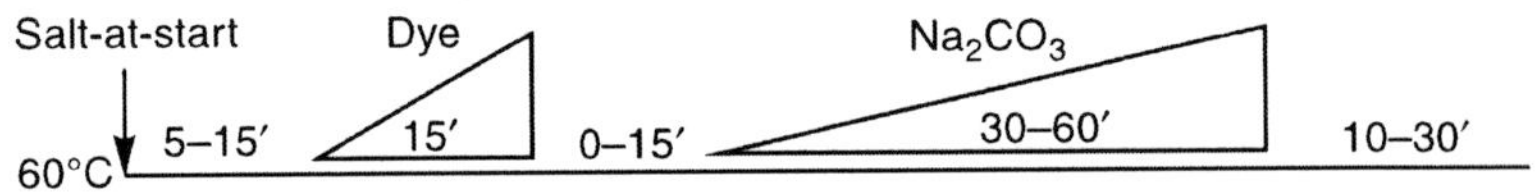

13.7 The alkali-controllable dyeing curve for dye with large (*E–S*) value.

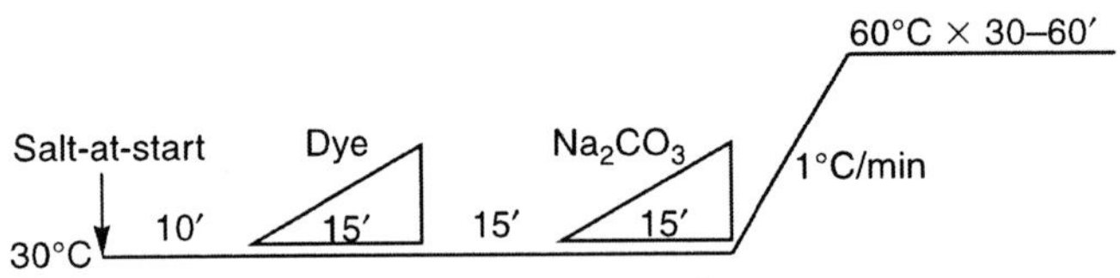

13.8 The temperature-controllable dyeing curve for dyes with large (*E–S*) value.

in this chapter denotes the linear dosing of chemicals and dyes. For alkali dosing, progressive dosing may be required for some dyeing processes in practice for achieving better dyeing results.

For dyes with low (*E–S*) value, indicating high substantivity, 'salt-at-start' is not recommended, because these dyes have the poor migration ability and unevenness is inevitable. As a result, controlled salt-dosing is necessary to control the substantivity. Salt-dosing combined with temperature increase is adopted to control the exhaustion rate (Fig. 13.9). Controlled dye-dosing can be adopted for dyes with moderate (*E–S*) value (Fig. 13.10). The other dyeing profiles are shown in Table 13.7.

When the dyeing process is ended, a wash-off process is necessary to achieve optimal colour fastness. The unfixed dye on fibre will be effectively removed under hot soaping treatment.

Table 13.6 The exhaust dyeing profiles of dyes with large (*E–S*) value

Process flow	Exhaustion → fixation → rinsing → soaping → rinsing → drying			
Dyebath recipe (liquor ratio: 1:5–10)	Ingredient	Shade		
		Light	Medium	Dark
	Reactive dyes[1] (%, owf[2])	<1	1–3	> 3
	Na_2SO_4 (g/L)	10–30	30–60	60–80
	Na_2CO_3 (g/L)	5–10	10–20	20–30

[1] Example of available dye: Remazol®.
[2] owf: on the weight of fabric.

Table 13.7 The exhaust dyeing profile of dyes with small to moderate (*E–S*) value

Process flow	Exhaustion → fixation → rinsing → soaping → rinsing → drying			
Dyebath recipe (liquor ratio: 1:5–10)	Ingredient	Shade		
		Light	Medium	Dark
	Reactive dyes[1] (%, owf)	< 1	1–3	> 3
	Na_2SO_4 (g/L)	10–45	45–70	70–90
	Na_2CO_3 (g/L)	10–15	15–20	>20

[1] Example of available dyes: Procion H-E for small *(E–S)* value; Procion XL+ or Procion H-EXL for moderate *(E–S)* value.

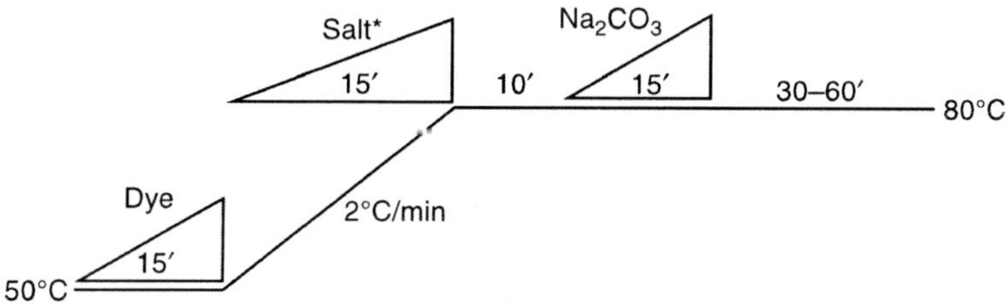

13.9 Dyeing curve for dyes with small (*E–S*) value. * Linear dosing of salt during temperature rising period for time saving.

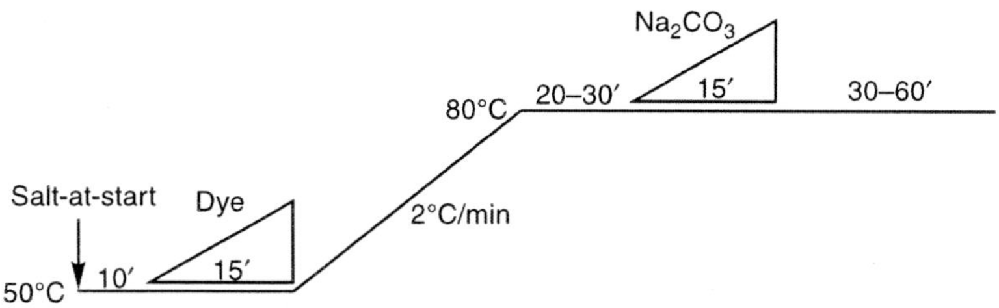

13.10 Dyeing curve for dyes with moderate (*E–S*) value.

Figure 13.11 shows a general batchwise wash-off process for products dyed with medium shade. More washing-off may be needed for deep shade, until the required colour fastness is achieved.

Reactive dyeing process control in continuous process

The continuous dyeing method for woven fabric, especially for cotton or cotton blend, is an important processing technology for reactive dyeing, especially when a large amount of fabric needs to be dyed. Unlike exhaust dyeing, in which dyes of different substantivity are properly all available, dyes with only low to moderate substantivity are preferable in continuous dyeing. The continuous dyeing process can generally be classified into four types: (1) pad-dry-pad-steam method (Table 13.8), (2) pad-steam method (Table 13.9), (3) Econtrol process (Table 13.10), and (4) pad-dry-cure method (Table 13.11).

Note:

1. Pre-drying with an infrared dryer is recommended to reduce the pick-up ratio to 30% before drying in order to minimise unexpected dye liquor migration.
2. The addition of salt to the alkali bath is to minimise colour-bleeding. However for dyes with low solubility, a high concentration salt might cause dye precipitate, therefore a dispensing agent (10 g/L) in the pad-bath is needed to ensure stability.
3. Saturated steam is preferred for reactive dye fixing. The steaming time may vary depending on the dye applied.
4. In order to prevent the dye migration, a migration inhibitor, such as sodium alginate (4%), can be added into the dye solution (20–40 g/L).

The process parameters of pad-steam method are similar to those of the pad-dry-pad-steam process. But more attention has to be paid to the

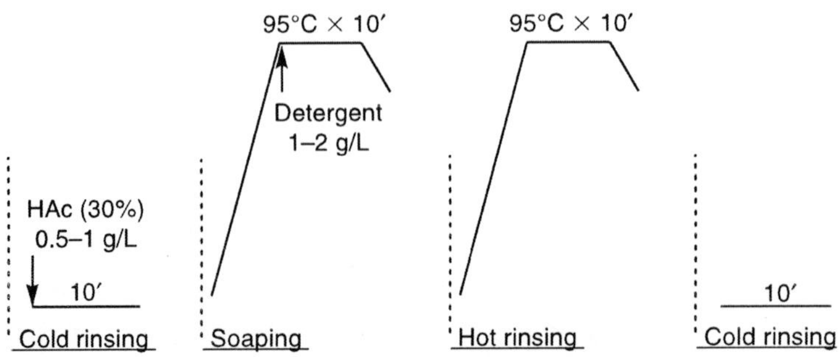

13.11 A general batchwise wash-off process for the dyeing of medium shade.

Table 13.8 The continuous dyeing profile of the pad-dry-pad-steam method

Process flow	Dye-padding → pre-drying (reduce pick-up ratio to 30%) → drying (120°C, 2–5 min) → alkali-padding → steaming (102–105°C, 30–90 s) → rinsing → soaping → rinsing → drying			
Recipe of dye-padding solution	Ingredient	Concentration (g/L)		
		Light	Medium	Dark
	Reactive dye[1]	< 20	20–50	> 50
	Salt	0–15	0–25	0–30
	Urea	0–20	0–40	0–50
	Migration inhibitor	0–20	0–30	0–40
	Sodium m-nitrobenzenesulfonate	0–5	0–8	0–10
	Wetting agent	1	1–2	1–2
Recipe of alkali-padding solution	Ingredient	Concentration (g/L)		
	NaOH (38°Bé)	10–20		
	Na_2CO_3	20		
	Salt	250		

[1] Example of available dyes: Remazol®, Levafix® E.

Table 13.9 The continuous dyeing profile of pad-steam method

Process flow	Padding (dye + alkali) → infrared pre-drying (reduce pick-up to 30%) → drying → steaming (102–105°C, 30–90 s) → rinsing → soaping → rinsing → drying			
	Ingredient	Concentration (g/L)		
		Light	Medium	Dark
Recipe of padding solution	Reactive dye[1]	< 20	20–50	>50
	Na_2SO_4	0–15	0–25	0–30
	Urea	0–50	0–80	0–100
	Migration inhibitor	0–20	0–30	0–40
	Sodium m–nitrobenzenesulfonate	0–5	0–8	0–10
	Wetting agent	1	1–2	1–2
	$NaHCO_3/Na_2CO_3$ (1:1) or Na_2CO_3	5–10	10–15	10–15

[1] Example of available dyes: Remazol®, Procion® H-E.

Table 13.10 The continuous dyeing profile of Econtrol method

Process flow	Padding (dye + alkali[1]) → fixation[2] → rinsing → soaping → rinsing → drying	
	Ingredient	Concentration (g/L)
Recipe of dyebath	Reactive dye[3]	X
	Urea	0–30
	Wetting agent	1
Recipe of alkali bath	Na_2SO_4	100–200
	Na_2CO_3	20–40
	NaOH	3–10

[1] The dye solution and alkali solution are mixed at the point when they were just introduced into pad-bath trough;
[2] Hot air with relative humidity of 25% (120°C, 2–3 min);
[3] Example of available dyes: Remazol®.

Table 13.11 The continuous dyeing profile of pad-dry-cure method

Process flow	Padding (dye + alkali) → pre-drying (reduce pick-up to 30%) → drying (120°C, 2–5 min) → curing (180°C, 1–3 min) → rinsing → soaping → rinsing → drying			
	Ingredient	Concentration (g/L)		
		Light	Medium	Dark
Recipe of padding solution	Reactive dye[1]	< 20	20–50	>50
	Na_2CO_3	10	15	20
	Urea	50–100	100–150	150–200
	Wetting agent	0.8–1.2	1–1.5	1.2–2
	Sodium m-nitrobenzenesulfonate	0–6	0–8	0–10

[1] Example of available dyes: Procion®H-E.

stability of the mixed dyebath. Generally Na_2CO_3 is more appropriate for dyes with relatively lower reactivity, such as MCT dyes; $NaHCO_3$ is properly used with dyes of relatively high reactivity, for avoiding dye hydrolysis. It is recommended to use the continuous dyeing machine with the special chemical mixing equipment, which can mix dye and alkali at the point when they are just introduced into the pad-bath trough.

In the Econtrol process, recently developed to eliminate the drying process for energy saving and low carbon production, the padded fabric is directly steamed at 120°C with hot air at a low relative humidity (25%). This process combines the advantages of short processing time and high fixation rate.

The pad-dry-cure process has the advantages of higher fixation rate, easy operation and good reproduction. The curing temperature is around 180°C. Because little water is present on the fabric when curing, large amounts of urea must be added to the dyebath to help fibre swelling and dye fixation. One disadvantage of this process is the risk of urea decomposition during curing, especially when the temperature is extremely high. Recently, thermal-stable chemicals performing the same functions, such as dicyanoguanidine, are more appropriate (even though more costly) to replace urea to avoid harmful by-products generated during curing.

In the continuous dyeing range, the wash-off unit is usually installed at the end of the production line. The washing unit may include a set of washing boxes. The continuous washing process may involve three stages: (1) initial rinsing, (2) soaping, and (3) final rinsing. Initial rinsing employs either cold or warm water to remove superficial loose colour (sometimes HAc is required for neutralisation); soaping uses hot water (95°C) containing detergent (1–3 g/L) to remove the unfixed dye inside the fabric; and final rinsing is to ensure that loose colour is cleaned thoroughly.

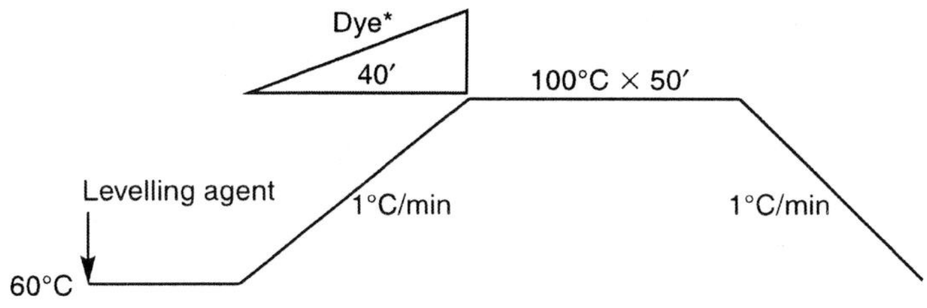

13.12 Direct dyeing curve for cotton. * Linear dosing of dyes during temperature rising period for time saving.

13.2.2 Other dyes available for cotton dyeing

In addition to reactive dyes, other available dyes for cotton include direct dyes, sulphur dyes, vat dyes and azoic dyes. The invention of direct dye, with substantivity for cellulosic fibres, greatly preceded the advent of reactive dyes. The weakness of direct dyes is poor colour fastness, especially the wet fastness. Although the wet fastness of direct dyes is generally not as good as vat dyes, due to easy application and a broader shade range direct dyes were the most important dye for cotton dyeing before the 1950s. Since then, the position of direct dyes in cotton dyeing has been replaced by reactive dyes. Direct dyes nowadays are only used in the applications where colour fastness is not important. An example of the direct dyeing process is shown in Fig. 13.12.

Sulphur, vat, and azoic dyes can be categorised together because they share the same dyeing mechanism, by which soluble forms of the dyes are first absorbed by cotton and then transferred into insoluble form in the fibre substrate. Vat dyes and sulphur dyes are still important for dyeing and printing (discharge and resistant style) of cotton. The dyed goods have very considerable fastness. Indigo dyes are the oldest vat dyes derived from natural sources. Indigo and sulphur dyes are the most important dyes applied in denim production. The dyeing application of vat dyes and sulphur dyes is similar, but the latter are particularly important for deep shades dyeing. The azoic dyes are of less importance in dyeing and printing because of their relatively low colour fastness.

13.3 Dyeing of synthetic materials

There are lots of synthetic textile fibres. However, the dyeing of polyester and nylon is discussed only below since polyester is the most popular synthetic textile fibre and nylon runs the second.

13.3.1 Polyester and its dyeing properties

Polyester (PET) is a synthetic fibre and has the properties of hydrophobicity, crystallisation and thermoplasticity. It becomes softer when being heated over glass temperature T_g (the transfer temperature of polymer from glassy state to a state of higher elasticity) and melts at temperatures above melting point T_m. When polyester fibre was discovered in 1947, none of the water-soluble dyes used at that time, such as direct, acid and basic dyes, could be applied for polyester dyeing, since there is no dyeing site on this fibre. Later, disperse dye, which was originally developed for the dyeing of cellulose acetate fibre, was found to be practical for dyeing polyester.

The diffusion of disperse dyes into polyester, including all the thermoplastic fibres, is described by free volume theory (Vrentas and Duda, 1977). According to this theory, disperse dyes are adsorbed on the fibre surface and diffuse into the fibre through transient passages, resulting from the segmental motion of polymer chains in material substrate when sufficient thermal energy is provided. When polyester reaches T_g, the lowest temperature at which the polymer chains begin to vibrate and slide past each other when a force is applied, disperse dye begins to diffuse into fibre by 'jumping' from one site to another. With continuous thermal energy provided dyes are gradually embedded into the fibre substrate, until the distribution in fibre and dyebath reaches equilibrium. When the system temperature is decreased after equilibrium, the vibration and sliding of polymer chains is weakened, and finally the dyes in fibre are blocked inside the fibre. Based on the free volume theory, dyeing temperature must be beyond T_g to generate enough free volume in fibre to accommodate the dye.

13.3.2 Disperse dye and polyester dyeing

Disperse dye, as indicated in Table 13.1, is the only dyestuff being available for polyester dyeing.

Properties of disperse dye

Unlike the anionic dyes for cotton, disperse dyes are relatively insoluble in cold water and have only limited solubility in boiling water. As their name implies, disperse dyes are present in the dyebath in a superfine aqueous suspension with dispersing agents. However, they possess substantivity for hydrophobic fibre, such as polyester and nylon, in which they are more 'soluble' than in aqueous. The slight water solubility of disperse dyes results from the presence of polar substituents in their molecular structures. It gives rise to the formation of dyes in their monomolecular forms in the true aqueous solution, which helps the disperse dye to penetrate into hydrophobic fibre during dyeing. Apart from a few bright pink and blue shades, anthraquinone disperse dyes (Fig. 13.13) are gradually

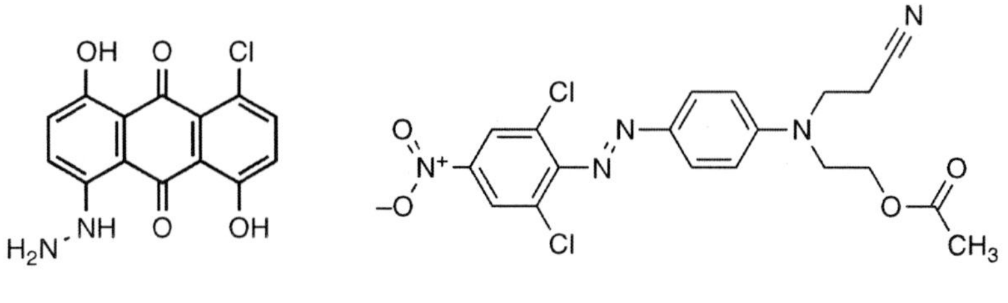

13.13 Chemical structures of disperse dyes.

being replaced by azo disperse dyes because of production costs and environmental threats in the synthesis process.

Disperse dyes are primarily marketed in the forms of powder or grain, and less frequently as paste or aqueous dispersion. All of these forms contain micro-fine dye particles and dispersants varying in dose and type. Dispersant is added, usually anionic surfactant, to clad each dye particle with a monomolecular layer of adsorbed dispersant. When the disperse dyes 'resolve' in the dyebath, hydrophobic chains in the dispersant structure embed into the hydrophobic dye particle, leaving the anionic groups exposed to the surrounding water. The overall negative charge on the surface of each particle prevents coalescence and aggregation. Consequently, the suspension of dye particles in the dyebath maintains stability and dyeing readiness (Broadbent, 2001).

Disperse dyes have distinctive dyeing properties on polyester. They are often classified in accordance with their dyeing rate and sublimation fastness, which are closely related to the polar group number and molecular weight of dye molecule. Table 13.12 shows the most common classification of disperse dye in the exhaust dyeing of polyester, as well as a set of the typical commodity dyes of three primary colours, red, yellow, and blue, applied for relevant shade. Disperse dyes with low molecular weight, classified as low energy disperse dyes, present more levelling and faster dyeing rate in the dyeing progress but poor sublimation fastness for its final products. On the other hand, those dyes with higher molecular weight, which exhibit weak migration and low dyeing rate but good sublimation fastness, constitute the high-energy disperse dyes. Low energy dyes are more suitable in the exhaustion dyeing, while high-energy types are preferred in thermofix dyeing.

Factors influencing disperse dyeing

The major influencing factors on disperse dyeing process are temperature, time, pH and auxiliaries. Polyester, in common with other synthetic fibres, is quite crystalline and very hydrophobic. Even at 100°C, which is greater than

Table 13.12 Classification of disperse dyes

Property	Low energy	Medium energy	High energy
Molecular weight	Low	Moderate	High
Polarity	Low	Moderate	High
Dyeing rate	High	Moderate	Low
Sublimation fastness	Low	Moderate	High
Dye example (C.I. disperse)	Yellow 54, Red 60, Blue 56.	Yellow 211, Red 73, Blue 183:1.	Orange 30, Red 167, Blue 79.

the T_g of polyester by about 20°C, dye diffusion is so weak that no satisfactory exhaustion can be obtained for most of the disperse dyes. As dyeing temperature increases from 100°C to 130°C under pressure, dyeing rate accelerates considerably, giving a better coverage of filament irregularities due to improved dye migration. For each particular dyeing, there always exists an optimum temperature/time profile, relating to the type of goods, dyes and the dyeing equipment involved. Dyeing time can be minimised by controlling the temperature to offer a reasonable exhaustion rate giving absorption uniformity. Conventionally, the dyeing time ranges from 30 to 60 min at 130°C to obtain sufficient equilibrium. The pH of dyebath is usually adjusted within 4.5–5.5 using either acetic acid alone or in combination with other acids. Acidic conditions could minimise reduction and hydrolysis of some disperse dyes with chemical groups that are sensitive to alkali at high temperature.

A carrier, an organic compound dissolved or emulsified in the dyebath, can be used to assist the dyeing of polyester around 100°C. The typical method is also called carrier dyeing. These substances enable the deep shades can be obtained at the boil within a reasonable dyeing time. Nowadays, carrier dyeing is declining for environmental and other considerations (Broadbent, 2001).

Disperse dyeing process control

Dyeing of polyester with disperse dye can technically be launched in either batchwise or continuous processing.

Exhaust dyeing

Batchwise dyeing for polyester is for dyeing polyester with batchwise dyeing machines, such as package dyeing machines for yarn, and jet or overflow dyeing machines for fabric, at high temperature (usually around 130°C).

Figure 13.14 shows the high temperature dyeing profile for polyester. Dyes selected must be subject to the same energy class. The build-up of colour on shade requires that the compatibility of dyes should be good, that is having about the same dyeing rate, similar to reactive dyeing. When the dyebath temperature reaches 90°C, at which dyeing initially begins,

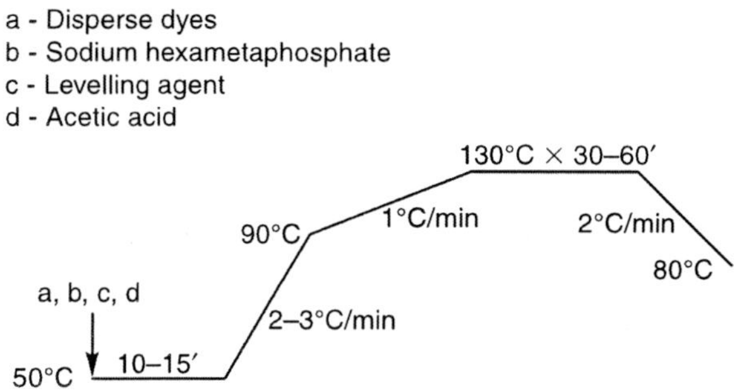

13.14 The typical exhaust dyeing curve of polyester with high temperature dyeing method.

Table 13.13 The polyester dyeing profile with high temperature dyeing method

	Ingredient	Shade		
		Light	Medium	Dark
Dyebath (liquor ratio: 1:5–10)	a. Disperse dyes (%, owf)	< 0.5	0.5–2.0	> 2.0
	b. Sodium hexametaphosphate (g/L)	0.5–1	0.8–1.2	1.0–1.5
	c. Levelling agent (g/L)	0.5–1	0.8–1.2	1.0–1.5
	d. Acetic acid (g/L)	0.7–1	0.9–1.2	0.9–1.2
	Ingredient	Concentration (g/L)		
Reduction cleaning bath	Sodium hydrosulfite	1		
	Na_2CO_3	2		
	Detergent	0.3		

the rate of temperature increase should be slowed down to eliminate dye agglomeration. When dyeing equilibrium is obtained, slow cooling is much preferred to minimise the generation of oligomer, which would lead to deterioration in appearance of the dyed goods. Reduction cleaning is important to get rid of the loose colour in fabric, and colour fastness can be enhanced. Thorough cleaning is necessary to ensure the dyed goods possess satisfactory fastness. In practice, reduction cleaning at 80°C for 15 min, followed by general cool rinsing, is usually enough to wash down the unfixed dye. To neutralise the dyed fabrics, acetic acid (60%) of 1–2 mL/L is required in the final rinse bath. The other related dyeing profiles are shown in Table 13.13.

Continuous dyeing

Continuous dyeing for polyester is usually carried out in the continuous Thermosol (thermofix) dyeing machine, which was invented by Du Pont in 1949. It involves first padding the disperse dyes into fabric, drying, and then

Nylon 6.6 | n = 60–80

Nylon 6 | n = 130–160

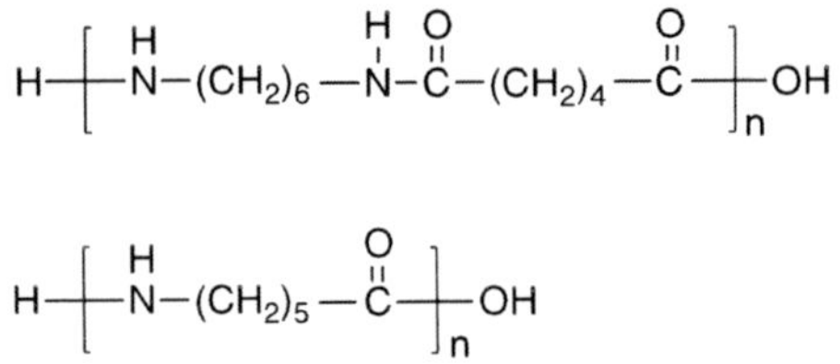

13.15 The chemical structures of nylon 6.6 and nylon 6.

Table 13.14 Classification of acid dye for nylon dyeing

Type	Molecular	Levelling	Wet fastness	Example (C.I. acid)
1	Small	Good	Poor	Blue 25, 78
2	Large	Moderate	Moderate	Blue 41, Yellow 172
3	Great	Poor	Good	Red 138, Blue 138

fixing the dyes by dry heating at a temperature of about 190–205°C. At this elevated temperature, the fibre molecular chains open up and the dispersed dyes vaporise and diffuse into the polymer. On cooling, the dyes are trapped within the fibre, yielding coloured fibres with good fastness. This process is rarely adopted for pure polyester dyeing, but is employed frequently in the dyeing of polyester/cellulose woven blends. The process control for the Thermosol process will be discussed later in Section 13.4.2.

13.3.3 Dyeing of nylon

As the second important synthetic fibre, nylon is becoming more and more significant throughout the textile industry. Nylon, with a high strength but low density, is a polyamide fibre with two kinds of chemical structure, namely nylon 6 and nylon 6,6. Their structures are illustrated in Fig. 13.15.

Acid dyeing

Nylon fibre is a hydrophobic polymer with an α-terminal amino group, so it can be dyed with disperse dyes as well as other water-soluble dyes such as acid dye and reactive dye. The exhaust dyeing process with batchwise dyeing machine is the most common method for pure nylon dyeing.

Acid dyes for nylon are divided into three types depending on their dyeing properties. These dyeing properties are shown in Table 13.14.

One of the most common defects of acid dyeing for nylon is the phenomenon of the barré result. Barré derives from chemical variation in manufacturing, oxidation of terminal amino in melt spinning and/or physical changes in spinning, drawing or heat-setting. It gives rise to the stripes on dyed goods. It is usual to add dyeing auxiliaries to control exhaustion to reduce the tendency of stripe formation and improve levelling. An anionic agent competes with an anionic acid dye for dyeing sites, while the cationic agent captures anionic acid dye first and then liberates it in proper condition, both of which decrease the dyeing rate to obtain levelness (Lewis and Marfell, 2004). A typical dyeing recipe is shown in Table 13.15, and the matched dyeing curves for different dye types are shown in Fig. 13.16 and 13.17.

It is difficult to obtain satisfactory wet fastness for dyed nylon with acid dyes for medium and dark shades; dye fixation is therefore a necessity. For example, tannin and tartar emetic are applied for acid dye fixation, and the process details are shown in Fig. 13.18. It is important to note that excessive tannin and tartar emetic will cause deterioration in handle.

Disperse dyeing

Disperse dye is another dye class that can be applied in nylon dyeing because of the hydrophobic property of nylon. The glass transition temperature (T_g) of nylon is 50–60°C and it is easier to swell than polyester, thus dyeing at boil

Table 13.15 Dyeing recipe for nylon 6.6 with acid dye

Ingredient	Amount (%, owf)	
	Type 1 and 2	Type 3
Dye	x	x
Anionic agent	1	1
Cationic agent	2–3	2–3
NH_4Ac	/	1–3%
HCOOH	3–5%	/

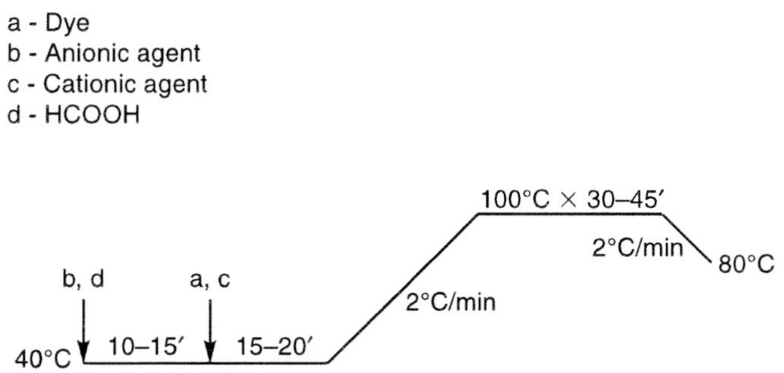

13.16 The nylon dyeing curve for the type 1 and type 2 acid dyes.

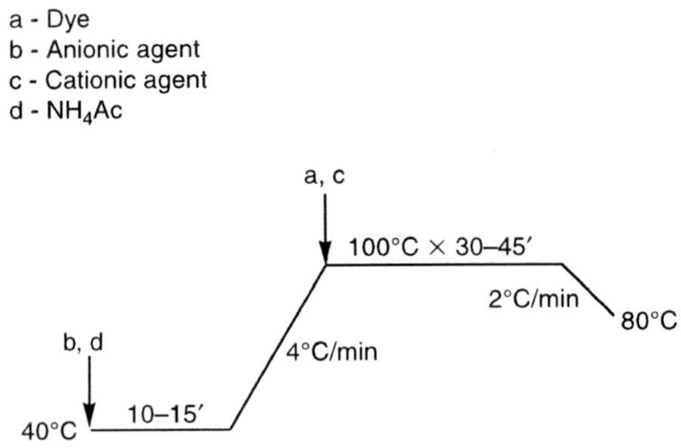

13.17 The nylon dyeing curve for the type 3 acid dye.

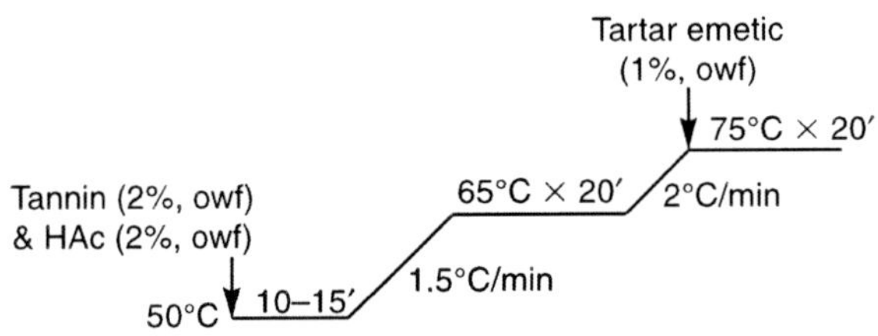

13.18 The curve of acid dye fixation.

under atmospheric pressure can yield satisfactory exhaustion. Disperse dye has better coverage of barré than acid dye, but it is difficult to obtain dark shades, and some disperse dyes exert poor washing fastness, especially for relatively deep colour. Nowadays, with the increasing requirement of higher colour fastness, dyeing nylon with disperse dye is declining. The typical dyeing curve is shown in Fig. 13.19 and the dyeing recipe is shown in Table 13.16.

In some cases, to obtain better levelling and dark shade, disperse dye and acid dye are combined in nylon dyeing. A cationic agent is not recommended in disperse dyeing, since it would be absorbed by disperse dye particles, decreasing the stability of dyebath. The non-ionic type is greatly preferred.

13.4 Dyeing of blends

Blends are composed of a mixture of two or more kinds of fibre materials into a single yarn or fabric. Polyester/cellulose blends occupy about 60% of the total production of all types of blends in textile industry, and are of more significant status than others.

When dyeing is involved in blend processing, the respective dyeing of each material in the blends is so different that solid dyeing mostly cannot

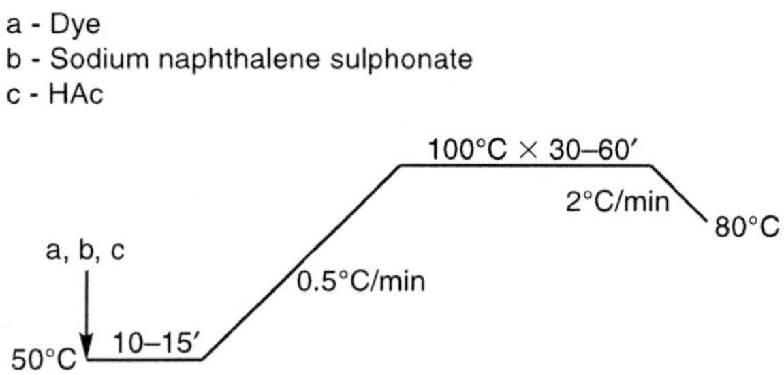

13.19 The dyeing curve of nylon with disperse dye.

Table 13.16 Dyeing recipe for nylon with disperse dye

Ingredient	Amount (%, owf)
a. Dye	x
b. Sodium naphthalene sulphonate	1–2
c. HAc (80%)	1

Note: liquor ratio: 1:5–10.

be completed with a single class of dye and/or in a single stage like its pure components. The dyeing property of blends follows that of its component fibres when they are dyed alone, although there are slight differences when varied processing methods applied. These differences will be discussed in the following paragraphs, along with process control.

Dye selection is closely related to the nature of the composite of the blends. Polyester/cellulose blends are dyed mostly with disperse dye for polyester part, and with reactive dye for the cellulose component, either in exhaust or continuous process.

13.4.1 Exhaust dyeing

Exhaustion dyeing processes for blends with disperse/reactive dyes are of two kinds. These are (1) the two-bath method, in which the two fibre in blends are dyed in two independent dyebaths separately with different dyestuffs, and (2) one-bath-two-stage, in which the dyestuffs are mixed together in a single bath, but the dyeing process is divided into separate stages for each fibre.

Two-bath method

In the two-bath method, usually the polyester is dyed first, followed by the cellulose part, which can avoid negative influences under the polyester dyeing conditions, such as high temperature, acidity and reduction cleaning on cellulose reactive dyeing. This process is similar to that of pure polyester and pure cotton dyeing. The two-bath process ensures the final goods have maximum brightness and colour fastness. Moreover, because of the independent dyebath for the different classes of dyes involved, there is considerable freedom in dye selection. There is no risk of interaction between reactive and disperse dyeing system. Nevertheless, the biggest shortcoming of such process is that the total dyeing time is greatly extended, which rapidly boosts production costs.

One-bath-two-stage method

The one-bath-two-stage method is an alternative method for blends dyeing. The term 'two-stage' means reactive dyeing for cotton is implemented in two steps. Exhaustion of reactive dye is carried out first and then fixation is activated by pH change (or pH-sliding) from acidic to alkaline. The dyeing curve for this process has some variation depending on the exhaustion sequence of each dyestuff. Fig. 13.20 shows two dyeing curves; in process (I) disperse dye and reactive dye are first exhausted by blends simultaneously, then the reactive dye is fixed by alkali, while in process (II) the pH adjustment is achieved by a pH-sliding agent. The pH-sliding agent is a specific auxiliary to gradually change the dyebath pH, either from acidic to alkaline, or from alkaline to acidic. In process (II), the pH-sliding agent presents alkaline character initially in the dyebath, and then it decomposes and liberates an acidic substance as the temperature increases. Reactive dye exhaustion is completed before the addition of the pH-sliding agent, and the dye fixation is carried out after the addition of the pH-sliding agent. When reactive fixation is accomplished eventually, the temperature is further raised and the pH-sliding agent releases acids to shift the dyebath from alkaline to acidic, meeting the requirements of disperse dyeing. In this process, the total processing time can be shortened to four hours. But there are further requirements for dye selection in this one-bath method.

In the above two methods, reactive dye and disperse dye are simultaneously present in one dyebath at the same time. As a result, each must be stable enough to withstand the dyeing condition for the other dyestuffs, that is reactive dye must withstand the acidic bath at a high temperature, and disperse dye must display sufficient stability to alkali as well as high electrolyte concentration. Furthermore, it is essential to choose reactive dyes and disperse dyes that do not react with each other. After dyeing, the reduction cleaning for unfixed disperse dye must be avoided otherwise there will be decomposition of reactive dye from the dyed cotton composite.

A more versatile dyeing process for blends is of dyeing in a neutral bath in a single stage with special reactive dye and disperse dye. The Kayacelon

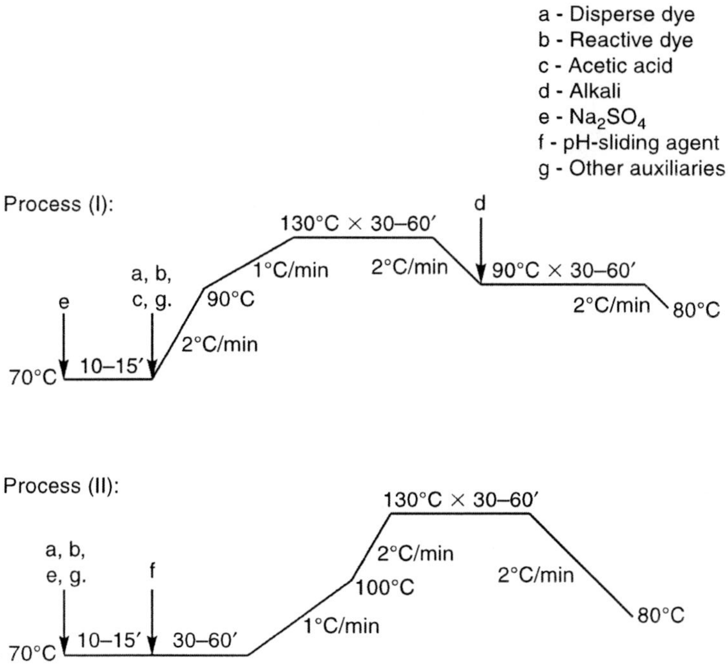

13.20 Dyeing curves of blends for one-bath-two-stage.

React CN® reactive dye, which can be fixed in a neutral condition as the temperature ranges from 110°C to 130°C, has admirable compatibility with disperse dye. In this dyeing process, pH adjustment is no longer a necessity and the total dyeing time is only about four hours.

13.4.2 Continuous dyeing

Continuous dyeing for woven polyester/cellulose blends is another significant process throughout the dyeing industry because of having large production capacity and high efficiency. There are three process flows as follows:

1. Two-bath-two-stage: padding (disperse dye) → pre-drying → drying → thermofix → padding (reactive dye) → pre-drying → drying → padding (alkali) → steaming → soaping → drying.
2. One-bath-two-stage: which is similar to two-bath-two-stage but reactive dye is padded with disperse dye together in one bath.
3. One-bath-one-stage: padding (disperse and reactive dye) → pre-drying → drying → thermofix → soaping → drying.

The first process avoids mutual interference between the dyeing of two fibres and is free of dye selection, but is more time-consuming. The condition for each stage is similar to that for pure polyester and pure cotton. One of the distinct characteristics of the continuous process is that the staining of disperse dyeing on cellulose is more severe than that in exhaust dyeing. To overcome this, an effective dispersing agent is necessary to transfer disperse dye from cellulose to polyester. High-energy dye is always preferred for the thermofix process although a complicated control is needed to obtain optimum dyeing results. Before the second padding of reactive dye solution, reduction cleaning of unfixed disperse dye is essential if high fastness is required. If the former disperse dye used is alkali-sensitive, such as those with vulnerable heterocyclic rings or hydrolysable ester groups, reduction cleaning may not be necessary, because these loose colours would be destroyed after alkali-padding and cleaned in the final rinsing, which is more economical. Reactive dyeing for a cellulose component can use either high reactivity dyes or low reactivity dyes following their characteristic processes.

The second process is relatively shorter than the first one, but there is a problem in dye selection, as in batchwise dyeing. The dyeing condition for this process is listed in Table 13.17.

The one-bath-one-stage employs disperse dye, reactive dye and alkali in a single pad-bath. The padded two dyes are fixed on fabric simultaneously in the thermofix unit in a continuous production line. The dyeing condition is shown in Table 13.18. Suitable disperse dyes for this process are those with low alkali-sensitivity and less cotton-staining tendency. The reactive dye with low reactivity and good thermal stability, such as MCT type, is thus suitable.

13.5 Process control in batchwise dyeing machines

Dyeing machines are designed in different shapes and with different loading capacities to accommodate textile materials in varied forms and qualities. The materials being dyed can be fibre, yarn, fabric, or even garments (Broadbent, 2001). In an exhaust dyeing, dyes in the dyebath are gradually transferred onto material, which is thought to be exhausted from the dyebath to the substrate (Shamey and Hussein, 2005). Generally, dyeing equipment intended for exhaust dyeing is designed in the form of manageable batches (Ingamells, 1993). Therefore, exhaust dyeing is also called batchwise dyeing, in which textile material is dyed from batch to batch. The representative machines designed for batchwise processing include: (1) hanks and package dyeing machines for yarns, and (2) overflow, jet, jig and beam dyeing machines for fabrics. All of these are based on (1) circulation of the dye solution through the material, (2) circulation of the material through the dye solution, or (3) circulation of the material and dye solution simultaneously.

Table 13.17 The continuous dyeing profile of one-bath-two-stage method

	Ingredient	Content (g/L)
a. Dye-padding	Disperse dye	X
	Reactive dye	Y
	Migration inhibitor	10–30
	Urea	0–100
	Wetting agent	1–2
b. Pre-drying (infrared)	Ensure the pick-up ratio is less than 30%	
c. Drying	120°C, 2–5 min	
d. Thermofix	180–210°C, 30–60 s	
e. Alkali-padding	NaOH (38°Bé)	10–20
	Na_2CO_3	20
	Salt	250
f. Steaming	102–105°C (saturated steam), 30–90 s	

Table 13.18 The continuous dyeing profile of one-bath-one-stage method

	Ingredient	Content (g/L)
a. Padding	Disperse dye	X
	Reactive dye	Y
	Migration inhibitor	10–20
	Wetting agent	1–2
	$NaHCO_3$	2–10
	Dicyandiamide	0–15
b. Pre-drying (infrared)	Ensure the pick-up ratio is less than 30%	
c. Drying	120°C, 2–5 min	
d. Thermofix	200–210°C, 1 min	

13.5.1 Package dyeing machines for yarn

In package dyeing, the most important dyeing method for yarn, yarns are wound on perforated cones. Before dyeing, yarn packages are first inserted onto vertical, perforated spindles in the dyeing machine. When dyeing begins, dye liquor can be pumped through the perforations in spindles and then forced to run into the package crosswise. Flow direction can be reversed from inside-to-outside to outside-to-inside automatically. A schematic of a package dyeing machine is shown in Fig. 13.21.

Evenness is always the predominant consideration in dyeing. For package dyeing, the evenness depends on good preparation of the yarn (proper package density, even winding), correct loading, right liquor ratio and appropriate liquor flow rate (good liquor circulation), and so on.

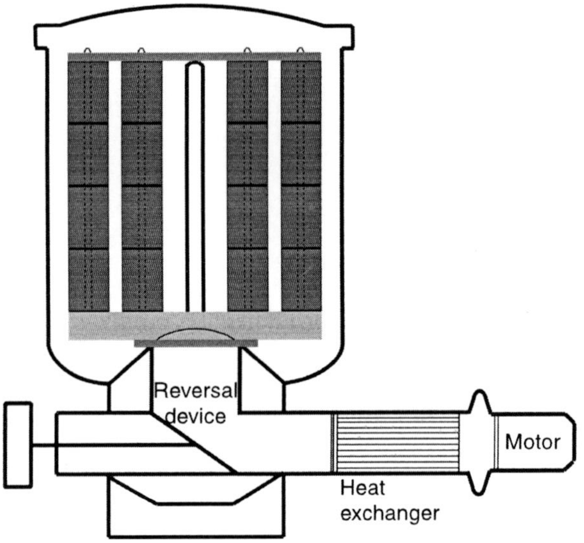

13.21 The schematic of a package dyeing machine.

Package density

The porosity of the package must be uniform to prevent the uneven dyeing caused by the liquor short-cut. The evenness of winding has great influence on dyeing quality. Good winding requires the packages are constant in density, including in any single package from inner to outer, as well as that from package to package in the same batch. It is very important to find the proper package winding density. Too low a density would cause the collapse of yarn from the perforated tubes, while too high a density could result in difficulty for dye liquor penetration. The correct density of the yarn package is associated with the nature of material, yarn texture, yarn count, and the liquor circulation of package machine. The examples of package densities are listed in Table 13.19.

Mounting of package

After packages are prepared, they have to be mounted on a series of hollow, perforated, vertical spindles in the dyeing machine. It is critical to fix the packages firmly and compactly on spindles, without any slip space. Gaps between packages, top spacer and top package, as well as bottom spacer and bottom package, should be eliminated (Fig. 13.22). Otherwise a leakage may occur between these spaces when dye liquor flows pumped from either inside-to-outside or in reversed flow direction during the dyeing operation, which can lead to unevenness in dyeing.

Table 13.19 Packages winding density as a function of yarn count and type

Yarn material	Density (g/dm³)	Yarn count
Cotton	340–400	40/1 (Nm)
Cotton	320–350	100/1 (Nm)
PET	340–380	/
PET/Cotton	340–380	/

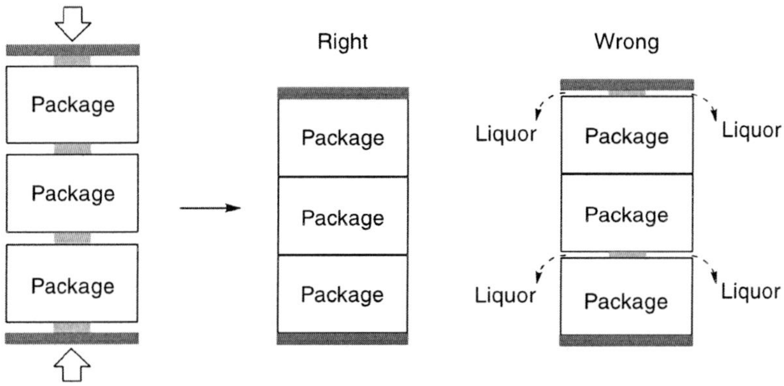

13.22 Load requirements for yarn package.

Liquor circulation

Liquor circulation, as the most fundamental and critical requirement, must be sufficient and uniform (Shore, 1995). A better dyeing levelness can result by changing liquid flow direction from time to time. Table 13.20 gives the recommended flow parameters for package dyeing.

Dyeing quality of a dyed yarn depends on both having dyeing evenness and maintaining the yarn in good physical quality. Generally, the higher the liquor flow rate, the better the dyeing evenness that can be achieved (Shang, 2002). However, undue turbulence in the dyebath can cause pilling which declining in physical quality of the yarn (Shore, 1995). Therefore, the flow rate of dye liquor should be optimally controlled so as to balance the dyeing evenness and the physical quality of the dyed yarn (Fig. 13.23).

13.5.2 Overflow dyeing machines and jet dyeing machines for fabric

The principles of the overflow dyeing machine and the jet dyeing machine are same, that is fabric being dyed runs cyclically, and dye solution at the same time also flows cyclically. Fabric loaded in the dyeing machine is in rope form, and

Table 13.20 Parameters of flow circulation

Flow direction	Time (min)	Differential pressure (Kpa)
Inside-to-outside	3–4	60–80
Outside-to-inside	4–6	40–60

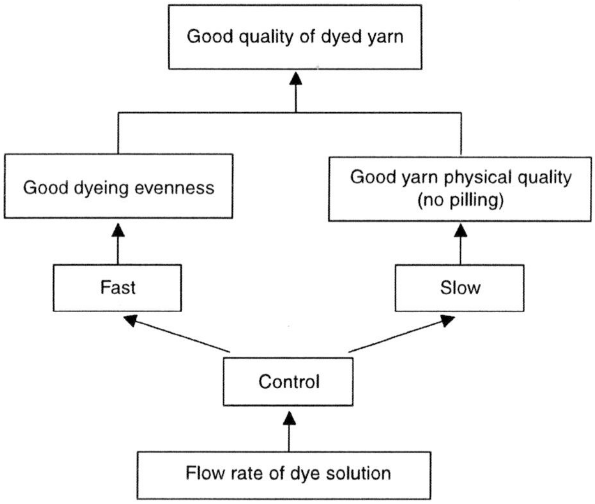

13.23 The influence of the flow rate of dye solution on both the dyeing quality and the yarn physical quality in package dyeing.

is moved by both the driving force from the lifting reel and the pushing force generated by nozzle liquor, which determines the fabric running rate. Dye solution is sprayed or jet-fed on to the fabric as the fabric passes through the nozzle, which is the main place that dye solution exchanges with fabric in terms of dye exhaustion. During repeated fabric circulation inside the machine, a dyeing process is carried out. Moreover, some machines are designed to operate above atmospheric pressure, which is particularly suitable for dyeing polyester at temperatures approaching 135°C (Broadbent, 2001).

Machine classification

The overflow and the jet dyeing machine are distinguished on the basis of the fabric running rate inside the machine. Generally, a fabric running rate below 350 m/min is classified as an overflow dyeing machine, and that over 350 m/min is called a jet dyeing machine. There are four main shapes of overflow or jet dyeing machines (Shang and Zhuo, 2003), that is O-shape (Fig. 13.24), U-shape (Fig. 13.25), up-L-shape (Fig. 13.26) and down-L-shape (Fig. 13.27). The comparison of these machines is listed in Table 13.21.

For the overflow dyeing machine, friction between fabric and machine, as well as that between fabric and liquor, is smaller than that of the jet dyeing

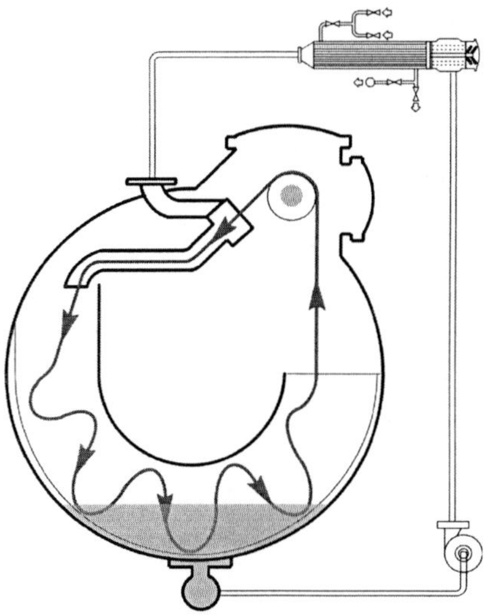

13.24 Schematic of O-shape dyeing machine.

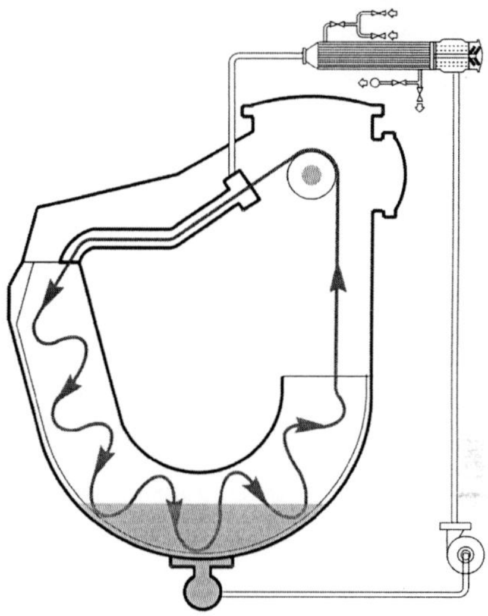

13.25 Schematic of U-shape dyeing machine.

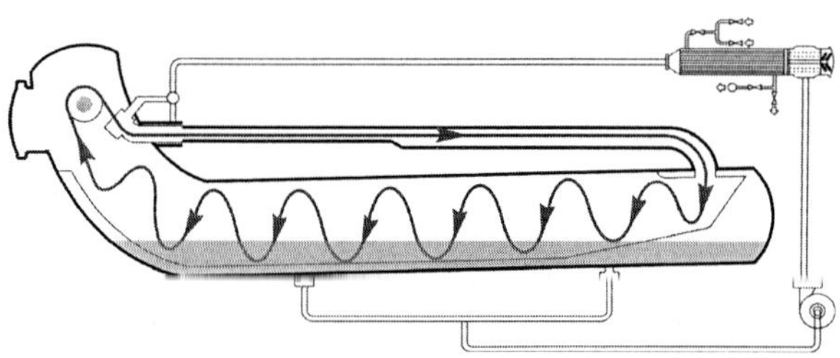

13.26 Schematic of up-L-shape dyeing machine.

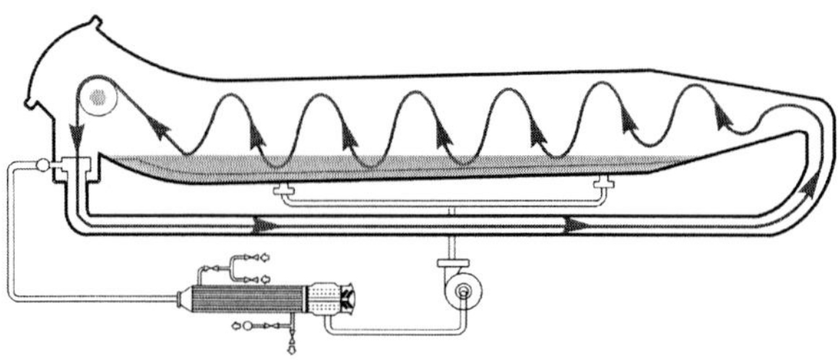

13.27 Schematic of down-L-shape dyeing machine.

machine, since the liquor pushing force generated by nozzle is gentle. The overflow dyeing machine is particularly suitable for those fabrics which cannot withstand high tension and high friction, such as cotton knits and their blends, as well as the elastic fabrics. Knitted fabrics should be dyed with overflow dyeing machines to reduce pilling and to retain its dimensional stability. This is because the nozzle liquor pushing force for fabric is moderate, and the slippage between fabric and lift reel is small. The jet dyeing machine is more suitable for pure synthetic fabrics, especially for thin fabrics and the easy crease-marked fabric.

Factors of dyeing process control

Overflow and jet dyeing machines are the most significant exhaust dyeing machines for fabric at present. Keeping the machines running under the optimal conditions has profound influences on the dyed goods. The important factors for dyeing process control include: (1) fabric loading weight, (2) liquor ratio, (3) fabric running rate in terms of the total actions of lifting reel speed and nozzle liquor pressure, (4) fabric cycle time, (5) chemical dosing rate, and (6) temperature/time profile.

Loading capacity

Generally, the maximum fabric loading capacity depends on the design capacity of the dyeing machine. However, the actual loading weight should be based on the fabric loading length. For a designed weight, the fabric length of thin fabric is longer, while the fabric length is shorter for thick fabrics. Since the requirement of shortening fabric cycle time for ensuring even dyeing, thick fabric could be loaded up to the designed capacity, while the loading of thin fabric should be discounted. The loading for easy crease marked fabric might also need to discount to avoid the un-removed crease marks happened. The actual loading capacity for a single chamber machine can be calculated as follows:

1. Determine the optimum fabric cycle time, usually no more than 3 min for cotton and its blends, and no more than 2.5 min for pure polyester.
2. Determine the optimum machine running rate, that is, no slippage between fabric and lift reel.
3. Calculate fabric length:
   Fabric length = machine running rate × cycle time

The elongation of fabric and the slippage between fabric and lifting reel should be considered.

Table 13.21 The comparison of overflow and jet dyeing machines

Machine shape	O	U	Up-L	Down-L
Overflow or jet	Overflow or jet	Overflow only	Overflow only	Overflow or jet
Maximum fabric running rate (m/min)	< 350 (overflow) or > 350 (jet)	< 350	< 350	< 350 (overflow) or > 350 (jet)
Nozzle pressure	High (jet) or low (overflow)	Low	Low	High (jet) or low (overflow)
Friction between fabric and machine	High (jet) or low (overflow)	Low	Low	High (jet) or low (overflow)
Liquor ratio	1:5–8	1:5–8	1:6–10	1:6–10
Maximum dyeing temperature (°C)	135	100	135	135
Fabric pile style inside machine	Plaiting (tight)	Plaiting (tight)	Waving (loose)	Waving (loose)
Availability for fabric	Jet for pure polyester; overflow for knitted cotton or cotton blend	For knitted cotton	For cotton, polyester, nylon, and blends; for both of knitting and woven	For pure (thin) polyester, nylon, blend, or the fabric easily having crease marks

4. Calculate the actual fabric loading weight:

Actual fabric loading weight = fabric length × fabric weight per metre

For a machine with multi-chambers, its actual total loading capacity can be calculated as:

Total actual loading capacity = the actual loading weight of a single chamber × chamber number

The development trend of the overflow and jet dyeing machines is towards lower liquor ratio and shorter processing time for environmental and energy considerations, so-called low carbon dyeing. Right-first-time and reproducibility in dyeing for all batchwise dyeing machines are other important issues needed to be improved. Various modified overflow and jet dyeing machines approaching these targets are now appearing in the market.

13.6 Process control in continuous dyeing machines

Continuous dyeing is mostly suitable for dyeing woven cotton and its blends in large amounts. Other advantages of the continuous process are the shorter dyeing process and the lower total water consumption than that of the batchwise method.

In continuous dyeing, the first piece of fabric to be dyed is fed into the dyeing range, and its end is stitched to the start of the next roll. Production goes at a constant rate without interruption until all the fabric is dyed (Broadbent, 2001). The dyeing range is assembled from pieces of basic equipment.

Figure 13.28 shows the complete continuous dyeing range for polyester/cellulose blends by the pad-thermofix-pad-steam method. It contains a dye padder, pre-dryer, dryer, thermofix unit, second padder, steamer, washing boxes, and dryer, etc. For particular dyeing processes, some equipment may be omitted or bypassed.

Factors of continuous dyeing process control

In continuous dyeing, the factors to be controlled include: (1) fabric moving speed, (2) dye liquor condition in padder trough, (3) padding pressure/liquor pick-up ratio, (4) evenness of padding pressure, (5) pre-drying, (6) curing, and steaming temperature, (7) evenness of drying, curing and steaming, (8) retention time in each unit, (9) rinsing units and rinsing conditions.

Fabric moving rate in dyeing covers a broad range. It can be as low as 10 m/min but also could often exceed 100 m/min depending on the fabric dyed and the efficiency of each unit in the dyeing range. Higher speed always benefits productivity, but it is more important to consider the completion status of equipment involved. In the continuous range, fabric threads through the pieces of equipment at a constant speed without interruption. When it runs through certain units, it is necessary to make sure the retention time

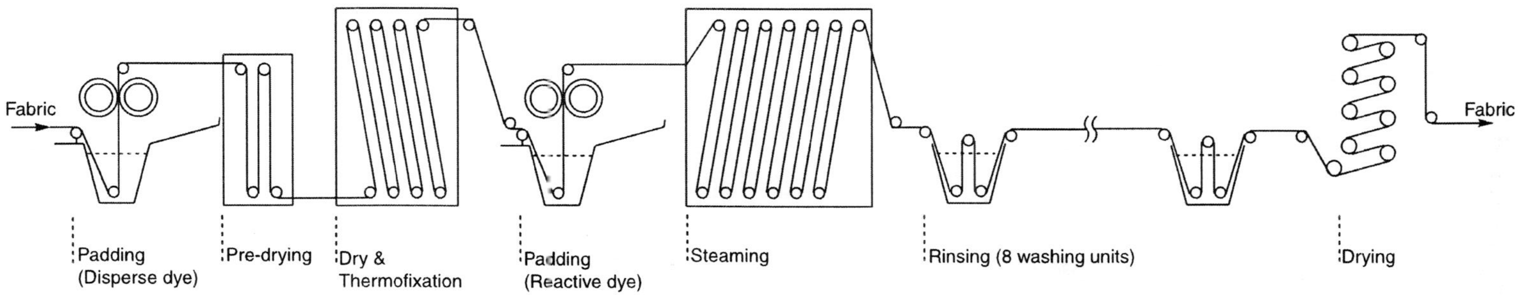

13.28 Schematic of continuous range for polyester/cellulosic blends dyeing with reactive and disperse dyes.

is sufficient to complete processing. If the retention time is less than 1 min in steamer, for instance, the movement has to slow down to meet the basic requirement.

With respect to dye liquor condition in the padder trough, there are factors to be considered, namely bath level and flow rate/dosing rate. The bath level must be kept constant to ensure the immersion time of fabric in the dyebath is without any variation, which would affect the consistency of shades. As fabric threads through dye liquor, a certain amount of dye liquor is absorbed by the fabric absorbency. Therefore the rate of padding solution complemented from a supply tank into the pad-bath should equal the rate of solution leaving the bath. The left solution from pad-bath is the fabric pick-up. If the processing loading is 15 kg/min, and the pick-up ratio is adjusted to 60%, the total flow supplied to pad-bath should be at a constant rate of 9 kg/min. In some cases the liquor would be complemented by two independent tanks, where for instance reactive dye solution and alkali solution are prepared separately. The complemented flow therefore equals the summation of the ingredients' flow. Moreover, it is better to keep the pad-bath at a low volume, in which leaves less waste residual liquor when dyeing reaches the end. More importantly the rapid turnover of the dyebath keeps the pad-bath consistent and fresh. Rapid replacement of padding liquor can also minimise initial colour tailing. Generally, fibre tends to absorb dye more than water in the padding process. The dye liquor squeezed out at the nip flows back to the padder trough with a lower dye concentration due to the preferential dye absorption by the fabric dyed. Consequently, while the dyeing lasts, the dye liquor in the padder trough is gradually diluted, and the drop in colour depth increases. This would be much more serious at the end of the process, causing the defect known as 'tailing' or 'ending'. To overcome this, rapid turnover of dyebath is effective, and selection of dyes with low substantivity is necessary.

Before padding, the fabric must be well prepared for good absorbency since the retention time of fabric in padding liquor might be as short as 0.5 s. Moreover, flat selvedges and crease-free surfaces are basic requirements for level dyeing. During padding, fabric goes into the padding solution under a roller submerged in liquor and then passes out of the bath towards squeeze rollers. Excess solution on fabric will be removed by pressured squeeze rollers, which are horizontal or vertical, the former being more common. This process impregnates dry fabric with dye solution mechanically. Impregnation must be as uniform as possible. The uniformity of the distribution is closely related to constant immersion time in the bath and a constantly even squeeze force, which gives constant pick-up on fabric. Constant immersion time depends on the fixed liquor level of pad-bath, as well as the constant moving speed of fabric as mentioned above. Even pick-up depends on the uniform pressure loaded by squeeze rollers. Generally, squeeze rollers are

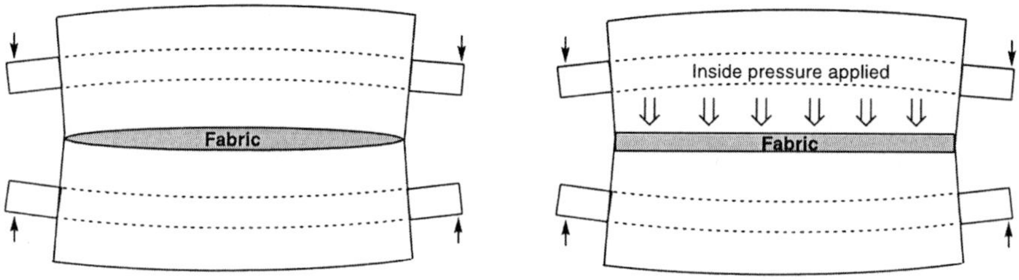

13.29 The comparison of conventional padder and the Küsters padder.

about 1.6–3.6 m long and 35–40 cm in diameter. They can squeeze a fabric with a linear pressure up to about 50 kg/cm of its length by a pneumatic system applying pressure at the two ends of one of the rollers for horizontal form, or the upper roller for vertical form. The higher the pressure exerted, the lower the pick-up. In practical application, the pressure at the two ends of the mandrel is usually higher than that in the centre due to the method of applying pressure. Such a non-uniform squeeze results in more pick-up in the centre, leading to paler depth at fabric selvedges known as 'listing'. To counteract this defect, modification of the pressure applying system is necessary to give level pick-up across the nip width under constant and even pressure. The swimming roller developed by Küsters is well known in that its linear pressure at the nip is more uniform across the width than that of other conventional equipment (Fig. 13.29).

Before going through any units in which the dye on fabric is being fixed, fabric must be dried completely to eliminate migration during fixation. The infrared heater, as a uniform drying unit, is used for pre-drying to reduce the pick-up, and migration inhibitor is added to the padding solution to increase viscosity, both of which are effective in minimising dye migration in fixation and for obtaining even dyeing.

Nowadays, thermofix dyeing is mainly for dyeing polyester blends, and very rarely for pure polyester. In the thermofix unit, polyester fabric is heated by air or metal rollers to 190–220°C for 30–120 s depends on the particular situation. When polyester/cellulose blends are padded in a single dyebath, the fixation of disperse dye on polyester and reactive dyeing on cellulosic fibre can be simultaneously completed in the thermofix unit. In an alternative process, fixation for reactive dyes can be achieved by steaming the padded fabric with saturated and slightly superheated (105°C) steam for a reasonable time. Good roller alignment in the steamer is necessary to get rid of lengthways creases. The steamer roof must be well insulated to prevent vapour condensing into liquid drops, which may fall onto fabric causing drop marks. It is better to keep the steamer free of air by cold water exit seal. No matter what kind of heating methods are applied in the fixation stage, uniformity is still fundamental, and much more crucial than any other process stages, because mistakes occurring in this step cannot be repaired.

The fixation ratio is never 100%, thus any unfixed dye must be cleaned from the fabric by rinsing and washing to give optimum fastness. Removal of unfixed dye is carried out on continuous washer assembled in the continuous dyeing range. In the continuous process, coloured fabric passes into washing boxes with the counter-current flow of the washing solution. A typical number of wash boxes is eight. The first three may be used for initial rinsing to remove loose colour; the next three are settled for hot soaping to clean embedded colour; the last two boxes are for the final rinse. The removal of unfixed dye should be as thorough as possible. After that, the wet fabric is dried by heated cylinders and the process then totally completed.

13.7 References

Broadbent, A. D. (2001). *Basic principles of textile coloration*, Bradford, Society of Dyers and Colourists.

Gordon, S. and Hsieh, Y. (2007). *Cotton: science and technology*, Cambridge, Woodhead.

Ibrahim, N. A. and Sayed, W. A. E. (1993). 'Low-temperature dyeing of cotton fabrics with monochlorotriazine dyes', *American Dyestuff Reporter*, **82**, 44–9.

Ingamells, W. (1993). *Colour for textiles: a user's handbook*, Bradford, Society of Dyers and Colourists.

Lewis, D. M. and Marfell, D. J. (2004). 'Nylon dyeing', in Hawkyard, C., ed., *Synthetic fibre dyeing*. Cambridge, Society of Dyers and Colourists, 82–121.

Shamey, R. and Hussein, T. (2005). *Critical solutions in the dyeing of cotton textile materials*, Manchester, Woodhead Publishing Limited.

Shang, S. M. (2002). 'Package dyeing: Low Liquor Ratio and Short Cycle', *International Dyer*, **187**, 23–5.

Shang, S. M. and Chong, C. L. (2002). 'Bleaching and dyeing cotton with a shorter process', *AATCC Review*, **2**, 31–3.

Shang, S. M. and Zhuo, H. J. (2003). 'Breakthrough in Overflow Dyeing Machine Promotes the Development of Eco-friendly Dyeing Technology', *China Textile Leader*, **4**, 28–33.

Shore, J. (1995). *Cellulosics dyeing*, Bradford, Society of Dyers and Colourists.

Vrentas, J. S. and Duda, J. L. (1977). 'Diffusion in polymer-solvent systems. I. Reexamination of the free-volume theory', *Journal of Polymer Science: Polymer Physics* Edition, **15**, 403–16.

Xi, D., Chen, J. and Ma, C. (2006). The importance of water resource on development of dyeing and finishing industry. In: *China Association of Dyeing and Finishing Industry*, Annual symposium on environmental protection in dyeing and finishing industry, Qingdao, 2006, Beijing: China Textile & Apparel Press.

Yeung, K. W. and Shang, S. M. (1999). 'The influence of metal ions on the aggregation and hydrophobicity of dyes in solutions', *Journal of the Society of Dyers and Colourists*, **115**, 228–32.

14
Process control in printing of textiles

S. M. SHANG, The Hong Kong Polytechnic University, China

DOI: 10.1533/9780857095633.3.339

Abstract: This chapter introduces the general printing technology for printing cotton, polyester, nylon, and blend fabrics with respect to dyes (reactive, disperse, acid, etc.) and pigments applications, printing methods (direct, discharge and resist, and heat transfer) and printing machines (roller, heat transfer, flat-screen, rotary-screen and inkjet). The corresponding printing processes and the process controls are discussed in detail.

Key words: printing applications of dyes and pigments, printing methods, printing machines, printing process control.

14.1 Introduction

Textile printing is considered as a controlled process to dye fabric in definite patterns or designs. It involves the transfer of colour paste on to the specified location of the fabric. Traditional textile printing techniques can be classified into several types, such as direct printing, discharge printing, resist printing and heat transfer printing.

14.2 Direct printing

Direct printing is the process by which colourants, containing dyes, thickeners, and the mordants or substances necessary for fixing the colour on the fabric, are printed in the desired pattern directly. This technique is widely applied in modern industry. Theoretically, the dyes used for dyeing a fibre are possible to adopt in printing on the same fibre as a fabric. However, there are more considerations arising from dyestuff selection than that in dyeing. The predominant consideration, apart from disperse dye and pigment, is the solubility of the dyes selected. They should have excellent solubility, in that they can not only be dissolved in the limited water in paste, but also can be redissolved in the condensed steam at the fixation stage. Moreover, dyes with rapid transfer ability from paste to fibre are preferred; otherwise the fixation time to achieve adequate diffusion will be prolonged.

Table 14.1 Profile of direct printing by all-in method with reactive dye on cotton

Process flow	Bleached fabric → printing → drying → steaming (102–105°C, 5–7 min) → rinsing → soaping → rinsing → drying	
	Ingredients	Content (g/kg)
Recipe of colour paste	Reactive dye	x
	Urea	50–100
	Sodium m-nitrobenzenesulfonate	10
	Sodium alginate (3–6%)	300–500
	$NaHCO_3$	15–30 (for dichlorotriazine or bifunctional dye with two reactive systems); or 8–15 (for dichlorotriazine dye)

14.2.1 Direct printing for cellulose

The full range of colours ensures reactive dye plays the most important role in direct printing for cellulosic fabric. Besides the common considerations for selecting dyes in dyeing, additional factors are the stability of the paste and the staining property in wash-off stage (Kanik and Hauser, 2004). To minimise staining in wash-off stage, low substantivity reactive dye is always the primary option. Nowadays, a lot of special reactive dyes for printing are available in the market, such as Novacron® P from Huntsman, Procion® PX and Remazol® P from Dystar.

A good preparation of cellulosic fabric before reactive printing is essential. Woven fabric must be thoroughly desized, and mercerisation is recommended, which gives fuller colour strength. Reactive printing can be classified as (1) the all-in method (widely used), in which the paste contains all the chemicals required, and (2) the two-stage method, in which the paste is without alkali, and fixation is brought about by padding alkaline solution after printing.

The profile of the all-in method is shown in Table 14.1. The addition of dye can be either by sprinkling dye powder into stock paste followed by high-speed stirring, or pre-dissolved in hot water and then poured into stock paste, with the latter being the most common method. The addition of urea may be necessary in pre-dissolving. It must be pointed out that the $NaHCO_3$ for fixation cannot be added into hot paste until it has cooled down to room temperature. This is to avoid premature decomposition of $NaHCO_3$, which liberates Na_2CO_3 and decreasing the stability of colour paste, as well as aggravating hydrolysis of the reactive dye.

If the reactive dye selected is not alkali-sensitive, Na_2CO_3 or even NaOH is preferred because higher colour yields will be obtained if the stable dyes are fixed at a higher alkali level.

Table 14.2 Profile of direct printing by two-stage method with reactive dye on cotton

Process flow	Bleached fabric → printing → alkali-padding → fixation → rinsing → soaping → rinsing → drying	
	Ingredients	Content (g/kg)
Recipe of colour paste	Reactive dye	x
	Urea	0–50
	Sodium m-nitrobenzenesulfonate	10
	Sodium alginate (3–6%)	300–500

Table 14.3 Typical recipe of alkali pad solution

Ingredients	Recipe 1 (g/L)	Recipe 2 (g/L)
NaOH	30	10
Na_2SiO_3	50	50–100
Na_2CO_3	–	50

Dye fixation can be completed by steaming, drying, or baking depending on the equipment available and the dye selected. Steam-fixation is the most common method employed. Insufficient steaming will decrease the colour fastness and aggravate the burden of wash-off, whereas too much steaming will reduce the colour yield.

In the two-stage method, the fabric is printed by the paste without alkali. This printing profile was shown in Table 14.2. The fixation can be initiated by applying alkali solution.

Typical recipes for the alkali pad solution are shown in Table 14.3, where Recipe 1 is suitable for dyes with relative lower reactivity, such as MCT dyes or bifunctional dyes with two reactive systems (MCT + VS), and Recipe 2 is more appropriate for dyes with higher reactivity, such as DCT and VS dyes. In some cases, it is necessary to add a large amount of salt in alkaline solution to avoid colour-bleeding from the printed pattern, similar to the wet-on-wet fixation process in continuous reactive dyeing.

The subsequent fixation process follows the same principle as that in the all-in method.

There is an alkali-shock method to fix the reactive dyes in which printed fabric goes through into hot alkali solution (100–103°C) to initiate fixation. Therefore, the process flow listed in Table 14.2 becomes: bleached fabric → printing → alkali-shock → rinsing → soaping → rinsing → drying. The alkali-shock method is more suitable for fabric printed with the reactive dyes having high reactivity. The retaining time of fabric in the alkali-shock process is about 10–20 s; the presence of a large amount of electrolyte may raise the temperature of the pad solution over 100°C without pressure.

Table 14.4 Profile of direct printing with disperse dye on polyester

Process flow	Pre-treated fabric → printing → fixation[1] → rinsing → soaping (reduction cleaning for deep shades) → rinsing → drying	
	Ingredients	Content (g/kg)
Recipe of colour paste	Disperse dye	x
	Dihydroxybutanedioic acid	2–3
	Oxidising agent	1–10
	Thickener	500–800

[1] (a) saturated steaming, 128–130°C, 20–30 min; or (b) superheated steaming, 175–180°C, 5–10 min; or (c) thermofix, 200–210°C, 1–2 min.

Table 14.5 Reduction cleaning recipe

Ingredient	Content (g/L)
Detergent	1
NaOH	0.002
$Na_2S_2O_4$	1–2

14.2.2 Direct printing for polyester

For polyester fabric printing, disperse dyes are the only dyes available. A thorough pretreatment of polyester is always the most basic requirement before printing. The residuals on the fibre, especially non-ionic types, have negative effects on fixation. The thickener selected must meet the requirements of having low-solid content, giving elastic film on fabric, and being easily removed in wash-off. Alginates and locust bean ethers are the most common thickeners used.

The printing profile is illustrated in Table 14.4. Depending on the fixation processes employed, suitable disperse dyes should be carefully selected. It is necessary to avoid using dyes with poor sublimation in superheated steaming and thermofix because of the decrease in colour yields and the staining on the unprinted area. To obtain medium to deep printings, adding ethylene oxide condensate as the fixation accelerator is usually necessary.

After fixation, prints are first rinsed in cold water to remove paste and unfixed dyes, and then rinsed in warm water (40–50°C) containing alkali and reduction agent (Table 14.5). Afterwards soaping should be carried out at less than 80°C, otherwise staining on the ground would take place. After final rinsing, the preferred drying temperature is below 120°C to prevent dye migration.

14.2.3 Direct printing for nylon

As the second important synthetic fabric, nylon can be printed with acid dyes, metal-complex acid dyes and direct dyes. The typical recipe for

Table 14.6 Typical recipe of polyamide printing paste

Ingredient	Content (g/kg)
Acid or mental-complex acid dye	x
Thiourea	30–50
Thickener	500
Sulfonated oil	20
Ammonium sulphate	30–60
Antifoamer	0–1

polyamide printing paste is shown in Table 14.6. Fixation is usually carried out in saturated steaming at 102–105°C for 20–30 min. Sometimes steaming with pressure is adopted to achieve enhanced colour depth. The final washing-off is very critical in polyamide printing with acid dye because this dye has high staining tendency on the ground. Unlike hydrolysed reactive dye to cotton and reduced disperse dye to polyester, the unfixed acid dye still has high affinity to polyamide. To prevent such a defect, maintaining an alkaline washing condition is necessary because the affinity of acid dyes in alkaline bath is very lower. Therefore, even the initial cold rinsing also needs sodium carbonate. Moreover, a cationic auxiliary is required in the subsequent warm rinsing to entrap unfixed acid dye in the bath to prevent staining.

14.2.4 Direct printing for blends

The printing for blends should consider all the fibre properties made up of the blended fabric.

Pigment printing

Pigment printing, different from the pigment dyeing for blends, is a thriving technology for blends printing at present, because of easy application and less pollution. They can be fixed on the fabric surface by a binder film which adheres to the fibres. Such film on fabric surface may stiffen fabric handle in the printed areas. However, such defect in the limited area can be ignored because the large unprinted area retains the original properties. Moreover, recently developed binders make pigment printed goods with enhanced handle. Because of no substantivity for all types of fibres, colourations with pigments for different fibres have almost no differences. Thus blends with two or more fibres can be printed by pigments in one-stage without variations for colouration property. Actually, pigment printing is increasingly applied for pure fabric printing, except for the cases in which very soft handle and much higher fastness are required in printed area. The process profile of pigment printing is shown in Table 14.7.

Table 14.7 Profile of direct printing with pigment on blends

Process flow	Pre-treated fabric → printing → drying → curing (150°C, 4 min) → after treatment	
Recipe of colour paste	Ingredients	Content (g/kg)
	Pigment dispersion (30–40 % pigment content)	x
	Binder (40–50% solid content)	100–150
	Synthetic thickener	15
	Urea	50

Table 14.8 Recipe of direct printing with disperse/reactive dyes on blends (one-stage)

Ingredient	Content (g/kg)
Disperse dye	X
Reactive dye	Y
Sodium m-nitrobenzenesulfonate	10
Sodium hexametaphosphate	0–3
$NaHCO_3$	5–10
Urea	30–50
Sodium alginate (6%)	500

The colour fastness of pigment printing is closely related to the binder content, as well as the curing temperature and time applied for the binder. Excessive binder content in the colour paste would cause poor fabric handle, such as stiff and sticky, while inadequate binder could lead to poor colour fastness of the printed pattern. The optimum control of binder content is required. Typically, 4 min at 150°C is sufficient for the film formation to fix pigment particles. Unlike printing with dyes, wash-off for pigment printed goods is not necessary.

Disperse/reactive printing

Polyester/cellulose fabric is the most significant blend, as mentioned before, in blend dyeing. This blend can also be printed with single class dye or two-class dyes such as the disperse/vat, the disperse/direct or the disperse/reactive system. The disperse/reactive system is thought to be the most appropriate combination, because it offers a wider colour range as well as fewer problems that might appear in printing than other combinations. There are two processes for the application of the disperse/reactive dyes combination, that is one-stage and two-stage, which are similar to those in pure cotton printing with reactive dye. An example recipe for one-stage printing is shown in Table 14.8. The application of single stage printing is much easier than the two-stage one, but more factors have to be considered in the dye selection: (1) reactive dyes must have high fixation on fabric and not react with the

Table 14.9 Profile of printing with disperse/reactive dyes on blends (two-stage, pad-steam)

Process flow	Bleached fabric → printing → disperse dye fixation → alkali-padding → steaming (102–105°C, 5–7 min) → rinsing → drying	
	Ingredients	Content (g/kg)
Recipe of alkali pad bath	K_2CO_3	50
	NaOH	12
	Na_2CO_3	100
	NaCl	15–30
	Thickener	100–200

Table 14.10 Profile of printing with disperse/reactive dyes on blends (two-stage, alkali-shock)

Process flow	Pre-treated fabric → printing → disperse dye fixation → alkali-shock (100–102°C, 10–20 s) → rinsing → drying	
	Ingredients	Content (g/kg)
Recipe of alkali-shock bath	NaOH	12
	Na_2CO_3	150
	NaCl	200

disperse dyes containing amino groups, and (2) disperse dyes need better stability in alkali paste. The dye fixation can be either at 200°C for 1–2 min or superheated steaming at 180°C for 5–10 min.

The two-stage method has more freedom for dye selection but more complicated processes and less reproductively presented. The disperse dye fixation condition is similar to that for pure polyester printing (thermofix). The reactive dye fixation has two ways to be carried out, that is the pad-steam (Table 14.9) method and the alkali-shock (Table 14.10) method, similar to that for pure cotton printing.

Removing paste and unfixed colour on blends printed with disperse/reactive dyes is more complicated than that for other rinsing processes. Generally it contains the following steps:

1. cold water rinsing,
2. warm alkaline rinsing (50°C),
3. hot alkaline rinsing (98°C),
4. hot soaping and rinsing (80–90°C),
5. cold water rinsing.

14.3 Discharge, resist and heat transfer printing

Different from the direct printing, the discharge, resist and heat transfer printing are called of indirect printing to some extent.

14.3.1 Discharge printing

In discharge printing, unlike direct printing, the fabric being printing is dyed by dischargeable dyes first. Next, the dyed fabric is printed with paste containing chemicals which have the ability to destroy the dyed colour. Colour on the printed patterns will be discharged after steaming. After rinsing, white patterns presenting on coloured fabric are obtained. If the printed paste contains discharge-resist dyes, the pattern will display the hue of such dyes. The former method is called white discharge printing, and the latter is known as coloured discharge printing. The colour around a printed pattern is called ground colour, because it is the background of the printed pattern. Relevantly, the pattern colour is described as the illuminating colour, in which it appears that coloured light illuminates the ground colour on the fabric. This effect is more impressive if the ground colour is deep.

The selection of dyes both for discharge ground (dischargeable) and for illuminating colour (discharge-resist) is critical. Generally, the dischargeable dyes for ground are azo dyes with an –N=N– group, which can be reduced by splitting the double-bond. However, different dischargeable characteristics still exist among the dyes with varied structure, although they have the common azo group. The selection of discharge-resist dyes for illuminating colour is closely related to the exact conditions in printing, such as printed material, colour shades, as well as discharging agent (reductant) selected. Vat dye resists discharging and is the dye class most applied in coloured discharge printing as the illuminating colour. Nowadays, there are the scales for the dischargeablity of dyes ranging from 1 to 5 provided by the commercial dye manufacturers for reference.

The most widely used discharging agents are sodium formaldehyde sulphoxylate ($NaHSO_2{\cdot}CH_2O{\cdot}2H_2O$), thiourea dioxide [$(NH_2)(NH)CSO_2H$] and tin(II) chloride ($SnCl_2$). The choice of reductant depends on the dyes involved, both the ground and the illuminating colour, as well as the fabric being printed. The diverse dischargeabilities of reductants and dyes can provide more freedom to combine reductants with dyes for different discharging scales. Unless specifically to develop special styles, $SnCl_2$ is not recommended for hydrophobic materials, because the capillary migration of the liquor gives haloing defects around the pattern reducing its definition. A suitable thickener for discharge printing must meet not only the requirement of direct printing, but also satisfy the extra requirement characterised in discharge printing. The thickener used should be stable to the reductant. Thus a non-ionic type is preferred. Moreover, thickener with a high-solid, low-viscosity property is necessary to ensure obtaining defined patterns successfully.

One of the problems arising in discharge printing is the invisible printed pattern on the ground colour. It is difficult to detect when the mistake occurs.

Thus, it is common to introduce marked-colour in the discharge paste, usually white pigment for white discharge. The addition of white pigment not only gives rise to the visibility of discharge printing pattern but also enhance the whiteness of the printed area.

Generally, the process flow of discharge printing is as follows:

Dyed fabric → printing → drying → steaming → cold rinsing → oxidation (if necessary) → hot soaping → rinsing → drying.

Steaming with saturated steam for 5–8 min is usually sufficient. In some cases, a mild oxidation agent is padded on dyed fabric to prevent unnecessary colour fading when it is in steam unit, which may be full of reductive air. Moreover, it may also minimise the haloes around the pattern.

When discharge printing is applied to polyester, the discharging agent can be either reductant as usual or alkali, because some of the disperse dyes containing an ester group can be destroyed under certain alkaline conditions. Strictly, the term of 'discharge-resist printing' is more appropriate than 'discharge printing' for polyester. because the fabric before printing is not virtually 'dyed', the disperse dye is only padded on to the fabric without fixation. The process flow of discharge-resist printing for polyester is as follows:

Pre-treated fabric → padding dyes → drying → printing → thermofix → rinsing → reduction cleaning → rinsing → drying.

When alkali is used in discharge-resist effect, it is better to replace the thermofix with the superheated steaming.

Discharge printing on blends is less common because a number of problems may arise in the operation. Moreover, few blends with discharge effect are observed anywhere in the textile fashion field. Pure cotton is thought to be more suitable than blends for discharge style.

14.3.2 Resist printing

Resist printing refers to the particular style in which the pre-treated fabric is printed with the resist agent first and then the printed fabric is dyed. The printed patterns do not accept the dyeing, leaving uncoloured patterns against a coloured ground. The resistants can be either physical substances such as wax, fat and resin, which prevent the pattern contacting the dye liquor, or chemicals such as oxidants, reductants, acids and alkalis, which resist fixation of ground colour. Sometimes a combination of pairs of resistants is adopted to meet special requirements. The styles of resist printing goods are similar to those of the discharge method, but the resist printing method has an advantage in the range of dyes that can be selected for ground colour, whereas dyes applied in discharge method only concentrate on those with instabilities, which can be destroyed easily.

14.3.3 Heat transfer printing

Heat transfer printing, also called sublimation transfer printing, is generally applied for pure polyester or for a blend with PET over 50%. It involves the pattern being first printed on the sublimation transfer paper with sublimation dyestuffs, and then the dye will sublimate and migrate to textile material by heat press with a hot press or hot calender. The printed patterns can achieve the features of vivid image, high definition and a rich gradation of shades. This method can be used for printing the natural landscape and artistic patterns. Its process is very simple, just pressing the transfer paper on the pre-treated fabric with a hot transfer printer. The temperature of the hot press or hot calender should be carefully controlled, and generally around 180–200°C in about 20–30 s for pure polyester. This printing method has more advantages, such as low production costs, no need of making and storing screens for a printing factory, low reject rates, being capable for garments printing, etc. The biggest weakness of transfer printing is its lack of flexibility in printing on different fabrics, which is an advantage of direct printing.

14.4 Process control in roller and screen printing machines

The printing process is mainly carried out by the printing machine. Based on the printing methods introduced above, there are five types of printing machines widely used in the textile industry. A comparison of these machines is shown in Table 14.11.

Besides the printing process, the process of fabric pretreatment before printing and the process of after-treatment for the printed fabric are also the important steps in printing. Therefore, in addition to the printing machine, the corresponding equipments for pretreatment and post-processing are also required for printing production and need to be well matched. But here the discussion is focused only on the major printing machines applied.

14.4.1 Roller printing machine

The roller printing machine, also called the engraved roller printing machine or cylinder printing machine, was once the most widely used machine for fabric printing. This machine can carry out the major printing methods, including direct printing, discharge printing and resist printing for cotton

Table 14.11 The comparison of printing machines

Features	Printing machine				
	Roller	Heat transfer	Flat-screen	Rotary-screen	Inkjet
Applicability for printing method	- Direct - Discharge - Resist - Pigment	Transfer	- Direct - Discharge - Resist - Pigment	- Direct - Discharge - Resist - Pigment	- Direct - Transfer - Pigment
Applicability for fibre	- Cotton - Cotton Blend	- Pure PET - PET blend	All kind of fibre	All kind of fibre	All kind of fibre
Applicability for fabric construction	Woven	- Woven - Knit	- Woven - Knit - Towel - Pile fabric	- Woven - Knits	- Woven - Knit
Fabric width	Medium	Medium	Medium to wider	Medium to wider	Narrow to wider
Applicability for garment	No	Yes	Yes	No	Yes
Machine/printing speed	Very fast	Slow	Slow	Fast	Very slow
Image definition	Better detail	Excellent detail	Improving detail	Better detail	Excellent detail
Image Size	Small to medium	Small to very large	Small to large	Small to large	Small to very large
Optimum production capacity for one pattern	Mass	Small to medium	Small to medium	Medium to Mass	Very small to small
Application advantages	High printing speed	- Simple printing process - Excellent image	- Machine investment is medium - Versatile for printing - Low production cost for medium order - Easy making screen	- Versatile for printing - Low production cost for large order - Easy making screen	- Mass customisation - Good for small order - Free of screen or roller - Printing process simple
Weakness	- High cost for engraving roller - Limitations in fabric type and width, and image size - Low level of automation	Lack of flexibility for fabric	- Relative low printing speed - Poor image definition for small pattern	Machine investment is large	- Low printing speed - High printing cost
Estimated market share (%)	10	6	25	55	4
Application trend	Declining	Stable	Stable	Stable	Increasing and developing fast

woven fabric and its blends. The main advantages of roller printing are high productivity, broad applications for different delicate pattern designs, and multicolour up to 12 colours. Due to the weaknesses of relatively smaller pattern size, complicated roller engraving process, especially unsuitability for pure synthetic fabric, the roller printing machine is gradually being replaced by screen printing machines, both the flat-screen printing machine and the rotary-screen printing machine.

14.4.2 Screen printing and screen printing machines

Screen printing is a printing technique that uses a patterned, partial paste-blocked mesh screen. The open areas (unblocked) of the mesh screen can transfer colour paste onto fabric by squeezing the paste with a roller or squeegee. Because of the simplicity of the application process, a wider range of dyes is available in screen printing than that in any other printing process. Therefore, the major printing methods, including direct printing, discharge printing and resist printing can be carried out by the screen printing method for most kinds of fibres, fabrics and their blends. The process control in screen printing is the same as the printing method (e.g. direct printing) being adopted. However, optimum control of the viscosity of the colour paste plays an important role in printing quality. Depending on the screen shape applied, screen printing can be classified as flat-screen printing and rotary-screen printing.

The corresponding printing machines are called flat-screen printing machines (Figure 14.1) and the rotary-screen printing machines (Figure 14.2). The widest printing width of these two machines can be 3.2 m or even more. Generally, the machine speed of flat-screen is lower than that of rotary-screen. For production cost, the flat-screen printing method is best suitable for a small batch; in contrast, the rotary-screen printing method is higher in production efficiency and is good for large volumes. Only one colour can be printed by each screen. A flat-screen printing machine usually has 8–12 screens and, as a result, 8–12 colours are available in a pattern. However, a rotary-screen printing machine may have up to 24 screens and, therefore, 24 colours may appear in a pattern. The rubber conveyer in the screen printer machine, to which the fabric is glued during printing, must be cleaned in continuous mode with water to remove excess adhesive and printing paste. At the end of each batch, or when changing colours, full cleaning of the printing machine is required.

Flat-screen printing

Flat-screen printing is the most versatile of all printing processes. It can be used to print on a wide variety of fibres (such as pure cotton, silk, pure synthetic or their blend), light or heavy fabrics (such woven, knitted, pile fabric, towel, etc.), and garments.

Flat-screen printing consists of four elements of the screen, the image carrier, the squeegee and the colour paste. The screen printing process uses a tightly stretched porous mesh over a frame made of wood or metal. The raw flat-screen is a combination of a frame and a piece of polyester fabric that is tightly stretched across the frame. When a detailed pattern is required, the fabric fixed on frame is entirely coated with a certain mixture of polymers that are water soluble but will become insoluble if exposed to a special light source due to self-crosslinkage. Not all the area of screen fabric will be exposed to the special light source, and the pattern area will be protected, where the coated substance can be washed off. As a result, the meshes on the pattern area are reserved but the rest are blocked by cross-linked polymer. Using a rubber squeegee, colour pastes are spread across the width of the screen frame and forced to pass through the pattern location only, producing a localised dyeing effect.

Flat-screen printing is a semi-continuous and reciprocating process. The screen is first moved into position over the fabric or garment, the squeegee is pressed against the mesh and drawn over the image area, and then the screen is lifted away from the fabric or garment to complete the process. The printing quality is influenced by many factors, such as composition, size and form of fabric, and angle, pressure, and speed of the squeegee.

Rotary-screen printing

Rotary-screen printing can increase production rates by its continuous production process, compared with the semi-continuous process of the flat-screen printing method, and is more suitable for mass printing production. Its typical speeds are between 50–100 m/min depending upon design complexity and fabric construction. The screen is made in rotary form and made from stainless steel, nickel usually. Nowadays, rotary-screen printing has become the most widely applied technology in the entire printing industry.

The basic operations of rotary-screen and flat-screen printing machines are very similar. An in-feed device, paste trough, rotating blanket (print table), dryer, and fixation equipment are used. The process involves initially feeding fabric onto the rubber blanket. As the fabric travels under the rotary screens, the screens turn with the fabric. Colour paste is continuously fed to the interior of the screen through a pipe. As the screen rotates, the squeegee device pushes paste through the design areas of the screen onto the fabric. The major disadvantage of rotary-screen printing is the high investment cost of the machine. As mentioned before, rotary-screen printing is most suitable for mass production, but generally not profitable for small batch

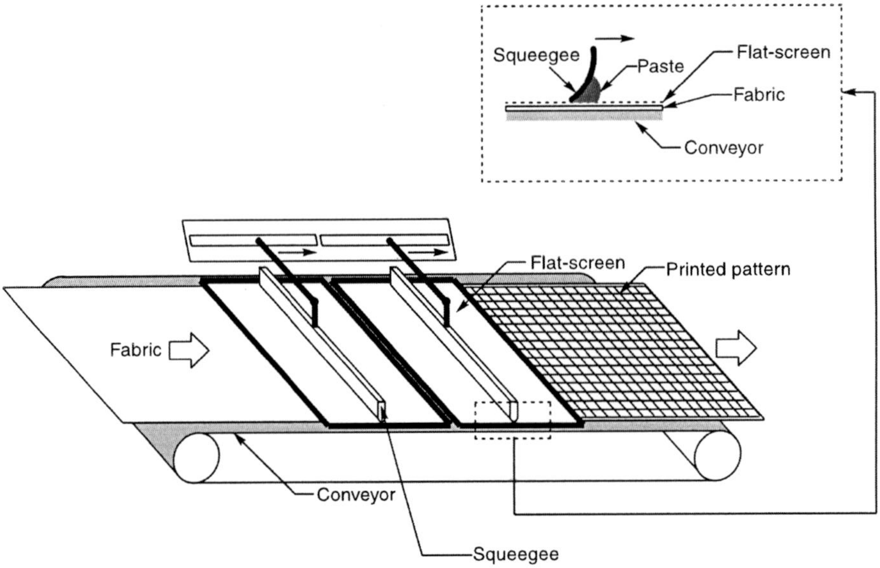

14.1 The schematics of the flat-screen printing machine.

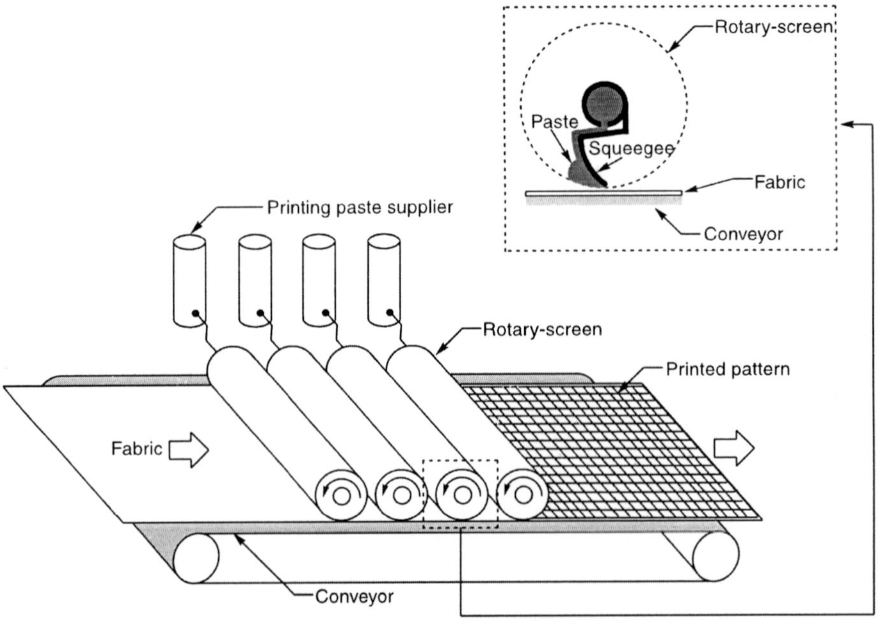

14.2 The schematics of the rotary-screen printing machine.

production, because of the cumbersome clean-up process for this machine when changing colours or patterns.

14.5 Inkjet printing and its process control

Inkjet printing of textiles is the method of direct printing of colourants onto fabric by a computer-controlled printer, which creates a digital image by propelling droplets of ink onto fabric. Therefore, this method is also called digital textile printing, that is solutions or dispersions of dyes (or pigments) are sprayed on the detailed region of fabric as droplets from printing heads. The printed fabrics then will be immediately steamed or cured for colourant fixation, followed by rinsing and drying if dyestuff inks have been applied, which is similar to conventional printing.

Even though inkjet printing has the serious shortcoming, i.e., its production speed is much lower than that of rotary-screen printing, it has still attracted worldwide attention in the past 30 years because of the advantages that inkjet printing can bring to the demands of printed goods in the future. Mass customisation replacing mass production and make-to-order with short supply chain will be the trend in the coming years. Inkjet printing can meet these trends since it has following benefits: (1) no limit in colour number and size of patterns printed; (2) precise registering and rich gradation of shades; (3) no need of screen production; (4) less pre-production sampling; (5) instant customer response; (6) no printing paste employed; (7) enhanced utilising efficiency of colourants; (8) fewer wastage of inferior printed fabric; and (9) less downtime for pattern change.

Most notably, inkjet printing is able both to print smaller designs onto garments (t-shirts, dresses, promotional wear) or to print larger patterns onto large-format rolls of textile. The latter is a growing trend in visual communication, where advertising and corporate branding are printed onto polyester media, such as banners, retail graphics, etc. Using inkjet technology in digital textile printing allows for single pieces, mid-run production and even long-run alternatives to screen printed fabric.

The following discussions are focused on inkjet processes, including the ink preparations, pretreatment of fabrics substrates, inkjet printing, defect prevention and after-treatment, and the relevant inkjet printing process control.

14.5.1 Ink preparations

The jetting ink is made up of colourant (dyestuff or pigment), solvent (water) and other additives (such as surfactant, salt, and binder for pigment ink, etc.). Inks are the main consumables for inkjet printing. All the ink

Table 14.12 Typical composition of printing inks (water-based type)

Component	Concentration (%)
Water	60–80
Dye or pigment	5–10
Surfactant	1–10
Other additives	1–10

formulated should meet the basic requirements, such as good jetting ability, good stability in storage and good performance when applied to substrates. Table 14.12 shows the typical ingredients of a jetting ink.

Compared to the ink used for paper printing, in addition to the requirements of colour purity, insoluble particle size, viscosity, surface tension, conductivity, stability, pH and foaming properties, the inks applied to textile fabric must have good colour fastness and good handle. There are four basic colours of cyan, magenta, yellow, and black (abbreviated as CMYK) for any kind of ink. In printing production, some more colours such as dark blue, dark red and orange are also needed. Currently, many ink producers can produce a set of 12-colour inks, which makes colour matching easier.

Reactive ink

Reactive ink is widely used for cellulose-based fibres, such as cotton and linen, and can also be used for silk, wool, and nylon to some extent. Reactive dyes react with nucleophiles such as hydroxyl and amino groups. Therefore, reactive inkjet printing has by-reactions as in reactive dyeing or in conventional reactive printing, which may reduce the colour yields unexpectedly. When preparing reactive dye ink, eliminating reactable ingredients is the most critical issue. However, by-reaction induced between reactive dye and water in ink is inevitable, because aqueous ink always provides hydroxyl groups to reactive dyes which might cause dye hydrolysis. To minimise hydrolysis of reactive dye, pH control is crucial and effective. In order to improve the long-term stability of reactive ink in storage, modification of commercially available dyes is an alternative approach. Three commercially available dyes (Drimarene Brilliant Red K-4BL, Drimarene Brilliant Blue K-BL and Drimarene Golden Yellow K-2R) were modified by the reaction of the parent dyes with 4-hydroxybenzensulphonic acid sodium salt (Clark *et al.*, 2009). The results show the inks based on modified dyes exhibit much better storage stability compared with the normal dye-containing inks. Generally, high quality reactive inks should possess the features of good fluency, perfect stability with shelf-life more than one year, high fixation (over 80%), and no harm to the printhead.

Table 14.13 Character of disperse dye used in inkjet printing

Printing method	Character of disperse dye
Direct printing	• Higher molecular weight • More hydrophilic
Transfer printing	• Lower molecular weight • More hydrophobic

Disperse ink

Disperse inks can be made into direct printing ink and transfer printing ink, based on the different properties of disperse dyes. Direct printing ink, as its name implies, can be used in direct inkjet printing similar to reactive inkjet printing. The transfer printing ink is used for printing the sublimation transfer paper for transfer printing as introduced before in Section 14.3.3. Table 14.13 shows the characteristics of two kinds of disperse inks for inkjet printing.

Unlike the jetting ink of reactive dye, disperse dye in ink is in the form of micro-particles, of mean size in the range of 100–250 nm. Hence disperse dye ink uses water-borne dispersion rather than solution, and the selection of proper dispersant for disperse ink should be considered more.

Pigment ink

Compared to dyestuff inks, pigment ink has shortcomings, such as difficulties in preparation of pigment dispersion, lower colour density and narrower colour gamut, etc. However, balanced with the advantages of being suitable for any fabric and of a much simpler printing process than that of dyestuff ink printing, pigment inkjet printing has been attracting much more attention over recent years.

Pigment inks are pigment dispersions prepared by a process similar to disperse dye inks. However preparations of pigment ink are much more complicated than those of disperse ink, different from the colouration mechanism of substrates with dyestuffs, pigment colouration bases on the fixation of pigment particles on fabric surface in the presence of binders. As a result, the selection of binder needs consideration for making pigment inks.

Low glass-transition temperature (T_g) and low viscosity are the basic requirements for the binder applied in pigment ink. Low T_g has the benefit of good handle for printed fabric, and more importantly it could reduce the risk of clogging the nozzles of the printing head. Low viscosity is essential because binder load needs to be large enough to maintain the colour fastness. A larger load of binder can cause the ink viscosity to increase, which can further result in the decrease of the ink jettability, and consequently to poor jetting reliability.

Although the preparations for pigment dispersion and disperse dye dispersion are similar, their colouration mechanisms are quite different. In direct printing or transfer printing of disperse dye, the colouration of printed area follows a dyeing principle in which dyestuff penetrates the internal molecules of the fibre. Therefore the particle size of dyestuff in ink has little effect on colour density and brightness. Most consideration focuses on the influence of dispersion size on jettability. Nevertheless, pigment inkjet printing is characteristically distinct, with no transformation of pigment particles. Hence design of size distribution of dispersion should take both ink jettability and colour quality into account. Generally, the mean size of pigments dispersed in ink ranges between 50 and 200 nm. Smaller particles have good light absorption ability, and bigger particles exhibit high colour density. Optimum particle size has to balance these two aspects for a given fabric.

Other ingredients in pigment ink include surfactant, wetting agent, defoamer, etc. These components are added to improve the applicability of ink, such as stability in storage/printing and ink reliability in the printing process.

14.5.2 Pretreatment of fabric substrates

Pretreatment for fabric substrates in inkjet printing is much more complicated than for conventional dyeing and printing. After the common pretreatment processes, including singing, desizing, scouring, bleaching and/or mercerisation, padding with thickener is essential for the fabric to be directly printed via the inkjet method. The thickener, acting as the printing paste in conventional printing to prevent dye migration, is not present in the printing ink but is padded onto the fabric prior to inkjet printing. Even though thickener in inkjet printing will bring some problems in practical production, such as instability of printing ink, clogging of jet nozzle, low printing speed, and some other less predictable problems (Leube *et al.*, 2000), pretreating fabric with thickener is a priority before it is printed with dye-based inks. Figure 14.3 shows the schematic difference of the shape of an ink drop landing on fabric substrate. It is obvious that the ink drop on non-pre-treated fabric shows much wicking and the drop size extends beyond the treated substrate. For fabrics printed with ink consisting of pigment, pretreatment with thickener might be an option depending on the particular situation.

To obtain good pretreatment results, the thickener should exhibit the properties of: excellent ability to hold water/moisture, easy removal ability, good fluidity, and good stability to pH, temperature, electrolyte and dyestuff.

Sodium alginate (medium viscosity), because of its inexpensiveness, is the most common thickener adopted in pretreatment to preserve the edge

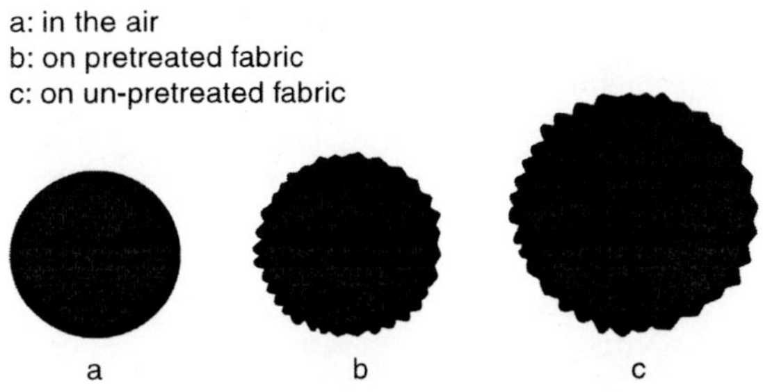

14.3 Vertical view of the shape of ink drop.

and outline sharpness of printed pattern. It also modifies the flow property of printing ink and holds moisture content to help dyes dissolving and entering into fibre during the steaming of printed fabric. The dose range of sodium alginate depends on the dye applied and the specification of fabric to print. Excessive sodium alginate on fabric would cause difficulties in after-treatment, such as in the wash-off process, while inadequate thickener could lead to poor definition of the printed pattern. In cases in which yellowing of the background of printed fabric is not acceptable, polyacrylic polymer is recommended to replace sodium alginate for a whiter and brighter background.

When preparing the pad bath, the formulation should ensure the other ingredients are totally dissolved before adding thickener, should eliminate impurities in pad bath, should prepare the pad bath in use, and is used shortly after preparation. With the development of inkjet printing, specialised products, recipes and processes are being developed by the printing ink manufacturers, which facilitates the printing mills obtaining optimum pretreatment.

Once the fabric is pre-treated, it becomes much more hydrophilic. Consequently, humidity of the storage condition for the pre-treated fabric has a profound effect on the moisture regain of the fabric. Controlling the moisture content to a proper level facilitates subsequent inkjet printing in terms of image definition and colour density. Printing on a fabric with high moisture content results in poor absorption of jetting inks, and substrate retaining low humidity leads to colour density reducing. The optimum moisture retention of the padded thickener on fabric depends on the substrates being pre-treated and the jetting inks being applied.

Cotton

The recipe for the solution for pretreating cotton consists of thickener, alkali and other auxiliaries. The typical recipe for the pad bath for pretreating cotton woven fabric is presented in Table 14.14.

Table 14.14 Typical profile of the pretreatment of cotton woven fabric

	Ingredients	Content (g/L)
Typical recipe of pad bath	Sodium alginate (medium viscosity)	80–100
	Urea	100
	Sodium carbonate	20–30
	Sodium m-nitrobenzenesulfonate	10
Padding and drying	Pick-up: 70–90%; Drying: 110–120°C , 2–5 min	

The functions of urea and alkali have been illustrated before in reactive dyeing and printing for cellulosic fibre. In inkjet printing, the difference is that these two ingredients are required to pad on fabric before printing. Selection of alkali depends on the reactivity of reactive dye in the printing ink. In continuous dyeing or conventional printing, in which alkali and reactive dyes are formulated in one bath, sodium bicarbonate as a relative weaker alkali is the first choice to reduce hydrolysis of dyes. But in inkjet printing, the stronger alkali, sodium carbonate, becomes the priority candidate. Other auxiliaries added in the pad bath should not react with the reactive dye in jetting ink, otherwise the colour yield of the final printed fabric would be decreased and severe staining might happen during washing process.

Polyester

The inkjet printing of polyester involves two methods, direct printing and transfer printing. Fixation of disperse printing requires high temperature, ranging from 180°C to to 210°C depending on the process adopted and equipment applied. Therefore the most critical requirement for thickener used in printing is excellent thermal stability at fixation stage, otherwise the quality of printed pattern will be reduced. Sodium alginate is a typical thickener, and alternatively, some new synthetic thickeners have also been developed for pretreatment of polyester. Urea as a multifunctional agent is not recommended for polyester pretreatment, due to its poor resistance to heat in fixation process. The fumes generated from decomposition of urea may cause pollution in the fixation stage.

New technologies in pretreatment

In addition to the regular pretreatment of cotton fabric, some researches and trials have been carried out to improve printing quality in terms of colour fastness and colour yield. These researches and trials focus on the modifications of fabric substrates. Plasma treatment (Zhang and Fang, 2009; Wang and Wang, 2010; Kan *et al.*, 2011) and cationisation (Chen *et al.*, 2004) are the most frequently used technologies. Plasma treatment aims to change the outermost layer of the fibre, in terms of modifying the physical morphology and chemical properties. As known to all, original polyester exhibits poor ability to hold inks and dyestuffs, due to its smooth morphology and stable chemistry. Therefore, patterns directly printed with pigment inks have poor colour yields and bleeding happens frequently. When original polyester is modified by plasma, the etching effect and polar groups induced onto the fibre surface bring about a colour quality improvement of the printed fabric. Cationisation of cotton is a regular technology in salt-free reactive dyeing, and is also an effective method in printing. The cationised cotton has a positive charge on its surface, which attracts the ionised negative dyestuffs in the ink. Consequently, dye uptake will be greatly improved. This principle is also available for pigment inkjet printing (Fang *et al.*, 2005), because the surface of pigment particles in dispersion is negative due to embedded dispersants, of which the anionic groups are exposed to the surrounding water.

14.5.3 Inkjet printing

An industrial textile inkjet printing machine is usually designed in roll-to-roll or roll-to-folder form. The printing machine is fitted with a rubber blanket and an online dryer and features different entry/exit configurations. A good inkjet machine should have such characteristics as high productivity (250 m^2/h or even higher), low ink consumption, suitability for a wide range of fabric (knitted, woven, non-woven, etc.) and a wide range of inks (acid, reactive, direct disperse, sublimated disperse and pigment), wide range of printing width (1800, 2400, 3400 mm, etc.), high resolution (600 × 600 dpi), and high reliability (embedded printheads maintenance system, unlimited printhead lifetime), and printheads adjustable to media distance. The appropriate software is also a necessary for an inkjet printer for image scanning, colour separation and colour management.

14.5.4 Defects prevention

When fabric is fed into an inkjet printer, one of the most frequent defects is banding, arising from wrong feed speed of fabric and/or non-uniform fabric structure. In practical printing, the causes for banding probably include slippage between the fabric and the feeder and/or conveyor belt, inappropriate or non-uniform tension applied on the fabric, and poor pretreatment of the fabric. To prevent slippage, the feeder should have strong friction on fabric and the conveyor belt should be able to maintain a constant adhesion level. Because of adhesion to the belt, fabric stretched distortedly can cause wrinkles. The tension applied on the fabric depends on the nature of the fabric. Generally, for elastic and stretchable material, such as knits or fabric blended with polyurethane, the tension should be as small as possible, otherwise deformed fabric will lead to a substandard printing image. Uniform tension on fabric requires the driving device from printer to apply a constant force across the roller axis. This can be achieved by automatic tension-adjusting instruments with a tension-monitoring device, which controls the tension to a minimum variation. Another non-uniform tension occurs in partial area of the fabric generated from wet expansion or shrinkage of the printed area by ink-wetting. In that case, increasing the adhesiveness of the conveyor belt is an effective technique to eliminate wet expansion or shrinkage. Furthermore, adjustment of the recipe of pad bath for pretreatment of fabric is necessary to improve its dimensional stability for ink-wetting.

14.5.5 After-treatment

After-treatment refers to the process required after inkjet printing. Curing is the only process required for pigment inkjet printing, and washing is not necessary. However, for dye inkjet printing, it includes dyes fixation and subsequent wash-off. Dye fixation, in terms of steaming or heating for dyes, helps to complete the dyeing process on inkjet printed fabric (Tyler, 2005), similarly to conventional printed products. The example profiles of fixation and after-treatment are listed in Table 14.15.

14.5.6 Future trends

At present, the main drawbacks of inkjet printing are slow production speed and relatively high production cost. Production speed primarily depends on the industrial printers used. The printing speed has been improved, and now reach 250 m^2/h or more. Lowering the ink price is considered to be the first objective to reduce printing expense. It is predicted that with the promotion of digital printed products, massive production of ink will be required, which could bring significant reduction in its manufacturing cost. A virtuous cycle is thereby created and a brilliant future for digital printing is coming.

14.6 Product safety and low-carbon production

In a dyeing or printing process, many chemicals and dyes must be applied. The safety of the dyed or printed products, especially with respect to the chemicals or dyes used, become an important issue. There are many product safety regulations in the EU, China, Japan, Australia, USA, Canada, etc. REACH is a new European Community Regulation on chemicals and their safe use (EC 1907/2006). It deals with the Registration, Evaluation,

Table 14.15 Profile of after-treatment for inkjet printed fabric

Colorant	Process	Condition
Reactive dye	Fixation	Steam at 102°C for 7–10 min; or thermofix at 150–180°C for 3–5 min.
	Wash-off	(1) Rinse 5 min with cold water; (2) Soap for 5 min at 98 °C with 2 g/L ERIOPON® R; (3) Rinse for 3 min at the boil; (4) Rinse warm/cold.
Disperse dye	Fixation	Direct ink-jet printing by high temperature steaming at 180 °C for 8 min; or thermofix at 180–210 °C for 1–2 min.
	Wash-off	(1) Rinse 5 min with cold water; (2) Soap 5 min at 40 °C with 1 g/L detergent; (3) Reduction clearing.
Pigment	Fixation	Heat 1–3 min at 160–210°C.

Authorisation and Restriction of Chemical substances. There is a legal obligation for dyeing and printing companies to conform to these regulations for selecting chemicals and dyes. Close attention should be paid to the substances of very high concern (SVHC) listed in REACH, the substances listed in the Restricted Substance List (RSL) of the American Apparel and Footwear Association (AAFA), and the limited or restricted chemicals and the limit values listed in Oeko-Tex Standard 100 etc. In the global textile markets, testing for harmful substances plays a significant role in the decisions to be taken when buying textiles.

Eco-friendly and low-carbon emission productions for dyeing and printing productions are the other important issues concerned in textile industry. All kinds of new dyeing and printing equipments, dyes and chemicals, and technologies related to the shortening process, lowering water and energy consumptions for increasing production efficiency and reducing carbon emission, have been launched on the market (Shang *et al.*, 2011). We should keep a close eye on the developments of new chemicals, equipment and technologies, which may bring about changes in dyeing and printing process control.

14.7 Sources of further information

In Chapters 13 and 14, the author has tried to cover as many technology principles and the related process factors as possible that need to be considered and controlled in dyeing and printing. Because of the limitations of text, it is difficult to describe all of these considerations in detail. The relevant references listed at the ends of these two chapters can provide more information. In addition, the two internationally respected professional organisations, *Society of Dyers and Colorists* (SDC) (www.sdc.org.uk) in

UK and *American Association of Textile Chemists and Colorists* (AATCC) (www.aatcc.org) in US, have published many articles in their journals (such as *Coloration Technology* and *AATCC Review*) and books on dyeing and printing technology, in which the technology principles and the process controls might be involved. More information can also be found in other related textile journals or books, and on the web.

14.8 References

Chen, W. G., Zhao, S. C. Wang, X. G. (2004). 'Improving the color yield of ink-jet printing on cationized cotton', *Textile Research Journal*, **74**, 68–71.

Clark, M., Yang, K. Lewis, D. M. (2009). 'Modified 2,4-difluoro-5-chloro-pyrimidine dyes and their application in ink-jet printing on wool fabrics', *Coloration Technology*, **125**, 184–90.

Fang, K. J., Wang, C. X. Zhang, X. (2005). 'Dyeing of cationised cotton using nano-scale pigment dispersions', *Coloration Technology*, **121**, 325–8.

Kan, C. W., Yuen, C. W. M. Tsoi, W. Y. (2011). 'Using atmospheric pressure plasma for enhancing the deposition of printing paste on cotton fabric for digital ink-jet printing', *Cellulose*, **18**, 827–39.

Kanik, M. and Hauser, P. J. (2004). 'Printing cationized cotton with direct dyes', *Textile Research Journal*, **74**, 43–50.

Leube, H., Ruttiger, W. Kuhnel, G., Wolff, J., Ruppert, G., Schmitt, M., Heid, C., Hückel, C., Flath, H., Beckmann, B., Brossmann, R., Söll, M., Sewekow, U. (2000). *Textile Dyeing*, Wiley-VCH Verlag GmbH & Co. KGaA.

Shang, S. M., Hu, E. L., Poon, P., Jiang, S. X. and Kan, C. W. (2011). 'Application of foam dyeing for developing the wash-out effect on cotton knitted fabric with pigment', *Research Journal of Textile and Apparel*, **5**, 44–51.

Tyler, D. J. (2005). 'Textile Digital Printing Technologies', *Textile Progress*, **37**, 1–65.

Wang, C. Y. and Wang, C. X. (2010). 'Surface Pretreatment of Polyester Fabric for Ink Jet Printing with Radio Frequency O-2 Plasma', *Fibers and Polymers*, **11**, 223–8.

Zhang, C. M. and Fang, K. J. (2009). 'Surface modification of polyester fabrics for inkjet printing with atmospheric-pressure air/Ar plasma', *Surface & Coasting Technology*, **203**, 2058–63.

15
Process control in finishing of textiles

A. K. ROY CHOUDHURY, Government College of Engineering and Textile Technology, Serampore, India

DOI: 10.1533/9780857095633.3.363

Abstract: The chapter discusses the importance of process control in textile finishing operations. Controlling parameters for basic finishing machines as well as for various mechanical and chemical finishing processes are discussed. Low liquor foam finishing and eco-friendly finishing processes such as enzyme and plasma finishing are explained.

Key words: process control, textile finishing machines, chemical and mechanical finishes, enzyme, plasma treatment, foam finishing.

15.1 Introduction

The textile industry is one of the oldest industries in the world. The earliest known textiles include scraps of linen cloth found in Egyptian caves dating from about 5000 BC. In the western world, textile manufacturing remained a family business until the early 1500s, when the first factories were built. In Asia, especially China, centralising and standardising textile production occurred as early as the Zhou Dynasty (eleventh to eighth centuries BC).

In the West, an early regulation for quality assurance in the textile trade reported during the fourteenth century in Germany named 'Tuchshau' that is showing of cloth, which involved expert inspectors along with an equal number of city council members. As in most other manufacturing industries, for many centuries quality in the textile industry was achieved through inspection of finished goods. This final inspection was often used to create different grades of quality, products of which were then sold at various prices. Manufacturers slowly began to add inspection and control to the quality of raw materials and the production processes.

Intense competition, both nationally and globally, has resulted in a new focus on quality management in textile companies. The first reports on statistical quality control of yarn-manufacturing products appeared during the late 1940s and 1950s. Dyeing a textile yarn or fabric is one of the most difficult, monitored and controlled processes in the textile manufacturing chain. The finishing process, by contrast, still has few controls and relies heavily on inspection and testing. The two main methods used in finishing are chemical and mechanical. Evaluating the results of both these methods is still highly subjective, and very little statistical process control has been applied in these areas (Clapp *et al.*, 2001).

15.1.1 Approach to process control

The choice of process condition for the given product is taken by the previous history and forming new norms without affecting quality. The optimum norms may vary from unit to unit and machine to machine. This is for various reasons, such as working condition, type of machine, lay-out of machine, provision of utilities and variations in the quality of the fabric. Therefore every process house has to carry out their own experiments to identify their own optimum processing levels. Once the processing conditions are standardised, the implementation of these conditions during the normal course of production is carried out by maintaining required documents. However it is important to make regular inspection checks to ensure that the particular process is going on according to the norms fixed.

For quality optimisation it is necessary to know about causes for quality defects. Based on the cause and effect diagrams published by K. Ishikawa (1985) there are four critical causes for quality defects:

1. machines/equipment,
2. methods,
3. man (i.e. personnel),
4. material.

With machines or equipment, quality deviations may occur because of inadequate maintenance or servicing. In the textile industry, a breakdown in quality is said to have occurred if, for example, the required washing effect, degree of whiteness or absorbency of the fabrics being finished is not achieved. These are defects in the specification of the system because the expectations of the customer are not fulfilled.

A known problem in wet finishing ranges is the wear of the roller rubber, even if this happens to be outside the influence of the machine supplier. This is a potential source of risk of a reduction in quality. Two questions must be asked about a potential source of risk to quality:

1. Is this component really necessary?
2. Does the component have to be modified to suit the operating conditions, or can the conditions be improved to suit the critical component?

Rubberised rollers are critical components in the textile finishing industry. High temperatures and large amounts of chemicals lead to rapid and uncontrollable wear. It is well known that washing efficiency at boiling temperature is considerably higher than in cold washing water. Washing processes at boiling temperature create unfavourable conditions for rubber rollers. Cold washing would improve the operating conditions but make the textile results worse. This means that the critical component, in this case the first washing unit in a washing process, will have to be modified to suit the ambient conditions. The simplest way to do this would be to avoid using rubber rollers in such situations (Ströhle, 2009).

15.2 Instrumental process control

The increased demands of the textile finishing industry on the machinery with regards to quality of the products, productivity, cost reduction and reproducibility are not satisfied by the conventional machines. The cost-effectiveness of textile machinery as a result of automation systems has been widely discussed in several publications. The purpose and the benefits of such systems have never been questioned. Considerable differences of opinion exist, however, about the design of automation equipment, about strategies, versatility, systems and functions with or without computer, whether with integrated organisational hierarchy or as a straightforward automation of machine functions (Schreiber and Maschen, 1992).

Automation and control can be used in dyeing, printing and finishing for the following processes:

- temperature–time programmes,
- running partial or fully automatic processes,
- pH control,
- flow pressure/differential pressure,
- dosing of liquid chemicals.

Automation of regulation and control processes in production equipment requires a programme that allows analogue and digital techniques to be mixed on the screen without any problems, and the microprocessors that are introduced to be designed so that they can be integrated easily into the bus system. The introduction of Expert systems as the process control technology (e.g. for stenters or continuous open-width washing ranges) has been especially relevant to automation in textile dyeing and finishing (Rouette, 2000)

Any control system may be divided into three components (Duckworth, 1983):

1. The sensor detects a change in a process variable and produces a proportional output signal.

2. The controller receives the signal, compares it with the desired set value and emits a correction signal proportional to the deviation from the set value.
3. An actuator and servo device – commonly a power-operated valve – which receives the correction signal and mechanically adjusts the process variable.

A process control system should be applicable indiscriminately to all finishing machines. Identical hardware and software should be capable of controlling all continuous textile finishing processes.

Five steps to automate a continuous textile finishing machine (International Textile Reports, 1991) are as follows:

1. Division of the machine into a number of functional zones mostly based on drive sections of the machine to make controls flexible.
2. Defining various control parameters (e.g. temperature, flow, etc.) for each zone.
3. Calculation of the number of controllers required in order to control each zone and thereby for the whole machine. Analogue inputs and outputs are used to control temperature, flow, pH, etc., whereas digital inputs and outputs are used for level control, drain activation, end switches, etc.
4. Connection of multiplexer to the system in order to achieve a correct communication in two directions between the computer and the controllers.
5. Connection of a standard personal computer or preferably its industrial version that presides over each machine and is connected to the multiplexer via an interface. Generally the computer is situated near the operating panel of the machine.

The software system should take care of the following points:

1. Parameter definition – each control parameter should be properly defined. A series of standard control algorithms should be available. The possibility of relating control parameters should exist. For simple definitions, the computer system should be menu-controlled.
2. Process library – a library of machine processes can be defined with chosen set points for each process parameter. When a process is chosen or changed on the status screen, the machine sets itself to the desired set points. Intervention in the process is possible on the computer and local levels.
3. Status screen – while the machine is running, an on-line status screen should be available. The status screen should display the actual status of the machine, set points, actual values as well as high and low alarms. In addition, the software should offer a number of user-selectable reports and graphs.

The standard control functions in a continuous textile finishing machine are as follows:

1. Bath temperature is measured by a PT 100 element and is controlled by a proportional valve to assure processing at the optimum temperature.
2. Cloth temperature is measured by pyrometer and is controlled to avoid over-drying.
3. Steam pressure may be controlled as an alternative to temperature control in the steamer.
4. Flow of chemicals and water is measured by an inductive flow-meter and is controlled by a proportional valve or through a measuring cylinder dosing system, in order to assure a constant flow, depending on machine production.
5. The pH is measured by a pH sensor and mainly controlled in washing of processes.
6. Washing efficiency is assessed by measuring and controlling conductivity of bath and/or cloth by a conductivity sensor.
7. Density (concentration) control is mostly used in mercerising range to achieve a constant mercerising concentration.
8. Compensator pressure control in order to run the cloth with correct cloth tension.
9. Squeezing unit pressure control, especially for heavy-duty squeezing units to assure reproducible liquor pick-up.
10. Level control in washing units and chemical tanks used for safety interlocks (temperature control), as a safety level and to fill the machine automatically.
11. Speed measurement and control as an indication for the machine efficiency in order to make production-dependent dosing of chemicals and water into the machine.
12. Control of consumption of steam, water, electricity and chemicals to maintain quality and cost-effectiveness.

15.2.1 Flow-meters

These are used to measure the flow rate of a fluid in a pipe, duct or open channel. Flow-meters are of two types – quantity and inferential. While the former measures fluid flow rate directly, the latter measures fluid velocity, and thereby fluid rate. The most commonly used inferential flow-meters are:

1. pitot tube,
2. orifice meter,
3. venturi meter,

4. nozzle meter,
5. weirs,
6. rotameters.

Valves are devices used to control the flow of a fluid by placing an obstruction in the fluid's path. There are various types of valves; the selection for a specific function depends on whether it will be used as a stop valve (only 'off' or 'on' service) or a throttling (regulating) valve. The later again may be of various types namely cocks, ball, globe, gate, diaphragm, pinch, etc. A typical automatic process control valve (which may be globe or diaphragm types) consists of four separate but integrated components – actuator, packing gland, body and valve plug. The sensing element of the valve measures a particular process parameter (flow, temperature, pressure, level, pH) and if the value deviates from the desired value, the gas (air) pressure on the diaphragm in the pressure casing is altered automatically to deflect the diaphragm to move the value stem and plug, thereby altering the flow area in the valve to change the fluid flow rate (Duckworth, 1983).

The logical objective of a process control system is to assure repeatable process quality and efficiency by controlling process variables to predetermined and preset values and to provide management information assuring proper equipment operation. To fulfil these objectives, the system should have the following characteristics (Fulmer, 1983):

1. Communication with plant personnel through CRTs, printers or alarm systems (light and/or horns).
2. Input to the system by plant personnel may involve card readers, push buttons, keyboards, etc.
3. Reliability of the system in the industrial environment involving heat, humidity, dirt, etc. should be assured.
4. Ease of use and maintenance are desirable. Standard instructions for each type of fabrics and finishes are to be set-up.
5. Back-up control is necessary to avoid loss of production due to system breakdown.
6. The system should be flexible to cope with new fabrics and finishes.

The important parameters in textile processing are the quantities of chemicals, pH value, water quantity, temperature and reaction time. For many products, especially knitwear, the fabric tension also plays a decisive role. For reputed textile machine manufacturers it is necessary to accurately measure the parameters and to record these values in order to ensure long term both internal (cost reductions, quality assurance) as well as external (reliability for customers, handling of complaints) process reliability.

The chemical metering system offers the greatest potential for savings. Inductive flow rate meters ensure that expensive additives are dosed with millilitre accuracy on their way into the impregnating bath. Reproducibility can also be increased with the aid of impregnation compartments with low liquor content and a rapid rate of liquor exchange.

Modern databases for recipes or formulas also make the day-to-day work of machine operators easier. Not only the corresponding recipes for all product groups are saved, but also these recipes can be extrapolated for the specific fabric weight resulting reproducibility and quality of the highest order. The water quantity, temperature, fabric tension and pH value remain a key factor in these recipes.

One of the key factors for data management is the sensor systems, which are integrated in the machine. The process can only be started and finished correctly if the produced data are totally reliable. If a temperature sensor indicates an incorrect value then no data management system in the world will be able to save the overall result. Therefore continuous monitoring of the 'eyes and ears' of a system is still the best insurance policy for ensuring that, once selected, the parameters are maintained in the long term. Temperature monitoring with an external thermometer (or alternatively a second sensor), regular determination of the pH value via titration or via the Morapex process, checking the flow-meters through capacity gauging and constant checking of bearings and drives, are all important factors for guaranteeing constant fabric tension. Consequently, once modern technology is in place and the monitoring capabilities and process reliability of a state-of-the-art data management system is correctly set up, it is possible to deliver top quality in a cost-optimised framework (Kehry and Uhl, 2010).

15.3 Textile finishing processes and process control in finishing

Any operation (other than preparation and colouring) that improves the appearance and/or usefulness of fabric after it leaves the loom or knitting machine is called 'textile finishing'. Finishing is the final series of operations that produces marketable textile fabric from grey goods.

15.3.1 Textile finishing processes and their classification

The word 'finish' means all the different treatments applied to a fabric to change one or more of the following:

- appearance,
- feel or hand,
- wearability or care requirements.

Finishing improves aesthetic aspects or serviceability of the fabric or imparts certain desirable characteristics.

Textile finishes are classified in several ways. According to function, these can be classified into:

1. Aesthetic finishes, which modify the appearance and /or hand or drape of the fabrics.
2. Functional finishes, which improve the performance properties of the fabric such as durability, strength, etc. Property-changing functional finishes provide the added qualities desired for a particular fabric or they may be used to change an undesirable property to a more desirable one. Many such finishes add more than one property to a fabric.

Examples of aesthetic finishes are:

- calendering,
- fulling,
- mercerisation (chemical),
- napping and seeding,
- shearing (mechanical),
- bio-polishing (chemical),
- softening (chemical),
- stiffening (chemical),
- weight reduction (chemical).

Examples of functional finishes are:

- crease resistant/durable press,
- flame retardant,
- shrinkage control, Sanforizing,
- soil-release,
- water-proof and water-repellent,
- antistatic,
- antimicrobial/antiseptic.

Some of the finishes could result in both aesthetic and functional effects. Mercerisation is considered as an aesthetic finish, as it increases lustre. It may also be considered as functional, as it increases strength. Sometimes it is even considered as a pre-treatment process, as it improves dyeability. The classification of finishes and class-wise discussion of finishing processes are, therefore, quite difficult.

Depending on the machines or chemicals used, both aesthetic and functional finishes can be further classified into:

1. Mechanical finishes – these usually involve specific physical treatment to a fabric surface causing a change in fabric appearance. These are also known as dry finishes.
2. Chemical finishes – chemicals are usually applied to fabric by padding followed by curing and drying. These are also called as wet finishes.

Mechanical finishing typically refers to finishing effects 'done by machine' as opposed to finishes achieved through chemicals. Mechanical finishes have some advantages over chemical finishes. Mechanical finishing can process the fabric at very high speeds. The repeatability is better over time, since there are no formulations to change, and changing chemical restrictions and prices are not an issue. The mechanical finishes are now more acceptable because they are environmental friendly.

Except shrinkage control or Sanforizing process, all functional finishes mentioned above are basically chemical processes.

According to the quality, finishing may be classified into:

1. Temporary: a finish which is not stable and goes off after the first wash is known as temporary finish.
2. Semi-permanent: if the finish is stable for ten washes but not beyond.
3. Permanent: if the finishing effect in the fabric does not disappear and remains unaffected through all the conditions of wear and washing treatments, then the finish is said to be permanent.

The types of finishing agents or finishing processes applied to various textile materials are indeed very large in number and they can be classified in various ways. It is, therefore, difficult to describe each of them systematically.

15.3.2 General process control in finishing

The number of finishes applied to various textile materials is very large and are selected as per properties desired for the end products. One or more properties are modified by each finishing agent and most of such properties can only be assessed subjectively. The situation is further complicated by the fact that many finishes are mixed or combined. Quality control during finishing is, therefore, very challenging and subjective.

Compatibility

It is necessary to check various properties of the finish mixture before and during application. There are quite a number of factors making the chemicals used in a recipe mutually incompatible – the ionic characteristics, solubility, emulsion stability, pH conditions, etc. Stabilisers used in an emulsion may be disturbed by new or alternates additive in a recipe, thus causing breakup of the emulsion.

When two or more finishing chemicals in a recipe are contemplated or when the recipe is changed for reasons of non-availability of one of the components, it is desirable to check their mutual compatibility in the laboratory before administering in bulk.

Strength of the finishing agent

It is desirable to apply the recommended level of the finishing agents. Commercial finishing agents come in different strengths and it is the active matter that is important. In some cases, when dilution needs to be done to a working concentration level from concentrated pastes/emulsions with the application level fixed, the new consignments should be evaluated for the active content and accordingly the recipe should be adjusted to match the standard recipe quantity.

Moisture regain of substrate

The moisture regain of cotton is around 8%. This plays an important role in the feel of the fabric. Over-dried fabrics tend to feel relatively harsh despite application of softeners. It is also necessary for the fabrics to have satisfactory moisture regain before finishing by padding methods, particularly in the case of resin finishing, facilitating better and even diffusion and penetration of the finish.

It is desirable to control the moisture regain level of the fabrics during drying using control instruments like 'Textometers' in a stenter operation or 'cool tumbler' in the drying of the garments. There are conditioning cycles in the drying operations (e.g. in yarn drying) in the modern batchwise machinery.

The following process technologies take place in textile finishing (Rouette, 2000):

- monitoring and control of work processes across all areas by process computer (process data registration),
- shortening of work processes,
- reduction of reaction temperatures,
- recovery of chemicals and energy,
- high-temperature steaming,

- increase of passage or turnaround speeds (e.g. continuous systems at 300–500 m/min),
- reduction of liquor ratios,
- use of chemicals having no adverse effect on the environment such as formaldehyde-free finishing,
- products that dye and finish simultaneously,
- use of electrical power as a finishing agent,
- complete recovery of products that are not fully exhausted or fixed,
- self-destructing and self-disposing finishing agents.

15.4 Process control in basic finishing machines

Two types of machines are used in finishing of textile materials. Some basic machines are used in all or most of the finishing processes while some other machines are used for specific finishing processes. The former type includes padding mangles, drying and curing machines, etc. The process control in some of the basic finishing machines is discussed below.

15.4.1 Padding mangle

Traditionally, padders are used to apply chemical finishes in low wet pick-up techniques. In recent years such techniques have become more cost effective as the cost of energy has escalated significantly in comparison with the cost of finishing agents.

A padder consists of a trough and a pair of squeeze rolls (mangle). In the padding process, the fabric passes under a submerged roll in a trough filled with a treatment bath and then through squeeze rolls. In the simplest arrangement, a single dip is followed by a single nip. Alternately, three mangles can be used and the fabric is impregnated twice, double dipped – double nipped.

The amount of finishing solution or emulsion applied is referred to as the '%wet pick-up' which is usually expressed as a percentage on the weight (wt) of the dry untreated fabric at room temperature and humidity.

$$\%\text{ pick-up} = \frac{(\text{wt of padded fabric} - \text{wt of dry fabric}) \times 100}{\text{wt of dry fabric}} \qquad [15.1]$$

where % conc. is the concentration of the finishing chemical in the applied solution or emulsion expressed as percentage by weight (wt/wt).

To determine the amount of supplied chemical or finish added to the fabric, the '% add-on' is given by Equation [15.2]

$$\%\text{ add-on} = \frac{\%\text{ pick-up} \times \%\text{ conc. of finish in the solution}}{100} \qquad [15.2]$$

$$\% \text{ conc. of finish in the solution (wt / wt)} = \frac{\text{conc (g / L)}}{10 \times \text{density (g / mL)}} \quad [15.3]$$

$$\% \text{ add-on (solid)} = \frac{\% \text{ pick-up} \times \% \text{ conc. of finish in the solution} \times \% \text{ solid content}}{100 \times 100} \quad [15.4]$$

Since most finishing formulas are given in terms of grams per litre (g/L), Equation [15.3] can be used to convert the g/L concentration to percentage by weight.

When the actual solid level added to the fabric is desired, the percentage of solids add-on can be calculated from % solid content in finishing chemical using Equation [15.4] (Schindler and Hauser, 2004).

Typically, for uniform application the wet pick-ups are maintained in the range of 70–100% during chemical finishing by padding. Subsequently it is necessary to remove of large amounts of water during drying. The evaporation of this water can lead to uneven finish distribution in the dried textile owing to migration of the finish to the fabric surface during drying. The high rate of evaporation at the fabric surface leads to movement of the finish solution from the wet fabric interior to the dried fabric exterior resulting in a higher concentration of the finish at the fabric surfaces with a corresponding lower concentration in the fabric interior regions.

The migration of finish may be minimised by reducing the amount of water applied initially. However, too low a wet pick-up can be equally problematic and also lead to uneven finish distribution. 'Critical application value' (CAV) is defined as the minimum amount of durable press finish liquid that can be applied to a given cotton fabric without producing a non-uniform distribution of cross-links after drying and curing. Cellulosic fibres, because of their inherent hydrophilicity, have CAVs in the range of 35–40% wet pick-up. Hydrophobic fibres like polyester can have CAVs of less than 5%, allowing much lower wet pick-ups than hydrophilic fibres.

There are two main types of low wet pick-up applicators. The first is the saturation–removal type where the fabric is completely saturated with the finish liquid and then the excess liquid is removed mechanically or with a vacuum before drying. With the second type, a precise amount of finish liquid is uniformly applied to the fabric using transfer roll, spray or foam techniques.

The factors controlling the amount of solution remaining on the fabric (i.e. %pick-up) after padding (Schindler and Hauser, 2004; Tomasino, 1992) and the relevant conditions for higher liquor pick-up (under bracket) are:

1. Fibre type (hydrophilic fibres).
2. Yarn construction (low twist yarn and open end yarn).
3. Fabric construction (loosely constructed fabrics).
4. Wettability (easily wettable fabrics).
5. Squeeze pressure (lower pressure).
6. Hardness of roll covering (softer covering).
7. Immersion time (longer immersion time).
8. Viscosity of the padding liquor (higher viscosity).
9. Surface tension of the solution or emulsion (faster wetting solutions).
10. Temperature of the solution (viscosity and surface tension change with temperature thereby changing pickup-values).
11. Fabric speed (slower fabric travel).

All fabrics have upper and lower wet pick-up limits. Within these limits, adjustments in wet pick-up can be made by increasing or decreasing the squeeze pressure.

15.4.2 Drying and curing

Drying is a process of evaporation of the embedded liquid from the fabric. While the concept of drying is simple, in practice it can be the source of unsuspected problems. For drying to take place, the liquid must be converted to vapour and the vapour must be moved away from the fabric surface. Some of the factors affecting drying rate are air temperature, relative humidity of drying air, and volume of air passing over fabric (air flow). The following factors are to be considered in controlling the cost of drying processes (Heywood, 2003):

1. Mechanical removal of water must be maximised and achieved as uniformly as possible.
2. Drying should not proceed beyond the equilibrium regain of the fabric.
3. Thermal efficiency of convection drying requires that optimum humidity of the drying medium be maintained.
4. For speedy drying, air temperature should be high, but the cost increases if the temperature is increased above 150°C.
5. Air velocity at the fabric surface should be maximised – through-flow drying shows a clear advantage.
6. Losses by radiation and in the exhaust should be minimised – steam supply losses should be eliminated by direct heating within the drying chamber.

Generally speaking, the machinery used to dry fabrics can be used for curing, provided the fabrics are capable of reaching curing temperature. In many finishing procedures, drying and curing are divided into two steps. Each step will have its own individual specified conditions. Sometimes however, no delineation between the two is made. Wet fabric enters the oven and cured fabric exits. It is well to remember that fabric temperature will not approach air temperature until all the moisture is gone. In many chemical finishing procedures, the fabric's actual time/temperature relationship is critical in order to activate the chemical reactions. Therefore, in one-step drying and curing operations, it is important to know when the fabric is completely dry so that the residual time matches the required curing time at that particular temperature.

The thermometer probes generally indicate the temperatures. However it would be prudent to counter check the temperatures with thermal papers periodically to confirm that the temperatures shown by the probes are accurate particularly where temperature parameters are critical like in curing machines. All instruments can go wrong and there must be some verification procedure to ensure that such failures do not affect performance. Similar checks on the temperature of stenter chambers need to be monitored particularly during heat setting and thermosoling. Such precautions are necessary for all finishes involving chemical bonding or thermo-fixations at elevated temperatures.

15.5 Process control in stenter machines

A stenter is the most universal and most expensive fabric drying and finishing machine, but is indispensable for a modern process house. It is also known as a 'tenter' in the woollen industry. The stenter consists of two endless auto-lubricated driven chains, typically 40 to 60 m in length carrying pins or clips to hold the fabric edges while passing through a number of hot-air chambers (3–5, each of about 3 m). Hot air is directed onto the fabric equally from above and below. A stenter has the provision for overfeeding the fabric to allow required shrinkage during heat setting of fabric while the width is increased to the precisely specified value by the chains. The use of clip stenter has declined because of the difficulty of applying overfeed. The stenter speed ranges from 10 m/min for heavyweight furnishing fabrics to 100 m/min for lightweight dress-goods. The speed also depends on the processes carried out in stenter namely:

1. drying,
2. heat setting,
3. weft straightening,
4. curing after application of finishes.

The controllers fitted in a modern stenter may monitor and control the following parameters:

1. chamber temperatures,
2. moisture retention,
3. stretch/shrinkage (over feed),
4. fabric width,
5. fabric weight,
6. padder pressures,
7. exhaust humidity.

Efficiency and uniformity of drying demand attention to the airflow. Powerful fans (one or two) fitted on one side of each chamber pushes air into two tapered (sideways) ducts fitted at the top and bottom of the chamber. The ducts possess precisely designed nozzles through which air is injected at high velocity on both sides of the fabrics passing between the ducts. The slightly cooler air passes through the filter and heating section before returning to the fans.

For optimal thermal efficiency, it is essential to monitor and control the humidity of the chambers. The energy cost increases dramatically when humidity falls below 10% (0.1 kg water per kg of air) (Heywood, 2003). It is possible to control humidity with a fluidic oscillator developed by Mahlo (Merritt, 1982) or by adjusting exhaust dampers. Energy costs can be reduced by 50% by using various heat recovery systems.

Skewing of the product during manufacture alters the structure of textile materials. A distorted product depreciates in useful value. The Orthopac RVMC (Mahlo) is just such a modular process control system for use on stenters. It is a modular system for both fully automatic correction of bow-and-skew and process control. A number of scanners, sensors and lamps are spaced evenly across the passage of the fabric. The structure of the passing weft and courses modulates the light intensity measured by the scanners. The repetitive nature of the passing picks or courses creates a regular light-dark pattern. This in turn generates a modulated signal in the scanner's receiver system. A centrally-pivoted, cylindrical lens in the scanner oscillates to a specific angle in relation to the passing picks. When the lens is parallel with the pick or course, signal modulation is at its maximum. It then decreases as the oscillating lens progressively cuts across the weft line. Unwanted signals that differ from the frequency of the picks, knitted courses or rows of tufts are filtered out digitally, so that the system can analyse the pure bow-and-skew related signals and, from those, automatically compute the weft or course configuration (www. mahlo.com, RVMC-12_84-010242-002_en.pdf). Mahlo also developed OPTIPAC® VMC-12, a modular system for measuring, logging and controlling critical process parameters such as dwell time, thread density, residual moisture, weight/m^2, etc. across the full width of the product, and exhaust air humidity (www. mahlo.com, VMC-12_84-010243-003_en.pdf).

Controlling parameters for the stenter (Shah, www.sulphurdyes.com) are as follows.

15.5.1 Nip pressure

It is examined by checking liquor pick-up. It should be uniform throughout the width and length. To obtain uniformity the necessary action is to check the surface of padding mangle and adjust the pressure (pneumatically or hydrolytically).

15.5.2 Bow and heading (skew) controllers

During processing visual checking is necessary. There should be no bow or heading in the fabric. The synchronisation of photo cell, heading and bowing rollers is to be checked. The hardness and alignment of bowing rollers are to be checked.

15.5.3 Chamber temperature

Checking is by dial or digital thermometer. It is to be set as per the process and quality. The necessary action is to regulate oil supply in the radiator. Proper functioning of solenoid valve and digital controls are to be checked and confirmed.

15.5.4 Dwell time

This may be checked during the finishing process with a stopwatch. Dwell time depends on the material quality and time recommended for the particular process and is to be regulated accordingly.

15.5.5 Overfeeding

Overfeeding is allowed during heat setting so that the fabric can recover from the stress applied on it at various stages of its production by shrinking and becoming dimensionally stable. During overfeeding fabric is fed to the stenter faster than the running speed of the fabric. If the shrinking is to be allowed in width-wise direction then the chains are kept closer and are not stretched.

Method of checking: to check % overfeed a certain length of cloth is marked before heat setting and the length between the marks is measured again after the process.

Action: the required optimum overfeed of synthetic and blended fabric of a particular variety during heat setting is to be assessed beforehand.

15.5.6 Underfeeding

During underfeeding speed of the fabric feed to the stenter machine should be less than the output. This is done to stretch the length of the fabric.

15.5.7 Expanders and uncurlers

During this process, usually the working of uncurlers is to be checked. Pneumatic uncurlers and mechanical uncurlers are generally used.

Standard: there should be no crease on the fabric.

Necessary action: the smooth working condition of the uncurlers and smooth revolution of the expanders are to be checked.

15.5.8 Blower

Proper functioning of the blower during the finishing process is to be checked.

Standard: proper air circulation.

Necessary action: for proper air circulation, the air filters are to be cleaned. The fan direction is to be checked – air is to be taken from out to in.

15.5.9 Width of the fabric

Width is decided by the distance between the chains and it is to be checked at the delivery end.

Proper working of width adjusting shaft is to be ensured.

15.5.10 Leakages of thermic fluid

There should be no leakages. If many small brown spots are seen on the fabric it means that there are small leakages where the fluid falls as a spray. Larger brown spots on the fabric may indicate bigger leakages.

15.5.11 Concentration of the chemicals

The chemicals and their respective concentrations are to be listed.

Necessary action: the required optimum concentrations are to be maintained. Higher concentrations of the chemicals will lead to white patches, called chalk marks, when scratched with nail. Chalk marks may be due to higher concentration of chemicals.

15.5.12 Temperature and viscosity of the finishing bath

Temperature and viscosity of the finishing liquor are to be kept constant throughout the process.

15.5.13 Drying efficiency

It is checked during and after the drying process with a conductometer that is with the help of transducers.

Standard: no over-drying of the fabric, which will lead to high-energy consumption and strength loss. Drying efficiency of 95–98% is expected.

15.5.14 Temperature of thermic fluid oil

The temperature of thermic fluid should be optimum and should be frequently checked during the process with digital thermometer.

Standard: regulate thermic fluid circulation.

15.5.15 Clips and pins

Random inspection of clips and pins is to be done before processing.

Standard: proper working of the pins and clips should be ensured.

15.6 Calendering process

Calendering may be defined as the modification of the surface of a fabric by the action of heat and pressure. It is mainly done to impart lustre and smoothness to the fabric. The finish is obtained by passing the fabric between heated rotating rollers when both speed of rotation and pressure applied are variable. The surface of rollers can be either smooth or engraved to provide the appropriate finish to fabric. The rollers may be made of various materials, from hardened steel to elastic thermoplastic.

A calender is a multi-roll machine (2–16 rollers or bowls). The bowls rotate under adjustable mechanical, pneumatic or hydraulic pressure. The principle of operation is based on at least two adjacent calender rolls, one or more being a:

1. Metal roller (usually steel or chilled iron), the surface of which may be ground, polished, chromium-plated or engraved and can be heated (by steam, gas, electricity) if required, with.
2. Resilient roller.

To prevent damage to the surface of the resilient rolls when particularly thick seams pass through, 'seam detectors' are provided, which cause instantaneous pressure relief. Calendering effects are varied by various factors such as (Rouette, 2000; Heywood, 2003):

1. number of bowls,
2. arrangement of the bowls,
3. bowl composition (e.g. metal or resilient),
4. bowl pressure (50–100 tonnes, varying according to fabric type and width),
5. metal bowl temperature (up to 300°C depending on fabric and effect),
6. running speed,
7. the degree of friction (e.g. up to 300% friction by having an inclined fabric-filled bowl or faster running metal bowl),
8. number of passages,
9. fabric construction (weave),
10. fabric state (dry, damp, wet) or moisture content,
11. fibre content,
12. finish applied.

Objects of calendering are:

- To change the fabric handle – softening or stiffening effect.
- To impart a smooth silky touch to the fabric.
- To compress the fabric and reduce its thickness and porosity.
- To reduce the air permeability by closing the threads.
- To change appearance – increase lustre of the fabric.
- To reduce yarn slippage.
- To increase the opacity of fabric.
- To embossing patterning on the surface.

Various types of calendering are discussed below.

15.6.1 Swissing, normal gloss or simple calendering

In this case, the fabric runs through the nips of several bowls having the same surface speed. As a result, the fabric achieves a lustrous paper-like finish depending on the number and composition of the bowls.

15.6.2 Chasing finish

The fabric is passed through the nips of the calender, over external rollers and back into the bottom nip of the calender. Multiple layers of fabric run through the nip resulting in a thready appearance with soft handle. This is mostly done for linen fabric.

15.6.3 Chintz, glazing or friction calendering

The main difference between a swissing calender and a glazing calender is the use of a gear system to drive the smooth metal roller faster than the softer resilient roller – the peripheral speed may be up to three times higher. The latter machine is also comparatively heavier. Apart from the lustrous effect, the cloth handle can become quite papery and thin. Incorrect fabric conditions (especially incorrect moisture content) can lead to unacceptable handle, which cannot be corrected.

15.6.4 Schreiner calendering

The problem of producing lustrous but paper thin handle by friction calender can be solved by using a 'Schreiner' calender. Schreiner finish, though not fast to washing, produces a fabric more attractive from marketing point of view, often called 'silk' finish. Extremely lustrous fabrics can be obtained with the correct type of cloth and engraving (V-shape or U-shape) of 500 lines per inch at an angle of 20° to the axis. The angle of engraving should follow the approximate angle of the line of the twist of the yarn (left or right inclination at angles of 15–25° from horizontal and 15° from vertical for weft-faced and warp-faced fabrics respectively). Imitation Schreiner finish can also be obtained using bowls with 150 to 200 lines per inch only. A usual production problem is the picking-up of lint and dirt by the fine engraving spoiling the optical effects. Schreiner finishes reduce tensile strength especially with deep V-type engravings. The darker the colour of a fabric, the better is the lustre of the fabric after Schreiner finish.

Process conditions for Schreiner calendering are:

1. Moisture should not be less than the standard regain, which is 9–15% for cotton and it is ensured by pre-damping.
2. Temperature should be between 120°C and 160°C.
3. Nip pressure should be between 3.5 and 5.0 Bar.
4. Speed should be between 2 and 10 m/min.

15.6.5 Embossed calendering

This usually consists of two bowls. The top metal roller is engraved with a pattern and the softer composition roller has a surface that will accept the embossing pattern. The embossed rollers are quite expensive to produce. Originally these calenders were used to produce imitation leather cloths and book cloths. Embossed crepe design rollers easily deformed viscose fabrics to give a family of creping effects.

15.6.6 Moiré calendering

The moiré effect is an optical effect produced when a tightly woven fabric with very fine yarn is subjected to surface pressure that distorts the weave structure by yarn movement or yarn self-compression. The moiré effect can be produced only when the fibres being treated are deformable – wool fibres are not suitable for the purpose due to their high resilience.

Some of the process controlling parameters in calendering are as follows (Shah, www.sulphurdyes.com).

15.6.7 Nip pressure

The traditional method of applying pressure to the bowl set-up was by a system of levers or direct screw loading. The modern method of pneumatic or hydraulic pressure systems has altered the calender set-up. Calendering effect depends on the nip pressure, which is generally 7–9 tons on the rollers. The pressure is checked prior and during calender. The method of checking is nip pressure reading. The compressor valve and pneumatic control valve are to be regularly checked.

15.6.8 Threading

The threading through various calendering rolls depends on the finish required. It is to be checked visually before running of the machine. Proper threading of the fabric is to be ensured.

15.6.9 Damping

Controlled damping of the fabric is done before calendering. It is checked with the feel of the fabric, or with the conductivity meter.

Follow-up action: adjust water spray intensity.

15.6.10 Speed of the roll

Speed is checked during the process with speedometer.

Standard: there is no standard speed, but 40–80 m/min is usually maintained.

Follow-up action: regulate speed.

15.6.11 Width of fabric

This is checked during the process by measuring with tape as per the sort or quality of the fabric. Follow-up action: ensure desired width at the delivery end.

15.6.12 Bowl surface

Prior to running it is checked visually.

Standard: there should be no crack, hole or rough surface or deposition on the roller.

Follow-up action: proper cleaning of calendering rollers is to be carried at the end of each process.

The following parameters are to be accurately controlled for consistency of the calendering process on a day-to-day basis:

1. Pressure and distribution of pressure across the nip – the traditional method of applying pressure to the bowl by a system of levers or screw loading has been largely replaced by pneumatic or hydrostatic pressure. The former gives a more resilient system, while hydrostatic systems are very close to the dead-set screw method. The effect of loading at the ends of the bowls causes the bowls to deflect, resulting in the nip pressure in the middle of the nip decreasing. The remedy is to have a bowl whose diameter varies along its length so that the lifting of the middle is compensated by its greater diameter there.
2. Temperature of the bowls – accurate control of the metal bowl temperature is vital for consistent results and this is achieved by gas, electric, thermal fluid or steam heating. The temperature can vary between ambient for a light smoothing to 190°C for full lustrous calender finish. Modern hand-held infrared pyrometers can assess temperature more accurately as compared with conventional surface probes.
3. Speed and relative speed of the bowls – modern thyristor motor controls give highly accurate speed control as compared to conventional large DC motors or expensive AC motor controllers. The production speed of calenders depends on many factors and varies from a very slow speed of 5 m/min to a high speed of 75 m/min.

Quality control test: shine or lustre may be measured by lustre meter.

15.7 Surface raising and pre-shrinking finishes

Raising, emerising and pre-shrinking are three important machines widely used in textile finishing. They are mechanical finishes and hence, more eco-friendly. Moreover, they are economic as no chemicals are used. The process control for the above mechanical finishing machines are discussed below.

15.7.1 Raised surface finishes

Raised surface finishes are mostly mechanical finishes. Raised fabrics are mainly of two types, namely:

1. Raised from staple-fibre spun yarn fabrics such as woollens, worsteds, or cotton winceyette and velours – it consists of pulling out a layer of fibres from the structure of a fabric to form a pile resulting lofty handle, subdued weave/ pattern and blending of colours.
2. Raising from continuous filament synthetic yarns to form loop fabrics used for nightwear or bed-sheets – in this case loops in the fabric structure are stretched, but are not usually broken.
3. The machine consists of a drum or cylinder, round the surface of which are mounted a number of wire-covered rollers. Most modern raising machines are double-acting having both pile and counter-pile rollers (generally 12 each), depending on the direction of the point of the bent wires. The fabric is transported over the wire points, which penetrate the cloth surface to a depth depending on the relative speeds of cloth and rollers.

The causes of defects in raised knitted fabric as described by Pehl (1991) are as follows:

1. Variation in temperature and humidity condition of the fabric. Cotton fabric processed in warm and dry conditions may be badly creased. Such fabric may be pre-wetted and redried.
2. The cloth tends to cling to the pile rollers if the pile action is much greater than the counter-pile rollers. The cloth becomes very tight on the feed side and slack on the backside resulting creasing. The machine should be reset to a more balanced action.
3. Bad setting of the cleaning brushes can do a lot of damage – uneven raising should be corrected by regrinding or replacing the wire.
4. Changing fabric width may produce lines due to wire damage at the selvedge of the previous cloth.
5. Lateral stripping can be caused by yarn variation, which may not show up before raising.
6. Streaky or patchy raising may be due to traces of finishing agents.

The construction of knitted fabrics has an important bearing on the effectiveness of the raising process. The loops should stand erect without twisting, so twist factor (English) must not be higher than about 3.8 mm. Loop height should be 3.0–3.8 mm unlike normal terry customarily having a loop height of less than 2.5 mm. Usually raising is carried out after dyeing and drying.

Emerising

Emerising, sueding or sanding is a raising process in which the fabric in open-width is passed over one or more rotating emery-covered rollers to produce a suede-like finish. The major effect of emerising the fabric is the production of a very low pile – that is, short fibres protruding from the fabric surface. The handle after emerising depends on the fibre(s) present, linear density of the fibre and the intensity of the emerising. The handle becomes much softer, especially when micro fibres and chemical softeners are used and to provide a peach-skin finish. Emerising of microfibre fabrics should be carried out prior to dyeing.

The most versatile emerising machine is multi-roller type having 4–8 rollers covered with emery papers and driven independently in clockwise or anti-clockwise directions. Surface character of the rollers can be varied widely.

Ideally, the emery-covered rollers function as a cutting tool, severing the protruding fibres surface to a velvet-like, very short pile or nap. The effect on the fabric may be fine or severe depending on the emery grade (or grain) size. Microfibre fabrics are usually emerised with fine grade emery paper, followed by emerising with coarse grade emery paper, while for many fabrics, the opposite order gives more satisfactory results.

In a multi-roller machine, an emerised effect depends on the following parameters (Heywood, 2003):

1. Number of rollers in operation.
2. Direction of rotation of the rollers (i.e. with or against the fabric) – generally first and third roller run in the opposite direction to fabric passage, while second and other even rollers run in the same direction.
3. Fabric construction and tension – a tight fabric construction in plain will be more difficult to suede or emerise than 2/1 or 3/1 twill where the long weft float can be used to enhance surface fibre development. The tension must be adjusted to suit the particular fabric being emerised or sueded.
4. Fabric wrapping angle on the rollers.
5. Fabric speed (12–15 m/min for micro fibre fabric, 15–25 m/min for woven fabric using spun yarn and 10–20 m/min for similar knitted fabric).
6. Grade of abrasive grit used in emery paper covered rollers – a relatively coarse grade of 80–100 allows the weft threads to be caught and lifted by

the rollers producing a dense, long pile. The grains having size of 280–320 produces short and dense naps on lightweight ladies' outerwear fabrics having weight of 100–180 g/m^2. The grain size is increased to 400–600 for emerising finer micro fibres of polyester and nylon. Still higher grain sizes of 600–800 are rarely used as they exert polishing rather than emerising action.

7. Additional device that may be fitted for sueding back side of fabric.
8. Single-roller emerising or sueding machines generally consist of one abrasive-covered metal roller (optionally water-cooled) and one rubber-covered compression or pressure roller. The single-roller emerising machine is less productive – typical operating speed for microfibre fabrics is about 7.5 m/min. The single-roll machine, however, is used especially on fabrics with terry loops on the face that must be broken and also on difficult styles where the fabric surface must be effectively shaved or polished.

Multi-roller emerising machines may be cylindrical or of slatted design.

15.7.2 Pre-shrinking finish

In the past relaxation drying was the only way in which fabric shrinkage could be reduced. In the so-called London shrinkage the fabric is placed in contact with a dampened wrapping material under tensionless condition for a few days followed by drying by hanging. For virtual elimination of length shrinkage, however, it is necessary to resort to the technique called 'compressive shrinkage'.

Sanforizing and Rigmel are two methods of producing anti-shrink cellulosic materials developed by Sanford Cluett (USA) and Wrigley and Melville (UK) respectively. The methods are based on the fact that in an elastic material sinuous waveform, the convex surface is extended and concave surface is contracted. If cotton material is placed on the extended crest of the wave formed by the elastic material and moves with it into the contracted portion, being held firmly in contact with the elastic material during the movement, then the cotton material will be contracted by compression. Both methods very judiciously utilise a thick blanket for a closing up of the warp. They both allow the weft closing up to take care of itself. The fabric is not stretched weft-wise or set to an unstretched width in the stentering operation as a final stage in fabric finishing. Both processes pay particular attention to warp shrinkage, as most fabrics shrink more in the warp than in the weft and in garment warp shrinkage is usually more important (Marsh, 1979). The 'Sanforized' mark indicates that a fabric has been treated so that it will not shrink or stretch more than 1% or 2% on wetting.

Some process controlling parameters in a Sanforizer as described by Shah (www.sulphurdyes.com) are as follows.

Damping

Damping is done by sprinkling water over the fabric prior to compress it. It is checked by the conductometer.

Standard: it should be optimum. The fabric should not be too wet, it should be just damped. The spray intensity is to be adjusted.

Temperature

Shrinking is induced at the higher temperature of a Palmer drier. The temperature is checked by thermometer. The standard range is 140–160°C. The steam supply is regulated to get required temperature.

Width of the fabric

After shrinking the width of the fabric is to be checked at the delivery end manually. Dwell time and % shrinkage are to be adjusted.

Speed

During pre-shrinking the speed is generally kept in range of 20–40 m/min. The speed should be uniform throughout the batch.

Belts and blankets

The Sanforizer belt is made up of rubber and the blanket, which is a part of the Palmer unit is made up of wool. The surface of belt is to be checked before and during finishing.

Standard: it should have smooth surface. Correct surface of the belts and blankets is be maintained by regular grinding.

Shrinkage

After pre-shrinking the residual shrinkage is measured by boiling water test.

Cooling of belt

Water is sprayed for cooling and protecting the belt. As the belt is in constant contact with the heated roller, it must be cooled. During running, the temperature of the belt should not be more that 15–20°C above room temperature and the belt should be cooled to room temperature after completion of the process.

Standard: it must be always cool. Chocking of sprayer nozzle heads is to be checked from time to time.

15.8 Finishing with alkali

Strong caustic soda solution is used for mercerisation of cellulosic materials as well as for weight reduction of polyester to impart silk-like softness.

15.8.1 Mercerisation

Mercerisation is a treatment that gives lustrous appearance to a cotton fabric or a cotton thread. The process is applied to materials like cotton and hemp. The process was devised in 1844 by John Mercer of Great Harwood, Lancashire, England, who treated cotton fibres with sodium hydroxide. Mercerisation alters the chemical structure of the cotton fibre. It results in the swelling of the cell wall of the cotton fibre, increase in the surface area and reflectance, and gives the fibre a softer feel.

The cellulosic materials in yarn or fabric form are treated with a concentrated solution of caustic alkali whereby the fibres are swollen and the following properties are modified:

1. The strength and dye affinity of the materials increase.
2. The handle is modified.
3. If the material is stretched during or after the treatment, the lustre of the material is enhanced.

Some controlling factors in mercerisation are described below:

Moisture control

Drying cylinders are kept before the mercerisation tank to have the same moisture content in the fabric throughout for uniform results. Anther technique is wet on wet mercerisation, in which fabric is pre-wet but it requires high precaution.

Standard: Free from moisture.

Method: By sufficient steam in drying cylinder.

Caustic soda in padding solution

Standard: 25% or 50–52°Tw.

Method of checking: Twaddle meter or titration.

Necessary action: Adjust the concentration according to the requirement. At a later stage the concentration in the padding trough may increase due to the impurities from the fabric such as thickening agent, etc.

Temperature of padding solution

Ideally it is carried at room temperature. If the temperature is higher, then it is because the moisture in the fabric is more. Water and caustic soda lead to an exothermic reaction that will increase the temperature. So dry the fabric properly. If the temperature still increases then check the water cooling line.

Dwell time

45–60 s.

Wet pick-up

Wet pick-up is generally 120–125% but it should be perfectly uniform throughout the width and length. Pick-up is examined by comparing the original weight of fabric with fabric weight after padding. Pick-up is checked randomly.

Washing

The first compartment after mercerisation tank is the recuperator. Here the concentration of caustic soda should not be more than 10°Tw. If the concentration is very low then take less water (because if more water is present then caustic soda associates itself more with water molecule and its size becomes bigger which is very difficult to remove from the core of the fibre while washing and the washing may not be successful).

Action: Adjust the flow of water.

Temperature of recuperator

Here the hot washing is carried out. Live steam is blown in water. The temperature should not be less than 90°C.

Souring

This treatment is carried out when bleaching is performed on the fabric. In grey mercerisation souring is not done. Removal of caustic soda is very difficult and hence acid neutralisation is very cheap and easy.

Removal of alkali by washing consumes a large amount of water.

Standard: Extract of fabric must be neutral after the souring treatment.

Fabric after washing: Check the extract of fabric by pH indicator or pH paper

Standard: pH must be neutral. If fabric is acidic then adjust flow of water, if alkaline in nature than adjust souring percentage.

Residual caustic soda

Removal of all caustic soda is very difficult and uneconomical. It should not be more than 1% on fabric and if the quantity is more than 1% then check the washing efficiency, efficiency of recuperator and adjust the flow of water.

The most important controlling factor is the concentration of sodium hydroxide solution, which is monitored with a hydrometer. Several hydrometer scales are in use. Mercer used the Twaddell hydrometer scale (°Tw), which is still used in England, USA and elsewhere, especially in the mercerisation of woven fabrics. The mercerisation of circular knits developed in continental Europe, where Baumé scale (B) has been traditionally used and most knit-good mercerisers are familiar with the system. The concentration may also be described in terms of specific gravity (S).

At 15°C, Degree Twaddell (°Tw), $T = 200\ (S - 1)$ [15.5]

The relationship between the specific gravity (S) and Baumé (°Bé) (B) may be expressed as shown in Equation [15.6].

$$B = 144.3 - \frac{144.3}{S} \quad \text{or} \quad S = \frac{144.3}{144.3 - B} \qquad [15.6]$$

The relationship between °Tw and °Bé is shown in Equation [15.7].

$$T = \frac{200B}{144.3 - B} \qquad [15.7]$$

The most important quality control parameters during mercerisation are:

1. Tensile strength (warp and weft way).
2. Tearing strength of the fabric.

15.8.2 Weight reduction of polyester

The weight reduction of polyester by chemical method confers special aesthetic properties on polyester (drape, suppleness etc.). Modification of polyester fibre has been necessitated to overcome several inherent shortcomings in the polyester fibre properties.

The drawbacks of polyester include very low moisture absorption (0.4% moisture regain) and poor dyeability due to high degree of 'hydrophobicity' of the fibre molecular structure and compactness of polymer structure.

The weight reduction of polyester is done by saponification of terephthallic ester by sodium hydroxide in presence of a suitable wetting agent. The partial alkaline hydrolysis causes progressive 'weight reduction' of polyester fibre in the surface, resulting in a small weight loss, whereas its other favourable properties remain almost unchanged or changed for the better. The weight reduction treatment is thus a major breakthrough to obtain 'Skin care' polyester fabric.

Important variables for weight reduction method of polyester are as follows:

- temperature – most critical,
- caustic soda concentration – amount of chemical and volume of bath are important,
- time – governed by temperature and concentration of caustic and weight loss desired,
- catalysts, fibre type.

To achieve 15% weight reduction, at the optimum conditions the treatment was carried with 5% caustic soda and 0.5% wetting agent at 90°C for 60 min (Satees, www.scribd.com/). It is reported (Vigneswaran and Anbumani, 2007) that the treatment of 100% polyester rotor spun yarn (twist multiplier 3.0–4.0) with 5–15% concentrated caustic soda at 60°C resulted in 3–12% weight loss and 7–19% strength loss and treatment at 100°C registered 6–53% weight loss and 17–72% strength loss.

The following points are to be managed (Hayakashi, 2008) during deweighting of polyester:

1. There are many variations in the processes, such as atmospheric pressure method, high pressure method, continuous process, hanging method – the most suitable method should be followed.
2. Caustic soda concentration should be around 4–20%.
3. Use of accelerator (quaternary ammonium salts) – control to be exerted on the type and amount of accelerator.
4. Treatment temperature should be around 95–130°C, and liquor ratio between 1:20 and 1:60.
5. Weight reduction rate is generally restricted between 10% and 30% and precautions should be taken to prevent excess weight reduction.
6. Uneven weight reduction at different parts of the fabric is to be prevented which may otherwise cause uneven dyeing.

7. The deterioration of strength of fabric should be within the tolerance limit.
8. Removal of decomposed polyester oligomers from the deweighted materials which may otherwise cause clogging during beam dyeing, staining on dyeing machine and breaking of knitting needle.
9. No yellowing of fabrics – yellowing may occur due to the presence of residual accelerator.
10. Troubles caused by accelerator such as interaction with disperse dyes causing poor shade reproducibility.
11. Waste-water treatment problems, especially caused by accelerators such as increased BOD, and the effect on activated sludge.

15.9 Softeners

A survey conducted by the Marimount University concluded that 56% of the polled people believed that a soft hand feel relates directly to quality and would be helpful in making their purchasing decision (Biancalani, 2009).

Textile softeners are used to vary the handle of fabrics (similar to filling, stiffening and weighting finishes). They should demonstrate a positive effect on the handle of treated textiles; many textiles require softer, smoother, more supple handle for the best sales potential. They may also serve to improve the processability and wear characteristics of the textiles. Most textile processes, according to their intensity, will more or less lead to the removal of oils and the embitterment of textiles. The softeners must restore the natural softness and suppleness. Often, a softener must also reduce the tendency of textiles to build electrostatic charge. A wide variety of chemical compounds of differing constitutions are used for formulating textile softeners. They often contain a hydrophobic molecular component. The hydrophobic part is usually an alkyl chain of 16–18 carbon atoms length, if it is to contribute a softening effect. The varied application possibilities for softeners are further complicated by the variety of subjective assessments possible with so many different chemical types.

The requirements profile for textile softeners are as follows (Rouette, 2000):

15.9.1 Influence on textile properties

- Textile characteristics: handle, volume, softness, fall, odour.
- Mechanical properties: stretch, elasticity, abrasion resistance, tensile strength, tear strength, smoothness, pilling tendency, sewability.

- Functional properties: moisture management (hydrophilic/hydrophobic), antistatic, flame retardant, dirt resistant, sewability, rope crease prevention, antimicrobial.
- Aesthetic properties: colour nuance, fastness, permanence, whiteness, thermal migration.

15.9.2 Influence on production parameters

- Environmental acceptability (manufacture and use): biological breakdown, toxicity, irritant potential, corrosiveness, bath exhaustion, transport.
- Resistant to: electrolytes, water softeners, acids, alkalis, suitability for jet dyeing (foam, shear forces), storage stability (heat, frost), drying and fixation processes.
- Handling: viscosity (suitable for metering), concentrate, stock emulsion, solubility.
- compatibility: bleach liquor, dye liquor, reductive post-cleaning, optical brighteners, synthetic resins, catalysts, chemical finishes.

15.9.3 Quality control tests

- Handle or feel may be tested by the Kawabata system.
- Flexibility may be tested by drape tester.

15.9.4 Cationic softeners

From an exhaust bath, the speed of exhaustion of the cationic finishing agent on to cotton fabric depends on the strength of the positive charge it carries. This, in turn, depends on the pH. At lower pH, cationic softeners carry relatively higher positive charge and therefore are exhausted more rapidly even in cold. At pH 4–5 the exhaustion is almost total. This high rate of exhaustion on cotton is also very undesirable, since it tends to develop uneven spots/stains on the fabric surface. This is due to rushing and exhaustion of the softeners into sites that are easily penetrated and relatively less or none is available for sites that are more difficult to penetrate.

Therefore for different softeners the optimum pH conditions are to be established considering temperature of application and M:L ratio. Fabric construction and geometry also will influence ease/difficulty to penetrate. Generally, weakly acidic conditions are recommended by the manufacturers/suppliers of the softeners to achieve uniform and even exhaustion. The exhaust time should normally be 20–30 min.

Table 15.1 Properties of various emulsion types

Property	Macro	Micro	Nano
Size of drops (nm)	150–300	50–150	10 nm
Appearance	Milky, cloudy	Clear, water-like	Transparent gel or easily flowable.
Emulsifier content	Normal (around 3% on the wt. of softener)	Very high (may be equal to wt of solid softener	It is more of emulsifier attached to base polymer so self emulsifiable, external emulsifier is less.
Distribution after application	Mainly on the fabric surface	Inside fibre bundle	Completely inside fibre bundle
Handle	Soft and greasy	Very soft, not greasy	Soft, good drape but dry touch
Stability to shear forces	Moderate	High	Higher than micro
Substantivity	Good	Moderate	Moderate

Source: Schindler and Hauser, 2004.

Secondly, the variation in pH (particularly falling on the alkaline side) on the fabric across the width/length can cause differential exhaustion on the fabric surface. Accordingly the performance in terms of the actual 'finish' characteristics such as softness, lubricity, feel, drape, etc. and wash fastness also will vary. It is recommended that both the fabric substrate and the bath are maintained slightly acidic with safe organic acids.

Incomplete removal of anionic soaps and detergents normally used in the earlier soaping operations results in the cationic finishing agent forming a complex with the anionic soap/detergent and causing precipitation and thus diminishing the softening effect. This point is often neglected. A proper rinsing cycle after soaping is required to minimise this problem.

Softeners are water insoluble and are marketed in the form of emulsions. Microemulsions of softeners, mainly amino-modified silicones, give special softening affects. Due to their high emulsion stability, they are very suitable for applications with high shear as in jet or package dyeing machines. Unstable emulsion deposits on equipment and fabrics are very difficult to remove. The properties of various types of emulsions are shown in Table 15.1. The macro softener has made the fabric less flexible and less hydrophilic in nature than the same with the micro. The micro softener has imparted maximum flexibility and the nano softener the best hydrophilicity (Roy Choudhury, 2012).

Some properties of softener emulsions are as follows (Schindler and Hauser, 2004):

1. Emulsion stability – highly stable normal (not micro or nano) softener emulsions cannot provide a high degree of softness. The emulsions of moderate stability give better softness probably because small softener droplets deposit on the fibre surface. Unstable emulsion causes stains.
2. Reactive softeners – some softeners have functional groups (e.g. N-methylolated amines) that can react with the corresponding groups of textile fibres (e.g. hydroxyl group of cellulose). They provide a highly durable finish combined with typical merits–demerits of cross-linking chemistry.
3. Thermomigration of dyes – some hydrophobic softeners are solvents for disperse dyes and thereby increase thermomigration causing poor wash, crock fastness and staining of adjacent yarns.
4. Yellowing – the oxidation of cationic softeners or amino-modified silicones may cause yellowing of undyed finished fabrics. Yellowing may also arise due to interaction of cationic softeners and anionic fluorescent brightening agents. Careful selection of softeners and use of dispersing agents are essential for finishing of undyed fabrics.
5. Effect on dyeing shade: the use of silicone results in deeper shades as in the case of wet textiles due to lower refractive index of silicone (1.43) as compared to those of cotton (1.56) and nylon (1.57). Due to high reflection from a polyester surface arising from smooth surface and high refractive index (1.63), a higher amount of dye is required for obtaining deep black shade. The problem is more critical with microfibres having larger surface (about double that of normal). Deeper black and other shades may be obtained on polyester micro fibres by applying silicone, preferably amino-modified finish. The increased thermomigration by silicones may be controlled by avoiding overdosing and drying over 120°C.

15.10 Resin finishes

For the formation of wrinkles/creases in fabrics, the forces distorting the fabric must be transmitted to the individual fibres. The forces must then place a strain on the individual fibres and distort them. Some of the physical factors for wrinkling are set out below (Tomasino, 1992).

15.10.1 Fibre properties

Fabrics made from fine cotton fibre do not wrinkle as badly as fabrics from coarse fibre. This is because the bending radius of curvature is greater for a thick fibre than for a thin one. The greater the radius of fibre, the greater is the stress on the polymer chains. Fabrics made from fine Egyptian cotton wrinkle less than those made from coarse fibres.

15.10.2 Yarn properties

Fabrics made from high twist yarns wrinkle more than those made from low twist yarns. For low twist yarns, the distortion stresses are dissipated by the physical rearrangement that takes place as adjacent fibres slip by each other. The stress is dissipated before it can affect the individual fibres.

15.10.3 Fabric properties

Tightly woven fabrics wrinkle more than loose structured fabrics. In a loosely constructed fabric, the yarns can move as they respond to the wrinkling forces. The individual fibre is spared. Woven fabrics wrinkle more than knits. The knit loops allow an even greater freedom of yarn movement again sparing the individual fibre.

Resin finishing improves wet and dry crease recovery and crease retention. The chemicals used are thermosetting resins and catalysts. The resins may be nitrogenous, such as dimethylol urea, melamine formaldehyde, DMEU, DMDHEU, etc., or non-nitrogenous, such as dichloropropanol. The catalysts used are mineral and organic acids, metal salts, etc.

Crease resistant finishes are applied to cellulose fibres (cotton, linen and rayon) that wrinkle easily. 'Permanent Press' fabrics have crease resistant finishes that resist wrinkling and also help to maintain creases and pleats throughout wearing and cleaning. Wrinkle recovery is dependent on the presence of cross-links that hold adjacent molecules together and pull them back into shape when they are distorted.

Resin finish improves the following properties of fabrics:

a) low creasing,
b) high crease recovery angle,
c) easy to iron/non-iron,
d) higher dimensional stability,
e) higher durable press rating.

The desirable properties of easy-care finishing are as follows:

1. high 'durable press' rating, high dry and wet crease recovery angles,
2. minimum shrinkage and minimum loss of abrasion, tensile and tear strength,
3. negligible effect on shade and fastness of dyed materials,
4. no yellowing of white materials,
5. negligible effect on absorbency of the textile materials,
6. low or zero formaldehyde release during application and storage,

7. good resistance to pilling,
8. acceptable fabric handle.

The disadvantages of resin finishing are:

1. loss in tensile strength,
2. loss in tear strength,
3. lowering of abrasion resistance,
4. retention of chlorine as chloro-nitrogen compounds resulting in the damage of fabrics during subsequent ironing. The problem can be overcome by using nitrogen-free cross-linking agents,
5. increase in stiffness,
6. increase in soiling and affinity for soil,
7. loss in moisture absorption,
8. Release of formaldehyde during application and use of finished fabric.

The release of carcinogenic formaldehyde during application and fabric use is highly objectionable. The liberation of formaldehyde depends on the following factors (Lewin and Sello, 1983):

1. Type of cross-linking agent
2. Amount of cross-linking agent
3. Type of catalyst
4. Ratio between cross-linking agent and catalyst
5. Curing temperature
6. Curing time
7. Type of fabric

The various ways of reducing liberation of formaldehyde are:

1. Use of selected cross-linking agent
2. After-wash of the finished fabric
3. Addition of formaldehyde acceptors such as cyclic urea, cyclic carboxylic acid amides and cyclic carbamates in the finishing bath
4. Treatment of finished fabric with formaldehyde acceptors. The BASF fog chamber technique involves a spray-mist (fog. application to the finished textile fabric with a solution of formaldehyde binding substances as mentioned above.

The important effects of resin finishing on fabric properties are:

1. Crease recovery – a linear relationship between crease recovery and the logarithm of resin content has been demonstrated for cyclic ethylene

urea and for a modified melamine at solid contents 2–10% on cotton (Foster, 1957). A similar relationship probably exists for dimethylolurea. For a certain amount of resin the improvement in crease recovery is higher on viscose than on cotton.

2. Strength – on application of resin, cotton is invariably weakened, while viscose always gains somewhat in dry strength and substantially in wet strength. It is probably due to differences in the fine structures of the two fibres. Cotton has an intrinsically strong, highly ordered structure in which less ordered regions, the fibrillar boundaries, act as areas in which slippage can occur, allowing transfer and release of strains. Cross-linking in these regions prevents such slippage and results in a less perfect distribution of the applied stress, leading to localised stresses and consequently early failure. Rayon, on the other hand, has an intrinsically weak, less ordered fine structure; the molecular chain is much shorter and the system is easily torn apart. Insertion of cross-links has the effect of increasing the molecular weight and hence the cohesiveness and strength increase as a whole. With the introduction of a very large number of cross-links (which may not arise during normal resin treatment), the structure becomes hard and brittle.
3. Elongation – due to resin treatment a large reduction in elongation (50% or more) occurs for both cotton and rayon with consequent loss in tearing strength. Simultaneous application of softeners and lubricants restores tearing strength completely for viscose, but restoration on cotton largely depends on fabric construction and resin content.

The three methods of application of resin on textile materials are as follows:

1. Dry cross-linking (classical, shock and low temperature)
2. Moist cross-linking
3. Wet cross-linking

Comparison of the processing parameters and the effects obtained by the above three resin finishing methods is shown in Tables 15.2 and 15.3, respectively.

The controlling factors in resin finishing are:

- Concentration of resin in pad bath (depends on type of fibre, fabric structure and application). More the resin better is the performance. But the restricting factor is the loss in strength. 30 to 45% loss of tensile strength has to be taken in account.
- pH of the pad bath (should be acidic).
- Wet pick-up (should be as low as possible).

Table 15.2 Comparison of resin finishing methods

Parameters	Classical dry curing	Shock curing	Low temperature curing	Moist-cure	Wet cross-linking
Fabric humidity (%)	0–2	0–1	0–3	7–10	Completely wet
pH of liquor	3.5–5	3.5–5	2–4.5	< 2	< 1
Amount of resin (g/L)	30–80	30–80	150–200	150–250	150–250
Curing temperature, °C	140–160	150–190	130	25–35	15–25
Curing time	3 min	30–10 s	3 min	20–24 h	16–24 h
Shape of fibre	Non-swollen	Non-swollen	Non-swollen	Partially swollen	Swollen
Strength retention	Standard	Standard	Good	Very good	Very good
Wash after curing	No	No	Yes	Yes	Yes

Table 15.3 The effects obtained by the three resin finishing methods

Effects	Dry cross-linking	Moist cross-linking	Wet cross-linking
Stability to shrinkage	Very good	Good	Moderate
Dry crease recovery	High	High to very high	Weak
Wet crease recovery	Moderate to good	Good	Very good
Wash and wear behaviour	Good	Very good	Moderate
Loss of strength	High	Moderate	Low

- Moisture content of dried fabric (should be 5–7%).
- Time and temperature of curing (depend on resin and type of catalyst).
- With strongly acidic catalyst – curing may be carried at lower temperature; with metal salts 5–3 min at 140–160°C respectively is required.
- For better performance the fabric should be made of good quality long staple cotton fibres.
- Performance will also depend on yarn construction and fabric construction (weaving and design).
- The residual alkalinity on the fabric before treatment should be less than 0.04 g caustic soda/100 g of fabric.

The efficiency of curing may be assessed by measuring the nitrogen content of a sample before and after washing (i.e. after removing unreacted resin). The efficiency should be at least 80%.

Quality control tests: crease recovery angle, tensile strength, tear strength, free-formaldehyde content.

15.11 Protection from fire damage and water penetration

Repellency to fire and water are two most important finishes for textile materials. Fire retardancy is an important characteristic of textile materials in order to protect consumers from unsafe apparel. Fire fighters and emergency personnel require protection from flames. Floor coverings, upholstery and drapery also need protection, especially when used in public buildings. The military and airlines industries have multiple needs for fire-retardant textiles. The term flame retardant is used to describe fabrics which will not support combustion and are self-extinguishing. In case of accidental fire, this type of fabric will not contribute to the spreading of flame. The resistance of fabric to water may be of two types – water-proof and water-repellent. In the former type both water and air cannot penetrate into the fabric, while in case of water-repellency, only air can penetrate into the fabric.

15.11.1 Fire retardant finishes

Properties that characterise burning behaviour of the products can be listed as follows:

- Ease of ignition
- Rate of spreading of flames
- Development of heat
- Shrinkage and melting
- Development of smoke measured in terms of smoke density or visual obscuration
- Formation of toxic gases
- Oxygen depletion.

The physical and chemical properties of the fabric that influence burning are weight, weave, density, yarn size, twist, and ply, degree of purification of the cotton, presence of certain dyes, moisture content, characteristics of the fibres in a blend, and type of finishing agent.

The factors which promote the rate of propagation of combustion of textiles (Rouette, 2000) are:

- Rate of the pyrolysis reaction
- Melting behaviour of the fibre material
- Oxygen requirement of the decomposition products during burning
- Weight per unit area of the material
- Nature of the surface
- Fit of clothing (tight or loose)
- Number and nature of the layers of under and outer clothing
- Finishes (e.g. oil-containing preparations).

The effect of dyes is of minor importance in this connection. It is possible that they (e.g. metal complex dyes) may occasionally have a significant wicking effect, which does not significantly increase the rate of propagation.

In testing and evaluating the burning behaviour of textiles, the following variables are important:

1) Minimum ignition time – the time that is required to ignite a test sample of the textile material under defined conditions, that is, to produce persistent self-supporting combustion.
2) Burning time – also known as the after-burn time or the after-burn period, that is, the time over which the test sample continues to burn independently after ignition and removal of the source of ignition.
3) Rate of flame spread – the distance that the flame travels on the burning test sample in unit time.
4) Glow time – also known as afterglow time or afterglow period, that is, the time over which the material continues to glow after the source of ignition is removed.

Quality control tests: inclined and vertical flammability testers, cigarette burning test.

15.11.2 Water repellent finishes

Many terms are used to describe the water-repellency of textile materials, particularly fabrics. However, they are mostly classified into two unambiguous groups – water-proof and water-repellent.

Treated or untreated textile materials that prevent the absorption and penetration into their structure are called water-proof. In practice, a water-proof fabric shows no penetration by water below hydrostatic pressure of 100 cm (10 kPa).

The main differences between water-repellent and water-proof fabrics are:

1) The pores in water-proof fabrics are filled, while they are open in water-repellent fabrics.

Table 15.4 Surface tension and surface energy of a few liquids and a few fibres respectively

Liquid	Surface tension, mN/m	Fabric	Surface energy, mN/m
Water	72.75	Bleached cotton	44
Water with wetting agent	25–35	Wool	45
Coconut oil	40	Polyester	43
Benzene/petrol	26	Polycot	43
Fluoro-chemicals	10–15	Nylon 6,6	46
Paraffin oil	31	Polypropylene	29
Polysiloxane finish	~ 24		

Source: Goyal and Prabhu, 2009.

2) Air and water vapour permeability are zero or negligible in the case of water-proof fabrics, while they are usually high in the case of water-repellent fabrics.
3) Water-proof fabrics are resistant to penetration by water, while water can penetrate water-repellent fabrics under external hydrostatic pressure.

Water-repellent finishes resist wetting. If the fabric becomes very wet, water will eventually pass through.

When a drop of oil in contact with a textile surface forms a contact angle with it, the following three cases are possible:

Contact angle	Property on surface
>90°	no wetting (formation of droplets)
< 90°	surface wets
0°	instant wetting of surface

The contact angle depends on the surface energy of substrate (SE) and the surface tension (ST) of the liquid. Wetting occurs only when ST is less than SE.

ST of some liquids and SE value of some fabrics are as given in Table 15.4.

Hence, it is necessary to design proper structure of water repelling agent keeping in mind the surface characteristics of the material. The structure of the fabric is also equally important from the point of view of the ability of fabric to shed water resting on its surface and its ability to resist penetration of water.

Water repellents commonly available in the market fall under different categories as follows:

- Wax based repellents (20–25% paraffin and 5–10% zirconium salt, aluminium based salts) are cheap and can provide good water-repellency and resistant to water-pressure but they are not wash-fast and cannot impart oil repellency. The breathability of the finished fabric is low.

- Resin based repellents – products of condensation of fatty compound (acids, amides or amines) with methylated melamines.
- Silicon repellents – aqueous emulsions of polydimethylpolysiloxane – the finished fabrics have good water-repellency, water vapour permeability, soft handle, but cannot provide soil- or oil-repellency.
- Fluorochemicals – mostly copolymers of fluoroalkyl acrylates and methacrylates – they provide good water and oil repellency, resistance to washing and dry cleaning, and good soil-repellency but the price level is high.

For water-repellency, the fabric needs to have a rough surface and a loose structure while ensuring best water shedding properties and the right construction for offering resistance to penetration. Fabric for water repellent finish must be free from sizes, detergents, alkali and acids.

There is considerable growth in the use of more expensive fluorochemical finishes which can be applied at low add-on, for example 0.15–0.3% (on the weight of fabric). Fluorochemical-based water repellents give durable finish. The carbon–fluorine bond is extremely strong and stable. Chemical substances containing fluorinated carbons, 'fluorochemicals', have special physical and chemical properties such as chemical stability, thermal stability, and low surface energy. Fluorochemical products provide performance benefits not achievable with other substances. For example, textile fluorochemicals provide oil and water-repellency and stain and soil resistance without affecting the hand, appearance, or breathability of a fabric.

High fabric repellency depends on the following factors:

1) Fine yarns and compact textile structure
2) A thoroughly prepared fabric free from impurities such sizes, lubricants, surfactants with rewetting action
3) Uniform application of a chemical finish to provide a low-energy surface having lower critical surface energy than the surface tension of liquids likely to be encountered.

Three main test methods are used to evaluate the water-repellency of fabrics, namely:

1) Spray tests, to simulate exposure to rain
2) Hydrostatic pressure tests, measuring water penetration as a function of pressure exerted by water standing on the fabric
3) Wettability tests – sorption water by the fabric when immersed in water

Oil repellency is assessed by hydrocarbon resistance test noting contact angle or collapse time of drops of solvents falling on the fabrics. Various solvents namely n-hexadecane, n-tetradecane, n-decane, n-octane, n-heptane etc. may be used for the testing purpose.

15.12 Anti-pilling finish

Experience has shown that generally pure cotton fabrics are not prone to develop objectionable pilling. Some cotton fabrics may become hairy and some pills may form, but the rate of pill break-off is greater than the rate of formation leaving the surface void of pills. The reason for this is because cotton anchor fibres are relatively weak. Experience has also shown that some fabric constructions made with polyester/cotton yarns will exhibit severe pilling. When these pills are magnified, the anchor fibres are seen to be polyester filaments. The strong polyester anchor fibres do not break off easily so pills continue to build-up and not wear away.

The following factors affect the formation of pills on the textile materials (Tomasino, 1992).

15.12.1 Fibre variables

1) Fibre fineness – yarns made from fine denier synthetic filaments (≤1.5 dpf) pill more than yarns made from coarse filaments (≥ 2.5 dpf). This is because yarn twist imparts greater cohesive forces onto larger diameter fibres than finer fibres. As the filament denier decreases, the total number of synthetic filaments in any given weight percentage of a blend increases. This creates many more fibre ends that serve as anchor fibres. Finer fibres, because they are more in number, will move more easily through a yarn assembly than stiff thicker fibres which are in fewer number.
2) Fibre strength – fibre tenacity is a major contributor to fabric pilling because fabrics made from weaker synthetic fibres pill less than fabrics made from their stronger counterparts. Synthetic fibre producers offer pill resistant varieties that are based on lower molecular weight polymers, resulting in lower tenacity and flex life. Low flex life alone is not enough to produce pill-free fabrics; fibre migration must also be controlled.

15.12.2 Yarn variables

1) Yarn twist – yarns with low twist will pill more than yarns with high twist. The degree of twist will influence the ability of fibres to migrate to the surface. The lower the twist, the easier it is for fibres to migrate.
2) Hairy yarns – hairy yarns pill more than smooth yarns. Low twist contributes to yarn hairiness.

3) Yarn spinning methods – open end spun yarns pill more than ring-spun yarns because the yarn structure is more uneven, allowing for greater fibre mobility. Air-jet spun yarns with low flex life fibres result in relatively pill-free fabrics. Air-jet yarns have wrapper fibres holding the yarn assembly together. These act to keep the body fibres from migrating to the surface.

15.12.3 Fabric construction

Tightly constructed knits and woven fabrics pill less than loosely constructed knits and wovens. Tighter constructions reduce the migration tendencies of the fibres within the yarns.

15.12.4 Preparation and dyeing

Some wet processes make pilling worse while others provide substantial improvement. Preparation and dyeing processes that overwork the fabric will cause excessive hairiness and lead to poor pill performance. Long preparation and dyeing cycles are especially bad. In some instances pills may arise from just these processes alone. High-temperature dyeing of fabrics containing low flex life, high shrinkage fibres improves pilling performance. Auxiliaries such as carriers also have a positive effect on these fibres.

1) Singeing and shearing: singeing and shearing are methods of reducing fabric hairiness. In many cases, pill ratings are improved because of the reduced hairiness. This improvement may last for some fabrics for many wash-dry cycles; however, for others the onset of pilling is only temporarily postponed. Pills will start to form after several wear-wash cycles.
2) Heat setting: heat setting of fabrics containing thermoplastic fibres is often beneficial in improving pilling performance. For some fabrics, the improvement may be temporary and deteriorate after multiple wash cycles. On the other hand, heat setting is definitely helpful with those fabrics made with synthetic fibres having higher heat shrinkage. When high heat shrinkage is combined with low flex life, heat setting can enhance pilling performance to the point where the fabric appears to be pill-free.
3) Certain softeners, which decrease fibre-to-fibre friction by internal lubrication, such as non-ionic organo-modified silicone microemulsions and amino functional polysiloxanes, result in a decrease in fabric pilling performance.
4) For polyester/viscose blended fabrics, Sanforizing treatment also results in an increase in pilling propensity (Hussain *et al.*, 2008).

15.12.5 Fabric finishing types

Finishing procedures have a pronounced positive effect on fabric pilling. Many fabrics can be improved by selection of proper finishing conditions.

1) Film-forming binders - finishing with film-forming latexes will improve the pilling performance of nearly all fabrics. If the final fabric hand is acceptable, these film-forming finishes may be used for a universal solution of pilling tendency. These finishes reduce fibre migration by bridging across filaments binding them together. About 1.5% solids deposited on the fabric is needed to do much good.
2) Latex binders applied to fabrics utilising low flex life (pill resistant) fibres produce a dramatic improvement in pill ratings. Whereas the base fabric alone is only marginally better than one made from conventional polyester fibres (still rated objectionable), the latex finished fabric is virtually pill-free. The performance of latex finished conventional fibre fabric is improved but not to the same degree as the low flex-life-fibre fabric.
3) Durable press reactants – cellulose cross-linking resins applied to low flex-life fibres also produce dramatic improvement in pill rating. However, they have very little effect on conventional polyester fibres.
4) Fabric softeners – materials that reduce the coefficient of friction between fibres will make pilling worse. Fabric softeners increase the pilling propensity of a given fabric. Those applied after dyeing and/or in the finish bath make matters worse. Laundry added softeners may also interfere with pill resistance. These materials operate on the fibre migration portion of the pilling mechanism making it easier for the filaments to move. Softeners co-applied with latexes tend to overcome the improvements noted with film former. Those softeners that provide a soft hand (silicones, ditallow quats) are the worst offenders.

Quality control test: by pilling tester or abrasion tester

15.13 Other types of finishing: antistatic, soil release, antimicrobial and UV protection

In this section remaining four finishes namely anti-static, soil release, antimicrobial and UV protection are discussed. While the first two finishes are inter-related and mostly applicable for synthetic textile materials only, the other two have increasingly gained momentum in recent years. Static charges are produced during friction in synthetic materials due to their hydrophobic nature which create discomfort to the user as well as helping soils to be retained by the substrate. Antimicrobial finishes are applied to protect against accumulation of microorganisms on textile material which can cause rot and odour formation on textile materials and may also spread bacterial infection. Sun-induced skin damage may be caused by increased exposure to UV light. UV Protection fabrics may save us from such damage.

15.13.1 Antistatic finish

Static electricity is generated in various ways; however, most static generation originates from triboelectric charge generation. When two materials come in contact, for thermodynamic reasons there is a redistribution of electric charges at each surface. Rubbing is not necessary, but it will generally increase charge transfer.

Triboelectric series is the ranking of materials in order of their ability to produce static charge. Air, asbestos, glass, nylon, wool, cotton, wood can produce positive charge while rubber, acetate, polyester, acrylic, polyurethane, Teflon can produce negative charge. When a material from the first group is brought into contact with a material from the latter group both acquire respective charges. The amount of static charge build-up depends on the electrical resistances of the contacting surfaces. Static charge generation is high if the resistances are high. If conductivity is increased the charges will run back along the separating surfaces to the point of contact. In the textile industry, especially in dry or low-humidity weather, electrostatic charges may generate causing repulsion between fibres, roller lapping, ballooning of yarn while running in the machine, discomfort while wearing charged garment or walking over a carpet.

Virtually all textile fibres, except metal and carbon fibres, have high electrical resistivity when completely dry. On increase in relative humidity, the resistivity as well as static charge generation decreases. Due to very low moisture, synthetic fibres are prone to static charge generation – the problem can be minimised by blending with conductive metal or carbon fibres or by finishing with suitable antistatic agents.

The antistatic agent should have the following properties (Heywood, 2003):

1) Easy to apply, readily dilutable and applicable by conventional methods such as exhaust, spraying, kiss-roll method and padding
2) No effect on dyeing, colour and fastness of dyed materials
3) Stable to decomposition and vaporisation and should not diffuse inside the material
4) Low toxicity
5) No yellowing of treated materials
6) No enhancement of soiling
7) Good resistance to wash and rubbing.

The chemical classes suitable for use as antistatic agents include polyglycols, fatty glycol esters, fatty amide glycol ethers, ethoxylated amines, amine

Table 15.5 Relation between half-life time for decay of charge intensity and antistatic property of textiles

Half-life time (seconds)	Antistatic property
0–0.3	Very good
0.3–1.0	Good
1.0–2.0	Satisfactory
2.0–3.0	Sufficient
> 3.0	Insufficient

Source: Schindler and Hauser, 2004.

oxide, quaternary polymers and ammonium salts, phosphoric ethers and esters, etc.

Antistatic effective chemicals are largely chemically inert and require thermosol or heat treatment for fixing on polyester goods. Numerous types of chemicals are used as antistatic finishes, including polyglycols, fatty polyglycol ethers, ethoxylated amines, amine oxides, quaternary compound and polymers and phosphoric esters. Thermostable antistatic agents also have a good soil-release action.

Fabrics are generally treated with antistatic agents by padding followed by drying in a stenter or a suitable oven.

Quality control test: conductivity meter (Honestometer). The surface resistivity is easy and relatively reproducible to determine. But it is a static test with no information about the important charging and discharging behaviour of textiles. Therefore, combination with a charge dissipation test is favoured.

A charge dissipation test was performed with static voltmeter. The fabric was fixed vertically and charged with direct voltage or by rubbing with a glass rod. Maximum charge generated, as well as decayed, was measured. The time necessary for charge to fall to one half of its maximum value, called field intensity half-life time, was recorded. The relation between half-life time and the antistatic property of the textile materials is shown in Table. 15.5.

15.13.2 Soil-release finish

The purpose of soil-release finishing for textile is to facilitate the removal of soiling matter during laundering. Soils deposited on textile materials have complex composition containing both oily and particulate matters, namely clay, soot, street dirt, etc. Electrostatic attractive forces are primarily responsible for the deposition of air-borne particulate soils.

The degree of retention after laundering of a mixture of clay and six fats representative of human sebum on bleached cotton sheeting has been studied, with the conclusion that the presence of fats leads to a build-up of soiling, possibly as a result of the fats acting as adhesives for the clay (Sontag *et al.*, 1970).

Work on the deposition of fatty soil on polyester-fibre fabric during washing resulted in the following recommendations (Grindstaff *et al.*, 1970):

1) Non-ionic surfactants should be used rather than anionic.
2) The surfactant concentration should be kept above the critical micelle concentration when appreciable quantities of fatty soil are present.
3) The temperature of the first rinsing water should be kept well below 60°C.
4) Long rinsing cycles should be avoided.

Factors affecting soiling and soil-release (Heywood, 2003) are as follows:

Nature of soil

One of the most frequently occurring difficult-to-remove soils is sebum (shirt collar soil), which contains mostly fatty acids, triglycerides and fatty alcohols. Normally it can be removed by washing below 50°C. But if allowed to remain on the fabric for long time, fatty acids tend to oxidise to produce viscous material that requires higher temperature for its removal.

Nature of substrate

Soil removal from hydrophobic synthetic fibres is far more difficult than from hydrophilic fibres. The extent of soiling largely depends on the area of contact between soil and substrate. Hence, fibre geometry, yarn and fabric parameters play important roles in soil-release.

Effects of finishing

A linear correlation was observed between the extent of soiling by oily soil and the hydrophobicity of the fabrics. Hydrophobic finishes such as durable press finish, silicone and fluorocarbon finishes increase soil-retention. Soiling increases with decreasing moisture regain of cotton, particularly in the region from 5% to 4.3%, by cross-linking with DMDHEU. This corresponds to increase in nitrogen content of 1.6% to 3.4% respectively (Fiebig and Rezk, 1973). Durable press finish decreases the ability of cotton to resist both aqueous and oily soiling.

Soil-release finishes are mostly amphiphilic polymers, containing both hydrophobic and hydrophilic groups. The appropriate hydrophilic-hydrophillic balance is necessary for soil-release activity.

15.13.3 Antimicrobial finishing

Microbes, namely bacteria, virus, fungi and yeast, are present almost everywhere. Whereas human beings have an immune system to protect against accumulation of microorganisms, material such as textiles can easily be colonised by high numbers of microbes or even decomposed by them. Textiles are carriers of microorganisms, such as pathogenic bacteria, odour generating bacteria, mould and fungi. If the environment is favourable, certain bacteria can grow in a short time from a single germ to millions.

Microorganisms are small living forms of life that we cannot see with the naked eye, such as:

1) Tiny bacteria having rod, spiral or ball shape
2) Primitive plants such as fungi
3) Single or multiple cellular plants such as algae
4) Insects such as dust mites

An ideal antimicrobial treatment of textiles should satisfy the following requirements (Vishnu and Bhagyashri, 2010):

1) Effective against a broad spectrum of bacterial and fungal species
2) Non-toxic to consumers, no irritation or allergy to the user
3) Durable to dry cleaning, laundering and hot pressing processes
4) No negative effect to the fabric quality (e.g. physical strength and handle) or appearance
5) Compatible with other finishing agents
6) Not destructive of the resident flora of non-pathogenic bacteria on skin of the wearer.

Antimicrobial finishes may be of three types:

1) Rot proofing finishes provide long term or short term material protection against physical deterioration.
2) Hygiene finishes are concerned with the control of infection and unwanted bacteria; a specialised development is the prevention of dust mites.
3) Aesthetic finishes are used to control odour development and staining.

Some examples of antimicrobial agents are cross-linked polyethylene glycol, quaternary ammonium compounds (chains of 12–18 carbon atoms), triclosan, silver based antimicrobials, etc. Chitosan, neem, alovera and clove oil are some examples of natural antimicrobial agents.

Antibacterial efficiency may be evaluated by two types of tests:

- Tests performed on solid medium (diffusion tests)
- Tests performed in liquid culture (suspension tests).

15.13.4 UV protection finish

In recent years, consumers have become increasingly aware of the need for sun protection, which is related to the incidence of sun-induced skin damage and its relationship with increased exposure to UV light. UV radiation can lead to acute and chronic reactions and damage, such as acceleration of skin ageing and sunburn. Billions of people live on the earth, and each has his or her own colour of the skin. In the human body the skin colour depends on the quantities of melanin, carotene and oxygenated or reduced haemoglobin combined in the skin, as well as thickness, water content, etc. Among other factors, the quantity of melanin distributed in the skin determines its fairness or darkness and greatly influences the human complexion; at the same time melanin plays an important role in minimising the damage that UV rays cause in the skin.

Protection of the skin against the action of solar radiation is a relatively new objective of textile finishing, since the textile does not always guarantee adequate protection. The unfinished fabric is limited as a guarantee of adequate protection. Thus, a special additional sun protection finish is applied in the form of UV stabilisers. Electromagnetic radiation of wavelength between 150 and 400 nm are classed as ultraviolet rays. Approximately 10% of sun's energy is in the form of ultraviolet radiation. Atmosphere absorbs most of the noxious radiations emitted by the sun, only 5% of the harmful radiations reach to the surface of the earth.

UV absorbers, such as benzotriazole and phenyl benzotriazole, molecules are able to absorb the damaging UV rays of sunlight. UV absorbers convert UV energy into harmless heat energy. This transformation is regenerative and can be repeated indefinitely.

Solar protection factor (SPF), can be defined as the ratio of the time taken for a patch of skin to develop erythema, or reddening, with and without protection. The larger the SPF, the more protective is the fabric to UV radiation. Typically, a fabric with an SPF of > 40 is considered to provide excellent protection against UV radiation. In Europe and Australia, SPF is referred to as the ultraviolet protection factor (UPF). The SPF of textile fabrics can be assessed by measuring transmission through a screen (a fabric or chemical) with the aid of a spectrophotometer, after necessary correction for the nature of the solar spectrum in comparison with instrumental characteristics and to compensate for the fact that not all portions of the spectrum are actually harmful.

Cotton and silk fibres offer little protection to UV radiation since the radiation can pass through without being markedly absorbed. Wool and polyester, on the other hand, have significantly higher SPF since these fibres will absorb UV radiation. Nylon falls in between these extremes. One factor influencing nylon and polyester absorbance is the presence of the delustrant TiO_2, a material that strongly absorbs UV radiation.

Tight microfibre fabrics provide better UV protection than fabrics made from normal sized fibres with the same specific weight and type of construction. Many dyes absorb UV radiation as well as visible light. An SPF of 50 or higher can be achieved by dyeing cotton fabric to a deep shade. Since fashion and comfort often dictate the use of lightly coloured fabrics for summer apparel, the need arose for UV absorbing materials that could be applied to fibres to provide the desired SPF in light shades. Dyestuff and auxiliary manufacturers have responded by developing a variety of materials suitable for use as UV protection finishes. Benzotriazole derivatives and oxalic acid dianilide derivatives provide UV absorbers for natural fibres, while phenyl salicylates, benzophenones, benzotriazoles, phenyltriazines and cyanoacrylates provide UV absorbers for synthetic fibres.

UV absorbers should have sufficient wash fastness and light fastness as dyestuffs. Laundering trials should be carried out with all new formulations to confirm that the claimed UV protection is actually active during the life of the garment. If a UV absorber is also present in the fibre, the brightening effect from the OBA can be greatly diminished or even absent. Proper choice of an appropriate OBA can minimise this problem.

15.14 Wool treatment and enzyme finishes

This section deals with decatising and crabbing of wool fabric and enzymatic finish called bio-polishing. Decatising, also known as crabbing, blowing, and decating, is the process of making permanent a textile finish on a cloth, so that it does not shrink during garment making. The word comes from the French word 'décatir', which means to remove the 'cati' or finish of the wool. Though used mainly for wool, the term is also applied to processes performed on fabrics of other fibers, such as cotton, linen or polyester. Crabbing and blowing are minor variations on the general process for wool, which is to roll the cloth onto a roller and blow steam through it.

Crabbing is a treatment used to set the cloth and yarn twists permanently in woolens and worsted goods. In the crabbing process the fabric goes through hot and then cold water in order to set the cloth.

Bio-polishing removes the protruding fibers of a fabric through the action of an enzyme. Enzymes, such as cellulase for cotton, selectively remove protruding fibers. These enzymes may be deactivated by an increase in temperature.

15.14.1 Decatising of wool

Pressure decatising is a setting process for wool and wool-mix fabrics that provides conditions for achieving high levels of set. The principal objectives are:

- To stabilise lustre
- To set the finished aesthetics of the fabric.
- To stabilise dimensional stability

The process involves winding the fabric under controlled tension onto a perforated beam previously covered with a wrapper) interleaved with a decatising wrapper. When the batch is complete, it is loaded into a steaming chamber, sealed and steamed under pressure. Typical conditions for wool fabrics are 0.8–1.0 bar pressure at a temperature of 120–125°C for two minutes, followed by cooling to arrest the setting process. High levels of permanent set can be imparted, producing a 'permanent' finish. This is important for stability of finish (to steaming during making up), for control of fabric dimensional properties, and for handle and lustre. It is important that wool fabrics are at the correct pH and moisture content prior to all pressing and decatising procedures, particularly pressure decatising. Fabric pH is ideally around pH 6 – stronger acidity reduces set, and stronger alkalinity may cause yellowing. Moisture content for all-wool fabrics is ideally 12–15%, since low moisture levels can result in low levels of set.

Decatising effects can be influenced by the following factors (Rouette, 2000):

- Pressure, generated by backcloth tension and pressure from the pressure roller during the winding of the decatising cylinder
- Steam pressure
- Treatment duration
- Fabric moisture during decatising
- Cooling after decatising
- Condition of the backcloth.

If these parameters act for an appropriate period they influence the surface pattern and the handle of the product according to the following rules:

1) Brilliant sheen (permanent) and solid handle: high steam pressure, medium steaming time, long cooling in tightly wound state, satin backcloth

2) Brilliant sheen (permanent) and soft handle: high steam pressure, long steaming time, short cooling in less tightly wound state, satin backcloth
3) Lower sheen (permanent) and solid handle: medium steam pressure, medium steaming time, long cooling in less tightly wound state, satin or molleton backcloth
4) Low sheen (permanent) and soft handle: medium steam pressure, long steaming time, no cooling, winding with low tension, satin or molleton backcloth

Modern pressure decatising machines offer a variety of steam cycles and choice of steam directions (in to out, out to in) to produce variations in handle, lustre and finish. Computers are used to store and retrieve data, including fault monitoring. Most modern decatising machines offer continuous processing. They generally provide lower levels of set than batchwise pressure decatising (Myers, 2002).

15.14.2 Crabbing of wool

Setting or crabbing is mainly applicable to worsted fabrics and is required to relax and set the strains introduced into the yarns and fabric during spinning and weaving. If some weave structures are not set in this way, they may be susceptible to the formation of distortions (e.g. 'crows-footing') during subsequent wet finishing. The crabbing operation is carried out in the presence of heat and moisture, during which the intermolecular bonds in the wool are broken and then reformed in a more relaxed configuration. Setting is arrested by shock cooling.

In a traditional batch crabbing machine, the fabric is wound onto a cylinder (covered with a cotton wrapper) which is rotated, whilst half immersed in hot or boiling water, for a predetermined time. To ensure even treatment, the fabric is reversed and the treatment repeated. The fabric may then be steamed and finally passed through a tank of cold water to arrest the setting process.

Two basic types of continuous crabbing machine are available. In cylinder types, the fabric is initially wetted through a trough of hot water and then passed around a large, rotating, heated cylinder. The fabric is pressed at high pressure against the heated cylinder by a specially engineered impermeable belt. Special seals resist escape of steam and entry of air at the edges of the belt. Fabric operating temperatures as high as 135–140°C are claimed and superheated steam is created *in situ*, setting the fabric. Setting is arrested by shock cooling.

Superheated water machines differ in design from the cylinder types in having no pressure belt to maintain fabric/cylinder contact and use superheated water to facilitate setting. A possible advantage for fabric quality is that yarns are claimed to fully swell with minimal fabric compacting.

The fabric enters and exits through barometric columns. The temperature of water is around 110°C, although a series of steam battery heaters situated around the main cylinder are claimed to elevate the fabric temperature during its contact time with the cylinder, promoting fabric set.

15.14.3 Enzyme finishing

Bio-polishing is a process to remove the protruding fibres of a fabric through the action of an enzyme. This cellulase enzyme selectively acts on the protruding cellulose fibres and ceases to work after finishing the job by a simple raise in temperature of the treatment bath.

Cellulases are multi-component enzyme systems that are commonly produced by soil-dwelling fungi and bacteria. The effects achieved by treating with these enzymes are pilling reduction, increase in softness and amelioration of handle, surface structure, fabric appearance and dyeing yield. The most important cellulase-producing organisms are fungi of the genera Trichoderma, Penicillium and Fusarium. The cellulases used are chemically complex and consist of at least three enzyme systems working synergistically (Holme, 1998).

Acid cellulases exhibit the greatest activity generally in the pH range 4.5–5.5 at 45–55°C, whereas neutral cellulases require a pH of 5.5–8.0 at 50–60°C. Generally a treatment of 45–120 min is appropriate, as prolonged treatment time may significantly increase the fibre loss. Where acid cellulase predominates, required washing time is short (20–45 min), whereas for neutral cellulase a longer wash time (45–120 min) is required. Excessive cellulose dosage and vigorous agitation may also increase weight loss.

For cotton, restriction of the enzyme to the fibre surface is easily achieved, because cellulose is a highly crystalline material with only a small amorphous region, making the diffusion of enzymes into the interior of the fibre nearly impossible. Thus, by regulating enzyme dosage and choosing the right type of enzyme, the catalytic action of the enzyme can be confined to the surface of cotton and to the amorphous regions, leaving the fibres, as a whole, intact. Surface modification of cellulosic fabrics conferring cooler and softer feel, brighter luminous colour and more resistance to pilling using cellulases, is often known as *bio-polishing* – a term created by Novo Nordisk.

Bio-polishing is affected by many factors. the major ones are enzyme type, type of fabric and process variables. The predominant process variables that control bio-polishing are temperature, pH, duration of treatment, material to liquor ratio, enzyme concentration and mechanical agitation.

The bio-polishing process requires:

Enzyme dosage: 1–5% on fabric weight (depending on the activity of the enzyme)
Liquor ratio: 5–15 L/kg of fabric.
Time: 60–120 min (depending upon the amount of hydrolysis required).
Temperature: 50–60°C.
pH: 4.5–5.5 (For acid stable cellulase)

After treatment, the enzyme must be deactivated either by alkali treatment at pH 9–10 or by increasing the temperature to 70–80°C and giving a treatment for ten minutes.

15.15 Low-liquor finishing

Foam finishing, a low liquor finishing technique, depends on the following factors.

15.15.1 Foam properties

Foam properties are governed to a large extent by the method of generation and by the treatment liquor composition, plus other factors, such as the purpose for which the foam is generated and the method of foam application. The principal properties of foam for the textile finisher are the foaming degree, foam stability, viscosity, wetting power, and the bubble size and size distribution (Elbadawi and Pearson, 2003).

15.15.2 Foaming degree

This is the measure of the extent of foaming. It is an indication of the volume of foam in litres that has been produced from one litre of liquor. The foaming degree is normally expressed in terms of the foam density (g/mL), which is calculated from weighing a known volume of foam. Care must be exercised during sample collection by inserting the foam-discharge hose inside the container to prevent any air pockets from being trapped within the sample (Equation [15.8]).

Foam densities of between 0.07 and 0.14 g/mL and between 0.20 and 0.33 g/mL have been used for foam finishing and foam printing, respectively (Turner, 1981). The selection of the foam density to be used depends on the fabric mass per unit area and is directly proportional to it. It is reported that low foam densities yield low wet-pick-up values (Weber and Tuschen, 1980).

Another expression of the foaming degree is the blow ratio (Equation [15.9]), which is defined as the ratio of the mass of a given volume of liquid before foaming to the mass of the same volume of foam.

The blow ratio varies with the density of the liquor, whereas the foam density is independent of the liquor density. The blow ratio is altered through control of the amount and pressure of injected air or alternatively through the liquor-pump speed. Although blow ratio is widely used as a measure of the foaming degree, it is more usefully considered as a foam generation parameter than a foam property. Moreover, it assumes 100% dispersion of the injected air in the liquor, which cannot be true, because air pockets are often seen at the foam exit.

15.15.3 Foam stability

Foam stability is defined as the time that foam will maintain its initial properties as generated. Foam stability is required during generation, transportation, and application to the fabric and has to be lost thereafter. Foams that are too stable are difficult to collapse; hence penetration into the fabric is poor. On the other hand, relatively unstable foams collapse before the point of application, resulting in an uneven distribution of chemicals on the fabric.

Foam stability can be expressed in terms of foam half-life, which is the time required for half of the volume of liquid contained in the foam to revert to the bulk-liquid phase. The shorter the time, the lower is the stability of the foam. It is reported that foam half-lives vary according to the type of process for which they are used: for instance, 30–45 min for finishing, 30–180 min for dyeing, and 8–10 h for printing (Jerg, 1979).

Another method is the drop-emergence test, in which the time taken for one drop of liquid to emerge from foam is measured. For the determination of the drop-emergence time, a given volume of foam is placed in a measuring cylinder. The measuring cylinder is then inverted and the time noted until the first drop emerges.

Other less-used methods of foam stability are determination of the change in foam viscosity and measurement of the electrical conductance of the foam between two fixed electrodes (Cooke, 1983). As the foam drains, the bubble lamellae become thinner and the electrical conductance decreases, giving a measure of foam stability.

Foam stability can be attributed to several factors, which include the following.

1) Surface viscosity – high viscosity of the liquid film that forms the bubble walls contributes to film strength and durability, owing to the increased resistance to deformation; that is, there is a reduced thinning and drainage rate. However, the higher the surface viscosity, the harder it is to remove the entrapped air and to break the foam.

2) Film elasticity is defined as the ability of the liquid film to resist mechanical disturbance. Film elasticity is influenced by change in the surface tension, which is affected by the type and the concentration of the surfactant. An explanation of this phenomenon is that the surfactant flows from regions of higher concentration to those of lower concentration when the bubble is deformed. This is a self-healing mechanism that helps to maintain more uniform film elasticity and leads to better foam stability.
3) Zeta potential of the film – this is an electric or entropic effect due to repulsion of charged ionic surfactants in the film surfaces. Ionic surfactants are adsorbed at the bubble wall, with charged groups oriented towards the liquid and the nonpolar sections protruding into the air. As the inner and outer bubble walls (which have similar charges) approach each other, because of the drainage of liquid between them, the electrical double layers can overlap, and the resultant repulsive forces prevent collapse of the bubble walls, retarding destruction of the foam.

The other factors that contribute to foam instability are:

1) Buoyancy of the bubbles, which causes them to rise at a rate proportional to the bubble radius and density of the surrounding liquid medium, and inversely proportional to the liquid viscosity;
2) Thinning of the bubble walls as the liquid drains down the inside the bubble walls through gravity and collects at the intersections of the bubble faces, known as Plateau's borders; the rate of drainage, and hence foam stability, is dependent upon the liquid viscosity, and variation in bubble size, which results in pressure differences between bubbles.

15.15.4 Foam viscosity

This is a measure of the foam stiffness or resistance to flow when the bubble walls are still in the liquid state. The viscosity increases as the number of air bubbles increases. It is also affected by the addition of stabilisers and other foam ingredients. Higher foam viscosities result in greater stability through reduced foam-drainage rates. In practice, viscosity modifiers are added to bring about foam viscosities in the range of 2000–3000 centipoise (cP) for textile utilisation. Measurement of foam viscosity is a delicate process, since the foams may be either thixotropic (i.e. they possess the property of showing a temporary reduction in viscosity when shaken or stirred) or pseudoplastic (i.e. they show an instantaneous decrease in the apparent viscosity with an increase in the shear rate) in nature (Shaw, 1980).

There are various methods available for the measurement of foam viscosity, the selection of which depends on the type of foam produced.

Foam viscosity may be measured using a rotary Brookfield viscometer operating at a low shear rate, or by using a coaxial Brabender viscometer or a U-type Ostwald viscometer in a temperature-controlled water bath.

15.15.5 Bubble size and size distribution

Bubble size and bubble-size distribution are important criteria of foam behaviour. Foam of finer bubble size is more stable than foam of the same density with coarser bubbles. As foam ages, the bubble size increases and the distribution of sizes broadens owing to diffusion of air from the smaller (high pressure) to the larger bubbles (low pressure) through the thin lamellae. Uniform bubble size in the range of 50–100 μm in diameter is a necessity for textile applications (Ghosh, 1988).

Observing and measuring the complex structure of foam is extremely difficult, as the nascent bubbles jostle one another to seek a stable arrangement before they burst and collapse. Techniques such as photomicrographs of a foam sample in a recessed microscope slide and freezing the foam have been used to determine the bubble size and distribution.

The size of the photographically recorded bubbles is measured and classified in different categories, and the distribution is characterised by the mean and standard deviation of the bubble size (Kroezen and Wassink, 1987).

A visual technique has also been reported in which foam with a narrow distribution of bubble sizes is selected by visual inspection. This is then laid on a board, frozen, and used as a standard for comparison. This technique is quite subjective.

15.15.6 Foam wetting power

Textile substrates are not readily wettable by foam. The foam has to spread out on the substrate first and then break, releasing the liquid to penetrate the fibres. The wetting power can be assessed by the spreading or penetration rate, which is governed by the foam-drainage rate.

Liquid can drain out of the lamellae films between foam bubbles owing to gravity and surface-tension differences between the least and the most sharply curved parts of the bubble. The drainage rate is affected by the foam density, viscosity, and temperature and by the film thickness.

Rapid wetting by foam, or rather by the liquid released on collapse of the foam upon contact with the textile material, is a very important property where application as textile finishing is concerned. The wetting power of the foam can be assessed by the measurement of either the spreading or the penetration rate.

Foam spreading time is evaluated by measuring the growth in the wetted area. In practice, a measuring cup is completely filled with foam and inversely placed on a standard fabric material. After a specified time, the area of the wetted-out fabric outside the circumference of the original measuring cup is recorded.

15.16 Plasma treatments

Plasma is defined as a state in which a significant number of atoms and/or molecules are electrically, thermally or magnetically charged or ionised. Plasma, in general, refers to the excited gaseous state consisting of atoms, molecules, ions, metastables, and the excited state of these, and electrons such that the concentration of positively and negatively charged species is roughly the same. Theoretically, plasma is referred to as a 'fourth state of matter'.

Plasmas create a high density of free radicals by disassociating molecules through electron collisions and photochemical processes. This causes disruption of the chemical bonds in the fibre polymer surface, which results in the formation of new chemical species. Plasma treatment on fibre and polymer surfaces results in the formation of new functional groups such as $-OH$, $-C=O$, $-COOH$, which affect fabric wettability as well as facilitate graft polymerisation that, in turn, affects liquid repellence of treated textiles and nonwovens.

Gases commonly used for plasma treatments are:

- Chemically inert (e.g. helium and argon)
- Reactive and non-polymerisable (e.g. ammonia, air and nitrogen)
- Reactive and polymerisable (e.g. tetrafluoroethylene, hexamethyldisiloxane).

The potential use of plasma treatments for finishing of fibres, yarns and fabrics is as follows:

- Removing the surface hairiness in yarn
- Hydrophilic enhancement for improving adhesive bonding, wetting and dyeing
- Hydrophobic enhancement of water and oil repellent textiles
- Antibacterial fabrics by deposition of silver particles in the presence of plasma

- Graft plasma polymerisation for producing fabrics with laundry-durable oleophobic, hydrophobic and stain-resistant finishes
- Atmospheric plasma-based graft polymerisation of textiles and nonwovens having different surface functional properties on the face and back side of the fabric
- Flame-retardant coating using monomer vapours (halogen and/or phosphorus) in combination with nitrogen and/or silicone
- Durable antistatic properties using PU-resin and plasma processing
- Shrink resistance of woollen fabrics and preparation of machine-washable wool by plasma processing without resin
- Electro-conductivity of textile yarns by surface plasma deposition.

The advantages of plasma treatment include the following (Chan *et al.*, 1996):

- Modification can be confined to the surface layer (a few hundred angstrom units) without modifying the bulk properties of the polymer.
- Excited species in plasma can modify the surfaces of all polymers, irrespective of their structures and chemical reactivity.
- By the selection of the feed gas to plasma reactor, it is possible to achieve the desired type of chemical modification for the polymer surface.
- The use of plasma can avoid the problems encountered in wet chemical treatments, such as residual chemical in the effluent and swelling of the substrate.
- Modification is fairly uniform over the whole surface.

However, the processing parameters are highly system dependent, that is the optimal parameters developed and optimised for one system usually need to be modified for application to another system. The process and its set-up are complicated. It is very difficult to control precisely the amount of a specific functional group formed on the sample surface (Shishoo, 2007).

The effectiveness of plasma treatment depends on different factors namely (Kan and Yuen, 2007):

1) Nature of gas used
2) Flow rate
3) System pressure
4) Discharge power
5) Duration of treatment
6) Ageing of plasma-treated surface
7) Temperature change during the plasma treatment.

15.16.1 Nature of gas used

The result of plasma treatment depends strongly on the nature of the gas or the vapour used in glow discharge. Most organic, organosilicone or organometallic vapours tend to form a thin film on the surfaces, which are subjected to glow discharge, and the deposition of these films is the main factor modifying the polymer surface. On the other hand, glow discharges of non-polymerising gases, for example noble gas, nitrogen, oxygen, hydrogen, ammonia or water vapour, modify polymer surfaces through processes, such as oxidation, chemical etching, cross-linking and perhaps grafting.

Helium, neon and argon are the three inert gases commonly used in plasma technology. Because of its relatively lower cost, argon is by far the most common inert gas being used. Inert gas plasma has been used for the pre-treatment of substrates for cleaning purposes before reactive gases are applied.

If a plasma reaction is to be carried out with a high-system pressure but at a low reactive gas flow rate, an inert gas can serve as a diluent.

Oxygen and oxygen-containing plasma are most commonly employed to modify polymer surfaces, (it is well known that the oxygen plasma can react with a wide range of polymers to produce a variety of oxygen functional group, including C–O, C=O, O–C–O and C–O–O at the surface).

Nitrogen-containing plasma is widely used to improve wettability, printability, bondability and biocompatibility of polymer surfaces.

15.16.2 System pressure

The system pressure used may affect the energy of the plasma species. If the pressure is high, the probability of collision between plasma species will be increased leading to the loss of energy of the species before interacting with the material.

15.16.3 Discharge power

The intensity of plasma is a combined factor of pressure and discharge power. The breakdown energy necessary to produce plasma varies from one gas to another. Normally, the higher the discharge power applied, the more kinetic energy the plasma species will carry, resulting in strong intensity of plasma action.

15.16.4 Duration of treatment

The duration of treatment plays an important role in the plasma treatment. Generally speaking, the longer the duration of treatment, the more severe the modification of the material surface (e.g. sputtering or etching) will be.

A longer duration will not only affect the material surface but also provide an opportunity for the plasma species to penetrate the interior region of the material. This may alter the morphology of the polymeric material. However, when the treatment duration is too long, this will adversely affect the material and therefore careful control of treatment duration is required.

15.16.5 Ageing of the plasma-treated surface

In general, the concentration of functional groups introduced to a polymer surface by the plasma treatment may change as a function of time depending on the environment and temperature. This is because polymer chains have much greater mobility at the surface than in the bulk, allowing the surface to reorient in response to different environments.

Surface orientation can be accomplished by the diffusion of low-molecular weight oxidised materials into the bulk and the migration of polar function groups away from the surface. Ageing of plasma-treated polymer surfaces can be minimised in a number of ways.

An increase in the crystallinity and orientation of a polymer surface enhances the degree of order and thus reduces mobility of polymer chains, resulting in slower ageing. A highly cross-linked surface also restricts the mobility of polymer chains and helps to reduce the rate of ageing.

15.16.6 Effects of environment

When a polymer is exposed to the oxygen-containing plasma, the surface will change to a high-energy state with increased surface tension, as a result of the formation of polar groups. Various surface studies indicate that when the treated surface is placed in a low-energy medium, such as air or vacuum, the decrease in surface energy is caused by the rotation of the polar groups in the bulk or the migration of low-molecular weight fragments to the surface to reduce the interfacial energy. When a low-energy surface formed by treating a polymer in fluorine-containing plasma is placed in a high-energy medium, such as water, the apolar groups will tend to minimise the interfacial energy by moving away from the surface into the bulk. This phenomenon is usually described as ageing of a treated surface.

15.16.7 Temperature change during plasma treatment

The thermal decomposition of fibres under plasma condition is a major concern during the plasma treatment. The temperature increase inside the plasma reactor is the possible way of the thermal decomposition of fibre as a side effect.

15.17 Future trends

The general trend in chemical finishing is towards the use of more sophisticated chemical finishes that are more environmentally friendly and are specifically formulated for ease of application on automated machinery and equipment. Higher quality chemical finishes for the more discerning and demanding requirements of the market will in turn create pressures on manufacturers of chemical finishes to engineer their products to provide right-first-time processing.

Major driving forces will be (Holme, 1993):

- The need for higher quality, higher added value products
- Right-first-time, right-on-time, right-every-time processing
- Increased levels of automation and process control in machinery and equipment
- The increased use of process integration
- Greater emphasis on cost reduction by minimising the use of water, energy and all utilities
- The challenges posed by environmental issues, e.g. minimising discharges of chemicals onto land, into air and particularly into water
- The greater interest in multipurpose chemical finishes.

Quality reveals itself in daily use and is defined by the user as the situation when the expected and the experienced quality are as close as possible. The important aspects of quality policy in textile finishing plan design are:

- Avoid potential sources of risk
- Built-in freedom from maintenance
- Monitoring of specified and achieved values
- Control of exceptional situations.

All these measures are intended to increase process reliability and reproducibility. Higher quality results in gaining customer trust in the supplier and develops loyalty towards him, mutually benefiting each other (Ströhle, 2009).

15.18 References

Biancalani spa (2009). The 'Natural Touch' Quality and sustainability in fabric finishing, *Colourage Supplement* **56**(1): 33–4.

Chan C M, Ko T M and Hiraoka H (1996). Polymer surface modification by plasmas and photons, *Surf Sci Rep* **24**(1): 1–54.

Clapp T G, Godfrey A B, Greeson D, Johnson R H, Rich C and Seastrunk C, (2001). Weaving a quality textile industry, *Quality Digest*, October, http://www.quality-digest.com/oct01/html/textile.html.

Cooke T F (1983). Foam Wet Processing in the Textile Industry, *Text. Chem. Col.*, **15**(5): 74–85.

Duckworth C (1983). *Engineering in textile coloration*, Dyers Company pub. Trust, SDC (Bradford, UK).

Elbadawi A M and Pearson J S (2003). Foam technology in textile finishing, *Textile Progress*, **33**:4, 1–31.

Fiebig D and Rezk A A (1973). The Effect of Crosslinking Conditions on the Wet-Soiling Behavior of Cotton, *Textile Research J,* **43**, 438.

Foster S H (1957). A Simplified Relationship between the Properties of Resin-Finished Cotton Fabric and the Solids Applied, *Textile Research J.,* **27**, 129–35, doi:10.1177/004051755702700206.

Fulmer T D (1983). Computer supported controls for finishing ranges, *America's Textiles*, August, Vol. **12**, p. 15.

Ghosh D K (1988). Wet processing by foam technology, Man-made Text. *India*, 31(11): 483.

Goyal R and Prabhu C N (2009). Water repellent finishes, *Colourage*, February, issue 2, 88–90.

Grindstaff T H, Patterson H T and Billica H R (1970). Studies of Soiling and Detergency: Part IV: Deposition and Transfer Experiments With Fatty Soils, *Textile Research J.* **40**, 35–42, doi:10.1177/004051757004000106.

Hayakashi M (2008). Dyeing Technology on Polyester Micro-Fibre, presentation at Aleppo University on 26th April, www.arabytex.com/forum/.

Heywood D (2003). *Textile finishing*, Woodhead (Cambridge, England).

Holme I (1993). New Developments in the Chemical Finishing of Textiles, *Journal of the Textile Institute*, **84**(94), 520–33.

Holme I (1998). Recent Advances in Chemical Processing, *Colourage Annual*, 41–56.

Hussain T, Ahmed S and Qayum A (2008). Effect of different softeners and sanforising treatment on pilling performance of polyester/viscose blended fabrics, *Coloration Technology*, 375–8.

International Textile Reports (1991). Process control system for textile finishing, *Melliand Textilberichte*, **72**, 144–5.

Ishikawa, K (1985). *What is Total Quality Control?,* Prentice Hall, Englewood Cliffs, New Jersey, USA.

Jerg G (1979). Foam Printing, Ciba Geigy Publication F-C.5. 1979. p. 24.

Kan C and Yuen C M (2007). Plasma technology in wool, *Textile Progress*, **39**(3) 121–87.

Kehry S and Uhl H (2010). Improving the efficiency of textile machinery with intelligent data management, www.benningergroup.com.

Kroezen A B J and Wassink J G (1987). Bubble size distribution and energy dissipation in foam mixers, *J. Soc. Dyers Col.,* **103**(11): 386.

Lewin M and Sello S B (1983). 'Functional finishes, part A, Handbook of fibre science and Technology: vol. II', Marcel Dekkar (New York and Basel).

Marsh J T (1979). 'An introduction to textile finishing', B.I. Publications, India.

Merritt D S (1982). Energy savings by dryer exhaust control, *Am. Dyestuff Rep.* **71**(9): 26.

Myers S A (2002). Finishing *in Wool: Science and Technology*, Simpsion W S (Eds.), Woodhead Cambridge, England, 314–332.

Pehl R (1991). Surface finish of knitted fabrics – faults and options in the event of damage during the raising process, *International Textile Bulletin (Dyeing/Printing/Finishing)*, **37**(1): 27.

Rouette Hans-Karl (2000). *Encyclopedia of Textile Finishing*, Springer (Aachen, Germany).

Roy Choudhury A K, Chatterjee B, Saha S and Shaw K (2012). Comparison of Performances of Macro, Micro and Nano Silicone Softeners, DOI: 10.1080/00405000.2012.654666, *Journal of the textile institute*, Available online: 03 Feb 2012.

Satees G (2012) www.scribd.com/doc/29869975/Alkaline-Weight-Reduction, accessed on 27 August 2012.

Schindler W D and Hauser P J (2004). *Chemical Finishing of Textiles*, Woodhead (Cambridge, England).

Schreiber W and Maschen S (1992). Process automation for continuous wet finishing ranges, *Melliand Textilberichte*, **73**(10): 835–6.

Shah D R, 'Process control and safety in chemical processing of textiles', www.sulphurdyes.com/knowledgewall.html

Shishoo R (2007). *Plasma Technologies for Textiles*, Woodhead (Cambridge, England).

Shaw D J (1980). *Introduction to Colloid and Surface Chemistry* (3rd Edition), Butterworth (London. England).

Sontag M S, Purchase M E and Smith B F (1970). A double-label radioactive tracer analysis of the retention of selected fats on cotton and its relation to yellowing, *Textile Research J.* **40**(6): 529–36, doi:10.1177/004051757004000607.

Ströhle J (2009). Increasing the Process Reliability of Textile Finishing Processes, *Supplement to Colourage* **56**(4): 28–9.

Tomasino C (1992). *Chemistry & Technology of Fabric Preparation & Finishing*, College of Textiles, North Carolina State University: North Carolina, USA.

Turner G R (1981). Foam Technology: What's It All About?, *Textile Chemist & Colorist,* **13**(2): 13–18.

Vigneswaran C and Anbumani N (2007). Effects of Weight Reduction Treatment by Partial Alkaline Hydrolysis of Physical Properties of Polyester Rotor Spun Yarns, *Institute of Engineers (India)* Journal, TX, 88, August issue, 14–18.

Vishnu A D and Bhagyashri K (2010). Antimicrobial finishing of textiles, Man-made Textiles in India, March, **53**(3): 89–95.

Weber R and Tuschen J (1980). U.K. develops new dyeing for the continuous dyeing of carpets, *ChemiefasernTextilindustrie (Eng. Edn),* **30/82,** 745.29.

16
Process control in apparel manufacturing

G. THILAGAVATHI and S. VIJU,
PSG College of Technology, Coimbatore, India

DOI: 10.1533/9780857095633.3.428

Abstract: The quality of apparel is not only affected by the quality of the fabrics used in its manufacture but also by factors determined by the technology and parameters of the apparel manufacturing process. The selection and use of the processes and parameters can ultimately affect the apparel quality. In this chapter, process control issues relating to apparel manufacturing, such as spreading, pattern making, cutting, sewing, fusing, pressing and packaging are discussed. Quality evaluation of apparel and accessories is also discussed.

Key words: apparel, spreading, cutting, sewing, testing.

16.1 Introduction

Fabric is the basic raw material of apparel industry. The main task for the apparel manufacturer is to produce shell structures out of flat fabrics to match the shape of human body. The conversion of fabric into apparel involves various processes, such as spreading, pattern making, cutting, sewing, fusing, pressing and packaging. The quality of apparel is not only affected by the quality of the fabrics used in its manufacture but also by the factors determined by the technology and parameters of the apparel manufacturing process. The selection and use of the processes and parameters can ultimately affect the apparel quality (Fan and Hunter, 2009). In this chapter, process control issues relating to apparel manufacturing, such as spreading, pattern making, cutting, sewing, fusing, pressing and packaging are discussed. Quality evaluation of apparel and accessories are also discussed.

16.2 Process control in spreading, pattern making and cutting

In addition to the quality of fabric used for apparel manufacture, selection of suitable apparel preparatory processes such as spreading, pattern making and cutting also influence the quality of apparel considerably. The following section discusses the process control issues relating to spreading, pattern making and cutting.

16.2.1 Spreading

Spreading is the process of stacking the material one layer on top of another to create a lay. The lay can be a single ply of fabric or several hundred plies. Various factors that can affect spreading should be checked, such as ply alignment, ply tension or slackness, bowing and splicing (Kothari, 1999a; Solinger, 1980).

The greater the variation in width or length alignment, the greater the waste in precision cutting because the ends and sides must be trimmed to the narrowest and shortest plies. A tight spread will contract after cutting, resulting in components that are smaller and skimpier than they should be. A slack spread possesses excess length within the stipulated end of the spread. Cut components from slack spread will tend to be oversized. Bowing is the distortion of filling yarn from a straight line across the width of a fabric. This will cause unbalanced stresses in the fabric, resulting in slackness and tightness in the ply that will lead to undersized components. Also, the component containing such a defect will tend to twist or distort in laundering or dry-cleaning. Splicing is the overlapping of two ends of fabric in a ply. A short or insufficient overlap will result in incompletely cut pattern sections and a long overlap will result in waste. Static electricity in fabric may cause a distorted spread, resulting in incompletely cut pattern sections. Static electricity can be eliminated by either increasing the humidity in the cutting room or using static eliminators.

16.2.2 Pattern making

Various pattern defects that may arise during pattern making process are as follows (Carr and Latham, 1994; Kothari, 1999a):

- *Pattern alignment in relation to the grain of the fabric:* Pattern pieces normally carry a grain line. When pattern pieces are laid down on the piece of cloth, the grain line should lie parallel to the line of the warp in a woven fabric or the wales in a knitted fabric. When pattern pieces are laid across the piece, the grain line should lie parallel to the line of the weft in a woven fabric or the course in a knitted fabric. In bias cutting, which is often used in large pattern pieces as part of the apparel style in ladies dresses and lingerie, as well as in small pattern pieces such as pocket facings and undercollars in menswear as a requirement of satisfactory apparel assembly, the grain lines will be at 45°C to the warp. The designer or pattern cutter may define a tolerance that allows the marker planner to swing the grain line a small amount from parallel. If the marker planner lays down a pattern outside the stated rules for grain lines, then the finished apparel will not hang and drape correctly when worn.

- *Mixed pattern parts*: Parts not correctly labelled in marker, therefore a marriage of wrong-sized parts
- *Line definition poor*: Poor line definition (e.g. chalk too thick, indistinctly printed line, perforated lay not fully powdered), leading to inaccurate cutting
- *Skimpy marking*: Either the marker did not use the outside edge of the pattern or the pattern was moved or swung after partial marking to squeeze the pattern into a smaller space in the interests of fabric economy. Alternatively, the pattern is worn around the edges and should be replaced.
- *Marker too wide*: Apparel parts at the edge of the lay are cut with bits missing

16.2.3 Cutting

Cutting quality is a prerequisite for quality in a finished product. In addition, cut work quality affects the case and cost with which construction is accomplished. The quality of work leaving the cutting room is determined by how true the cut fabric parts are to the pattern; how smooth or rough the cut surface is; material or fabric defects in the cut fabric parts; shade differences between cut fabric pieces within a bundle. In addition, various factors in cutting that can affect the subsequent quality should be checked, such as under or overcut, size, placement and sequence alignment of notches and drill holes, ripped or pulled yarns, etc. Here are some defects that may arise in cutting (Anon, 2010; Carr and Latham, 1994; Kothari, 1999b):

- *Frayed edges*. May impede cutting time by clogging the knife action and/or spoil the fabric with rips or pulled yarns. The amount of fraying depends on fabric construction and finish. Improper cutting tools or dull knives cause excessive fraying in a pattern as the section is cut.
- *Fuzzy, ragged, or serrated edges*. The result of poor cutting implements. Such edges will impede sewing and/or diminish sewing quality. Such a condition is caused by faulty knife edges such as burrs, chips, or dullness.
- *Ply-to-ply fusion*. More common and troublesome. Adjacent plies in a block are fused together, which makes it difficult for the sewing machine operator to pick up a single ply quickly. Fusion occurs due to heat created by excessively high speed of cutting or by the friction of a dull knife. To prevent fusion, check knife speed, keep knives sharp, place wax paper between fabric plies, and lubricate the cutting blade.
- *Single-edge fusion*. Consists of a single ply whose cut yarn ends are fused to form a hard brittle rim on the cut edges. Sometimes, this is desirable to prevent fraying; however, hardness and brittleness are undesirable if

they impede sewing manipulation or may result in seams uncomfortable to the consumer.

- *Pattern precision*. Misshaping or distortion of the pattern perimeter as cut. Whether it is under or overcut is due to poor manual control of the cutting machine and poor lines on the marker. To assure precision in a pattern, check markers before cutting, use tensionless spreading, or allow time for the fabric to relax. After a cut, check the top, bottom and middle plies against the pattern.
- *Notches*. Notch size refers to the depth of a notch. If the depth is too great, the notch may show after sewing. If the notches are too small, sewing operators may have difficulty locating them quickly, resulting in decreased efficiency. Misplacement of a notch may be due to an improperly spread marker, poor control of a cutting machine with the cutter's notching tool stroking diagonally instead of vertically, incorrect marker in that the notches formatting parts do not coincide. Check notch placement against matching pieces. Quality control in stitching may be a problem if notches are not aligned (Fig. 16.1).
- *Drilling*. The drill hole may be too small in diameter. In addition, a drill may become too hot due to high speed or wrong size, causing the plies to fuse at the drill hole. The drill must stroke vertically to the table for uniform placement throughout the bundle. Sometimes fabric properties are such that the slight movement of yarns in a fabric would close a drill hole. In such cases, it is necessary to drill holes with a marking fluid. The drill used for such a purpose is hollow and carries marking fluid (ink) that is deposited at the drill point on the fabric as the needle is withdrawn. Such marks should last long enough so that further processing can be finished without difficulty, but should be easily removable after processing or in case of an error.

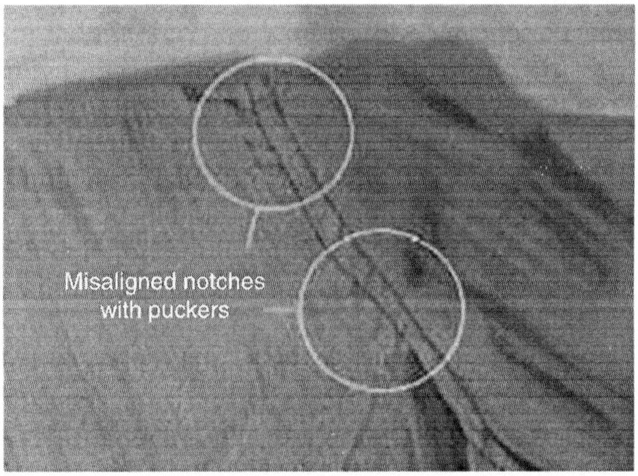

16.1 Misaligned notches.

16.3 Process control in sewing

Sewing damage is a problem in apparel production, causing minor appearance problems. The problems are most conveniently divided into (Carr and Latham, 1994, 2000):

- Problems of stitch formation, giving rise to poor seam appearance and performance;
- Problems of fabric distortion known as pucker, also giving rise to poor seam appearance;
- Problems of damage to fabric along the stitch line.

16.3.1 Stitching defects and control

The main problems that arise from the actual stitch formation include skip or slipped stitches, staggered stitches, unbalanced stitches, variable stitch density and needle, bobbin or looper thread breakage (Anon, 2010; Carr and Latham, 1994, 2000).

Skip or slipped stitches (Fig. 16.2) arise from the hook or looper in the machine not picking up the loop in the needle thread. In a lockstitch, a slipped stitch will create a small gap in a seam or a poor appearance in top-stitching but it will not cause seam failure. In a 101 or 401 chain stitch, it produces a weak point in the stitching that will run undone during use of the

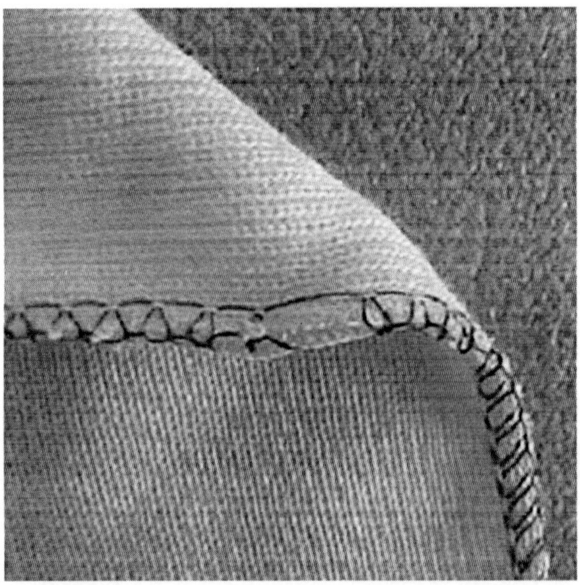

16.2 Skipped stitches.

Table 16.1 Causes and solutions for skipped stitches

Causes for skip or slipped stitches	Possible solution
Bobbin book falls to pick up loop	Check timing
Looper falls to pick up loop	Check timing
Thread does not form a good loop	Check thread tension and check spring Use better quality thread
Bent needles, damaged looper	Replace damaged needles
Deflection of needle	Replace needle
Incorrect needle type	Replace needle
Incorrect needle size	Replace needle
Incorrect thread tension	Adjust tensioner
Poor material control	Throat plate hole too large Presser foot control is poor Needle is inappropriate for fabric
Clogged needle	Replace needle, avoid clogging

garment. Thread elongation and recovery properties are very important in determining the thread loop forming properties. Threads which form large, consistent loops are much more safely picked up by the looper, even if the timing is imperfect or the needle is badly deflected in passing through heavy material. Other causes of slip-stitching are bent needles, incorrect needle size or type for the thread size or type, incorrect thread tensions or poor material control arising from a large throat plate hole and poor presser foot control. Table 16.1 summarises the causes and solutions for skipped stitches.

Staggered stitches (Fig. 16.3) can be caused by yarns in the fabric deflecting the needle away from a straight line of stitching, giving a poor appearance. In some hard, woven fabrics, really straight stitching will only be achieved at a slight angle of bias. The causes and solutions for staggered stitches are listed in Table 16.2.

Unbalanced stitches in lock stitching can reduce the potential for stretch in a seam in a knitted fabric. Unbalanced seams are often recognised by low extensibility leading to cracking. Bobbin tension should be adjusted until a full bobbin in its case will just slide down the thread when held by the end of the thread. Needle thread tension should also be adjusted to minimise the thread breakage. Where maximum stretch is required in a lockstitch seam, careful adjustments of needle and bobbin thread tension may be needed. Table 16.3 summarises the causes and solutions for unbalanced stitches

Variable stitch density (Fig. 16.4) may arise from insufficient foot pressure in a drop feed system, causing uneven feeding of the fabric through the machine. It can occur particularly with materials with a sticky or slippery surface. Pressure must be adequate to enable even feeding, but with many fabrics and sewing situations, specialised feed systems are necessary to achieve even feeding of all the plies of the material. Operators may be able to contribute to consistency in stitch density by sewing at a constant

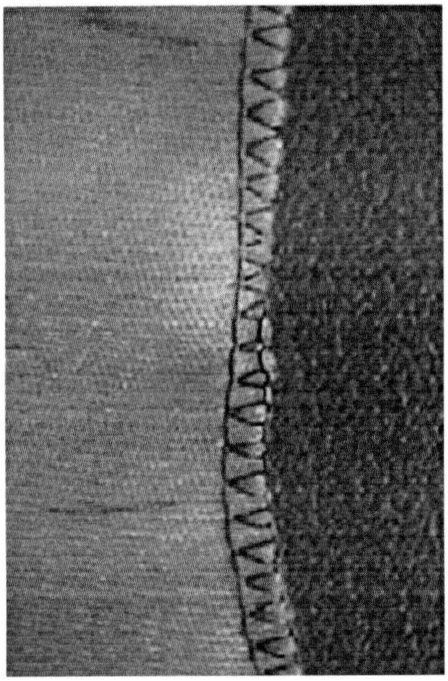

16.3 Staggered stitches.

Table 16.2 Causes and solutions for staggered stitches

Causes of staggered stitches	Possible solution
Inappropriate size needle	Go up a needle size
Inappropriate needle point	Select sharper point
Inappropriate needle blade	Select tapered blade
Damaged needle	Replace needle
Poor material control	Check current settings Use alternative feed mechanism
Inappropriate seam for selected fabric	Redesign to sew at a slight angle of bias

Table 16.3 Causes and solutions for unbalanced stitches

Causes of unbalanced stitches	Possible solution
Low extension, bobbin thread breakage	Reduce bobbin thread tension Check needle thread tension
Low extension, needle thread breakage	Reduce needle thread tension Check bobbin thread tension

speed, rather than in bursts. Table 16.4 summarises the causes and solutions for variable stitch density.

Breakage of thread (Fig. 16.5) and breakage of needle (Fig. 16.6) arises largely as a result of smooth metal surfaces in the machine becoming burred,

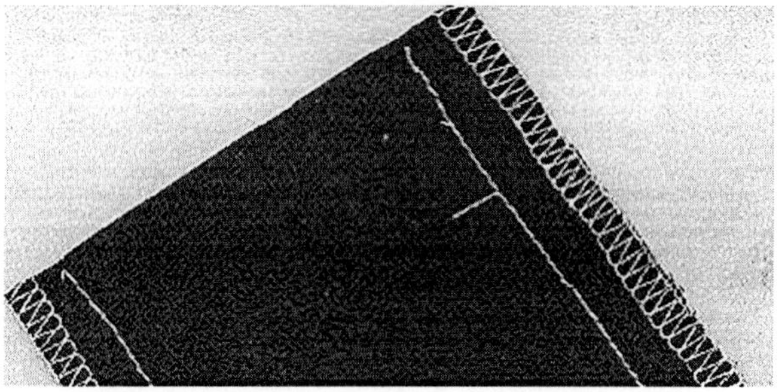

16.4 Variable stitch density.

Table 16.4 Causes and solutions for variable stitch density

Variable stitch density	Possible solution
Poor material control using drop feed	Increase foot pressure
Check feed dog setting	
Avoid sewing bursts	
	Change to better feeding mechanism foot pressure
Poor material control using specialised feed system	Check feed dog setting
Avoid sewing bursts	
Change to walking foot mechanism	Introduce puller

16.5 Breakage of thread.

chipped or damaged and then causing damage to the thread, needle and bobbin or looper. Other causes include improper settings, inadequate maintenance, and poor quality materials. Thread breakage during production is time consuming for the operator, especially if a join in the stitching is not acceptable and he or she must unpick it and start again. Both the cause of the damaged machine parts and the cause of the needle deflection must be investigated. Table 16.5 lists the causes and solutions for thread breakage.

16.6 Breakage of needle.

Table 16.5 Causes and solutions for thread breakage

Thread breakage	Possible solution
Damage to thread guiding mechanisms	Polish or replace all damaged parts
Damage to throat plate	Polish or replace throat plate
Incorrect thread tension	Check and adjust thread tension
Incorrect settings for stitch forming	Check timing and adjust
Poor thread quality	Change thread
Poor cone of thread	Adjust guides, remove faulty cone
Wrong needle size for thread	Match thread with needle size
Needle overheating	Reduce speed of sewing
	Reduce needle size
	Improve sewability via fabric finish

16.3.2 Causes and solutions for seam pucker

The visual appearance of the apparel is a principal factor deciding its value. Seam pucker influences the apparel appearance to a considerable degree. It has been well recognised that elimination of seam pucker completely is almost impossible and the common practice is to accept a small amount of pucker. Seam pucker may occur in apparel due to reasons such as structural jamming, differential feed, sewing thread tension, sewing thread shrinkage, fabric shrinkage and mismatched patterns.

General

Seam pucker (Fig. 16.7) is a wrinkled appearance along the seam, which influences the apparel appearance to a considerable degree. Seam pucker, identified as a sewability problem about 70 years ago, has been regarded as one of the most important parameters of quality control in apparel manufacturing industries (Anon, 2010; Hati and Das, 2011). It is usually caused by improper selection of sewing parameters and material properties, which results in unevenness on fabrics being stitched together, thus impairing their aesthetic value. In severe cases, seam pucker could appear like a wave front, originating from the seam and extending through the entire garment. In less severe cases,

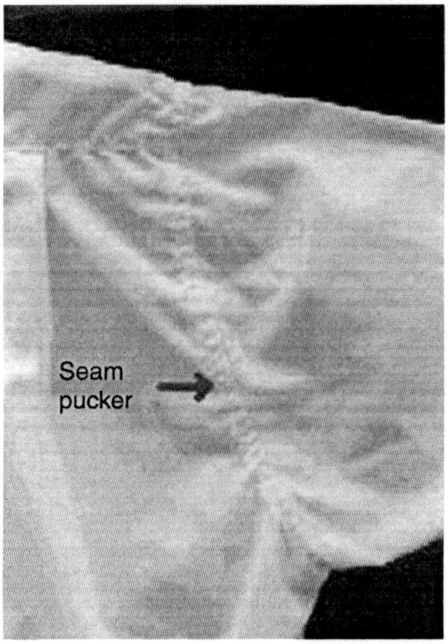

16.7 Example of seam pucker.

the wave formation is less pronounced, but nevertheless discernible. Indeed, apparel exhibiting pronounced seam puckers is certainly unwelcome by customers. Table 16.6 summarises the causes and solutions for seam pucker.

Structural jamming

During seam formation, stitches are made by interloping of bobbin and needle thread. These sewing threads displace the fabric yarns from its original position. Fabric yarns attempt to return to original position and they are prevented from doing so by the sewing threads. This causes the fabric layers to displace in a plane perpendicular to fabric plane and results in seam puckering (Fig. 16.8). This kind of pucker is visible mostly in tightly constructed fabrics that do not have enough space to accommodate sewing threads. The problem can be avoided or reduced by using a thinner thread and a small needle point; a needle plate and a presser foot with a smaller needle hole; fewer stitches per unit length; and stitching in a direction which allows the different yarns in the construction of the fabric to be displaced. This kind of pucker is visible on both sides of the fabric (Crum, 1983; Dorkin and Chamberlain, 1961).

Differential feed

Differential feed of the feed dog produces feeding pucker (Fig. 16.9). During sewing operation, bottom layer fabric is moved forwarded by the feed dog positively. But the movement of the top layer fabric is effected through frictional contact between top and bottom fabric. Thus movement to top fabric

Table 16.6 Causes and solutions for seam pucker

Type of pucker	Possible solution
Structural jamming	Select finer thread Reduce stitch density Select small needle size Cut and sew on bias Use needle plate with small needle hole Use presser foot with small needle hole
Differential feed	Check presser foot and feed dog setting Explore use of PTFE coated feet Use alternative feed mechanism
Sewing thread tension	Reduce thread tension Change to thread with reduced extensibility Ensure correct feed timing
Sewing thread shrinkage	Select thread with good lubrication Use a synthetic thread or core spun cotton
Fabric shrinkage	Refinish fabrics to reduce shrinkage effects Use a synthetic thread or core spun cotton
Mismatched patterns	Ensure patterns are correct Ensure cutting is accurate Ensure sewing ease is correct

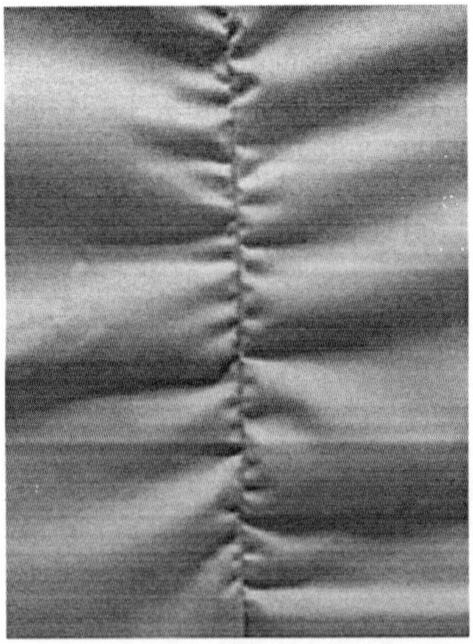

16.8 Pucker due to structural jamming.

is not a positive one. The velocity of the top layer fabric is generally lower than that of the bottom layer fabric. This causes accumulation of bottom layer fabrics and produces feeding pucker, which is visible at one side only. Use of an appropriate feed dog, presser foot pressure and sewing speed may help to reduce this kind of pucker (Hati and Das, 2011).

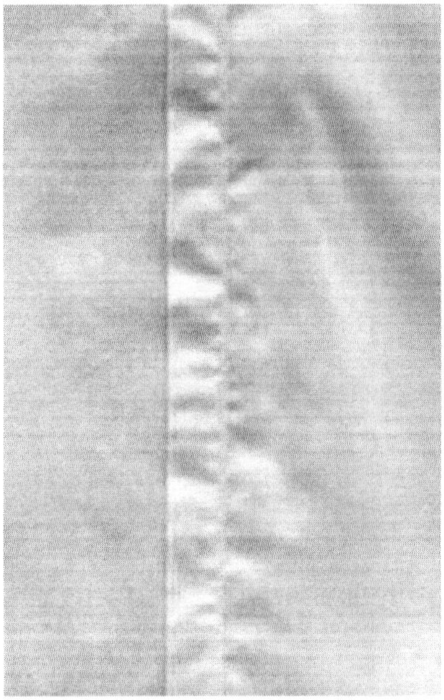

16.9 Pucker due to unequal feeding of machines.

The feed dog should have the optimum number of teeth and rows of teeth for the operation and fabric being sewn. Sometimes, pucker occurs when the material is not held down flat as it is being fed through the machine, creating a rippled appearance as the plies conform to the feed dog teeth. Generally, light weight, wrinkle-resistant fabrics should be sewn with feed dogs having 20–24 teeth per inch. Medium and heavy weight fabrics should be sewn with feed dogs having 14–18 teeth per inch and 8–12 teeth per inch respectively. The pressure exerted by the presser foot against the needle plate causes the fabric to spread. Therefore the dimensions of the seam produced on a fabric have been momentarily altered. When the fabric returns to its normal dimensions, waviness occurs. Reducing the pressure of the presser foot can prevent or at least reduce waviness. High sewing speed requires high presser foot pressure in order to control the feed and prevent the presser foot from bouncing under the impact of the feed dog movement. The bouncing of the pressure foot disturbs the fabric feed, leading to a puckered seam and uncontrolled sewing conditions (Fan and Hunter, 2009).

Sewing thread tension

Tension pucker occurs when the sewing thread is under very high tension (Fig. 16.10). Sewing thread extends due to very high tension, and afterwards

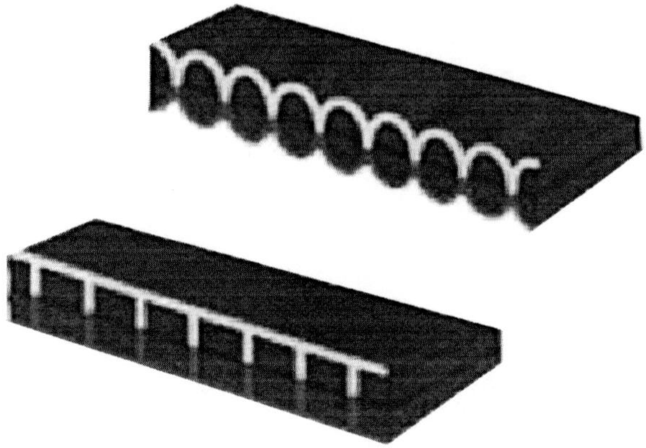

16.10 Tension pucker due to high sewing thread tension.

attempts to relax. If elastic recovery of thread and shrinkage of fabric coincides, pucker does not occur. But in most situations recovery of thread is more than fabric shrinkage causing fabric to pucker. This is inevitable from the point of view of the sewing operation, which requires higher needle thread tension than bobbin thread tension to snatch the latter for stitch formation. Although sizing or resin on the fabric can prevent or reduce this problem, the seam will still become puckered after laundering. Thread tension pucker can be minimised by reducing the thread tensions on the sewing machine or using a thread with a low elongation or high initial modulus to minimise stretching during sewing. If structural jamming does not occur, it is beneficial to increase the needle size or use a needle with a ball eye to open up a larger hole in the fabric so the stitch can be set. The take-up spring, thread control guides and eyelets should also be properly set to avoid undue tension of the sewing thread during sewing (Fan and Hunter, 2009; Hati and Das, 2011).

Sewing thread shrinkage

High shrinkage potential of fabrics is another source of seam pucker. Threads made of cotton and other natural fibre often shrinks when wet. This may cause seam pucker when fabric shrinkage and thread shrinkage differs. This problem can largely be overcome with the use of synthetic sewing threads. These threads are normally dry heat stabilised to withstand up to 110°C. They may also cause seam pucker during pressing, if the pressing temperature is above heat-set temperature (Dorkin and Chamberlain, 1961).

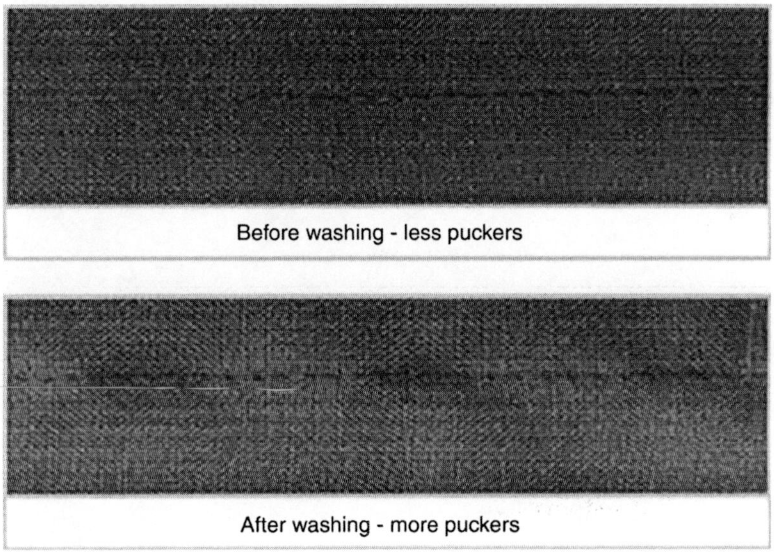

16.11 Pucker due to fabric shrinkage.

Fabric shrinkage

Different shrinkage of plies of fabric causes seam pucker that is visible only after few washing cycles (Fig. 16.11). The component that has the lower shrinkage rate will buckle, resulting in a shrinkage pucker. They will never occur during the sewing operation. They are concerned with dimensional stability of plies of fabric. To prevent or reduce this type of pucker, it is necessary to use compatible components that have similar dimensional stability (Anon, 2010; Hati and Das, 2011)

Mismatched patterns

If the patterns of fabric that are to be matched are of unequal length, the experienced tailor will feed more length of the longer pattern. This compensates for extra length, but fabric accumulates in the form of wrinkles and seam pucker appears (Fig. 16.12).

16.4 Causes of damage to the fabric during sewing

Damage to the fabric during sewing can be divided into two categories. One is sewing damage due to thermal problems and the other is due to mechanical problems. Some of the causes of thermal and mechanical problems are discussed here.

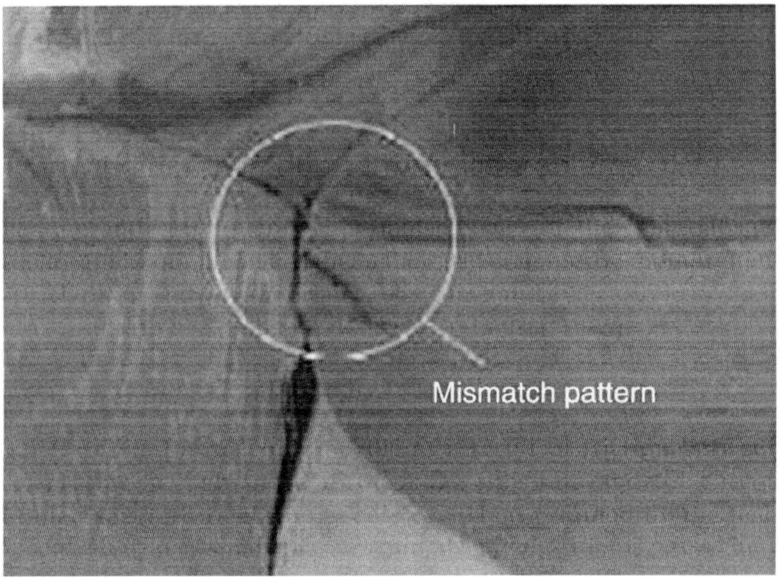

16.12 Pucker due to mismatched patterns.

16.4.1 Sewing damage due to thermal problems

Needle heating and consequential thermal damage of the material and thread used to stitch-join them have been the issue of numerous comprehensive reports (Dorkin and Chamberlain, 1963; Hurt and Tyler, 1971, 1972a,b; Laing and Webster, 1998; Sondheim, 1953; Worthington, 1975).

Generation and loss of heat in the needle

Heat is generated in the machine needle as a result of friction between the needle passing at high speed through the substrate and the thread passing at high speed through the needle eye. Heat retained by the needle is concentrated in a small mass of metal and the temperature may reach 300–350°C. During the stitching operation heat is also dissipated from the needle. Heat is transferred by conduction through the needlebar to other machine parts, to fabric surrounding the needle and to thread in the needle eye. Heat loss by conduction may be higher when materials of high thermal conductivity are used and when the needle size (diameter) is also increased. Conduction is the most important means of heat loss from needles. Movement of air around the needle allows heat loss by convection, with air currents created by moving machine parts and by movement of the sewing thread. Radiation has only a minor role in heat loss from needles. Conduction, convection and radiation are heat-loss processes which occur naturally but, by altering conditions of sewing, heat loss can be increased, though modifications to machine parts may be necessary (Laing and Webster, 1998).

Use of a bulged-eye needle provides for only a single line of contact between needle and material at the needle eye. Using a bulged-eye needle, as compared with a normal needle, results in a drop in needle temperature. Heat loss is also increased by using an air-cooler on the needle. Lubricants applied directly to needles may lower friction between the needle and thread, or they may be applied to thread. Absorption of energy by a vapour spray (i.e. a coolant atomised or directed straight to the needle, with heat being removed by vaporisation of the solvent), offers further scope for temperature reduction. Water may be a satisfactory coolant, but a rust inhibitor is required. Temperature losses achievable in this way can be up to 90°C.

High needle temperatures and their effects

Few fabrics were damaged seriously by needle heat prior to the 1940s. Since then fabrics and sewing threads manufactured from thermoplastic fibres have become common, and the maximum stitching speed of sewing machines has increased markedly from around 1000 to 6000 stitches/min. Thermoplastic sewing threads, such as nylon and polyester, may be heated by the needle, soften, melt, and break. Nylon and polyester threads melt at 240–260°C. Cotton and silk are not thermoplastic, but degrade at around 400°C. Thread breaks decrease production levels because of discontinuity in stitching, assuming that the partly-formed seam is not unpicked and fully restitched (Laing and Webster, 1998).

Softened polymer from either the sewing thread or the fabric being stitched (or both) may be deposited in the needle eye. The needle eye cools when machining stops, and the polymer hardens, thus clogging the eye. On restarting the machine, the sewing thread cannot move though the eye and so breaks. During the down stroke part of the stitch formation cycle a crease may form in the sewing thread at the needle eye. If the needle temperature is sufficiently high, this crease may be set in the thread by heating followed by rapid cooling. Any loop subsequently formed in the sewing thread (required during formation of ISO 301 for example), is improperly shaped when the thread is creased and the rotary hook may not enter the loop. This would result in missed or skipped stitches and seam performance is adversely affected. More generally, any thermal damage to the fabric may affect the properties, performance or appearance of a seam. Table 16.7 summarises the causes and solutions for thermal damage (Carr and Latham, 1994, 2000).

Table 16.7 Causes and solutions for thermal damage

Thermal damage	Possible solution
Excessive sewing speed	Reduce sewing speed
Inappropriate needle point	Select sharper set or finer ball point
Needle size too large	Reduce needle size
Inappropriate fabric density	Ensure fabric density is correct
Inadequate fabric finishing	Increase softener content on fabric
Inappropriate sewing thread	Select proper sewing thread

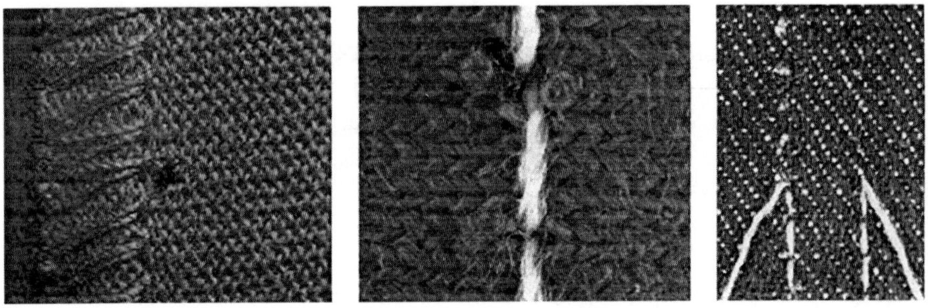

16.13 Examples of fabric defects produced by wrong needle choice.

Machine variables affecting needle-heat

Sewing speed has greater influence on needle temperature than any other single factor. Keeping other variables constant, the higher the speed the higher the temperature, up to an equilibrium point for any particular combination of variables. Equilibrium is reached when the heat generated is equal to that lost by conduction, convection and radiation. The actual temperature at equilibrium is specific to stitching conditions, for example the number of fabric plies, fabric properties, needle size and thread type. The time and number of stitches formed before the equilibrium may be quite short (Laing and Webster, 1998).

Needle size (Fig. 16.13) has been identified as an important variable. With lower temperatures typical of fine needles, less work is required for penetration of the cloth. Where needle diameter is larger, greater force is required for penetration of cloth and higher temperatures are observed. Needle temperature increases with increasing needle size but the magnitude of the effect is small, other variables being more significant (Hurt and Tyler, 1971, 1972b).

Needle-point shape (Fig. 16.14) and surface finish are also significant variables. Sewing with a heavy-ball needle usually results in temperatures at the higher end of the range. Bulged-eye needles operate at lower

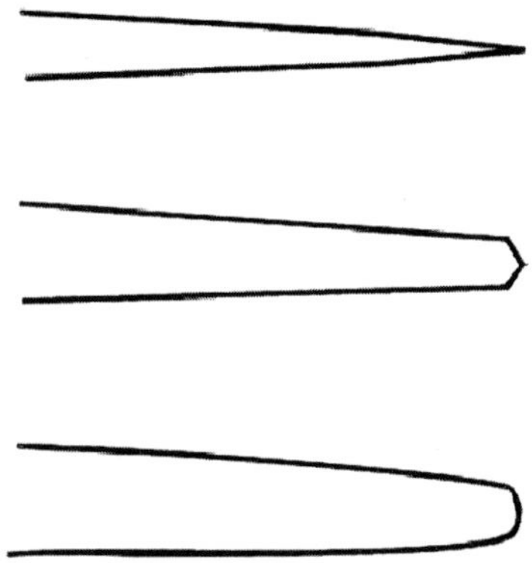

16.14 Different types of needle points.

equilibrium temperatures than cloth or set point needles, the reduction being 11–30°C. The surface finish of the needle also affects generation of heat. The highest temperatures are with dark matt surface needles, nickel, plain steel and chrome finishes giving progressively lower temperatures. The main parts in contact with material being stitched, the top of the blade and the needle point appear to have most effect on heat generated.

Any form of restraint on the movement of fibre and yarns in fabric might be expected to lead to an increase in the force required for the needle to penetrate the fabric. Machine parts such as the presser foot and throat plate affect the extent to which fabric is restrained. With respect to the throat plate, its hole size appears to be a critical variable. Movement of the fabric during the stitching operation is three dimensional with a glove of yarns/fibres forming around the needle as it passes down through the throat plate. If the throat plate hole is too small the fabric cannot be depressed and yarns are displaced temporarily. This may cause some fibres in the yarn to break. It has been shown that increasing the diameter of the throat plate hole decreases the equilibrium needle temperature (Laing and Webster, 1998).

Material variables affecting needle-heat

Needle heat generated during stitching depends on fabric-needle surface characteristics, fabric frictional characteristics, and fabric tightness (Laing and Webster, 1998).

Fabric density is a critical determinant of needle temperature. Any reduction in the ease with which fibre and/or yarns can be deflected from the path of the needle increases the force required for needle penetration. For example, woven fabrics with high cover factor and knitted fabric constructed of short stitches will be associated with elevated needle temperatures (Hurt and Tyler, 1971, 1972b; Laing and Webster, 1998).

Fabric finishing treatments (cleaning processes and solvents used to extract fats/waxes from the cloth, dyeing for colour change, resin finishes used to impart a different hand to cloth or to alter creasing characteristics) because of their effect on the frictional properties of the fabric, are the most important material variable influencing thermal damage. The amount and type of finish determines whether the needle temperature is raised or lowered. A decrease in needle temperature occurs with addition of lubricants or softeners – commonly long-chain fatty acid radicals that are hydrophilic. Softeners are applied primarily to improve the handle of fabric, but simultaneously may improve its sewing characteristics. Some softener is transferred from the fabric to the needle, the needle thus being lubricated, which reduces the friction thereby lowering equilibrium needle temperature. Fibres and yarns are stiffened by the application of many forms of resin and the handle and some other properties of the finished fabric are usually altered. These effects increase the resistance of yarns and fibres to movement by the needle, resulting in higher frictional forces and equilibrium needle temperatures (Hurt and Tyler, 1971; Laing and Webster, 1998).

The influence of sewing thread on needle temperatures is complex. Thread is able to cool the needle by conducting heat away (transfer occurring between the hot needle and the cold thread) and by increasing convective losses due to the cooling air stream created by the moving thread. The influence of both effects is greatest at the eye of the needle. Heat is also generated by the thread passing rapidly through the eye of the needle and against machine parts. Equilibrium needle temperature in a test run without sewing thread was lower than that obtained when subsequently stitching with a thread. A needle carrying thread, but without interacting with any fabric, heats by about 50°C when running at high speed (Hurt and Tyler, 1971, 1972b; Laing and Webster, 1998).

The type of thread appears to affect the equilibrium needle temperature. Continuous filament thread results in higher needle temperatures than staple fibre thread. With continuous filaments, there is a greater area of contact between the thread and needle, and in addition less air is carried along with the thread on account of its smooth surface. Sewing threads of bulked or textured yarns have both a reduced surface area for contact between thread and needle, and facilitate air movement through the needle eye. Staple sewing threads also transport more cooling air to the needle, allowing the sewing speed to be increased by up to 10%, relative to that with comparable continuous filament thread, without adverse needle-temperature effects (Dorkin and Chamberlain, 1961; Laing and Webster, 1998).

16.4.2 Sewing damage due to mechanical problems

During stitch formation the cloth may be damaged mechanically rather than thermally. When the damage is characterised by fragmented or broken fibres, filaments or yarns, it is referred to as mechanical damage. Such damage may be apparent immediately after stitching but frequently will not appear until after the product has been used, when seams have been subjected to some form of tension, stress, strain, deformation or after successive cleaning. In a woven fabric, damage of this sort may be masked by the sewing thread, whereas in knitted structures the visual effects of mechanical damage are more severe (Fig. 16.15). Fabric breakdown may extend well beyond the area adjacent to the seam. Seams damaged in this way are unacceptable from the perspective of both aesthetics and performance. Table 16.8 summarises the causes and solutions for mechanical damage (Carr and Latham, 1994, 2000).

The damage process

Inclusion of a seam in fabric generally reduces the fabric breaking load. Mechanical damage of yarns adjacent to the stitching may be a contributing factor. Rapid rupture of a yarn during stitching may occur as a result of extension to its breaking extension, or by cutting. The needle or the sewing threads may cut through the yarns of the material, but the mechanical damage outlined below appears largely independent of the presence of sewing threads.

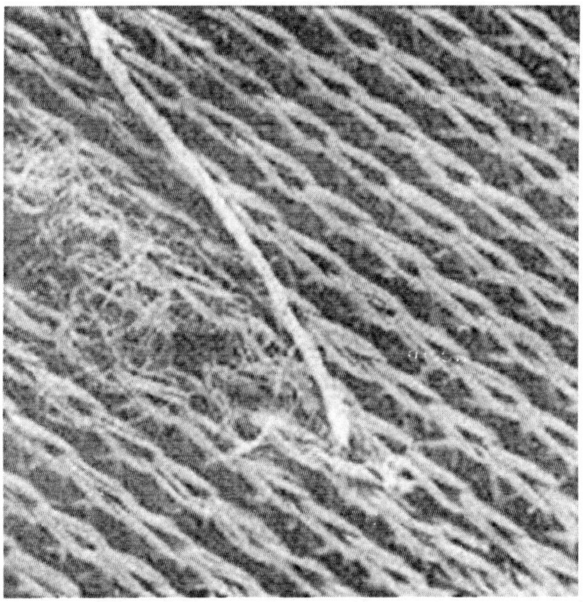

16.15 Knitted fabrics damaged during stitching.

Table 16.8 Causes and solutions for mechanical damage

Mechanical damage	Possible solution
Needle size too large	Reduce needle size
Inadequate fabric finishing	Increase softener content on fabric
Inappropriate needle point	Select sharper set or finer ball point
Damaged needle point and throat plate	Replace damaged parts
Mismatch between needle size and throat plate hole	Reduce tightness of fit

When the needle point strikes a spun and/or multi-filament yarn and passes through its centre, a diamond-shaped opening is formed and some fibres/filaments are extended to accommodate the thickness of the needle. Twist will be displaced. Where the yarn has a low twist, fibres/filaments can move easily, but as twist level increases, twist displacement become increasingly problematic. Fibre/filament failure will occur if the fibres are stretched to their breaking extension (Laing and Webster, 1998).

Many fabrics, especially those made from cotton and/or viscose, are treated with a stiffener, starch or crease-resistant resin finish. With the addition of starches and dextrins, the extensibility of individual fibres may not be affected but the fibres adhere to one another laterally. Adhesion prevents fibres slipping. With a resin finish, the extensibility of the fibres generally decreases. The effect of any finishing agent depends on the yarn's initial twist level, with low-twist yarns not being adversely affected by more severe finishing processes.

Interstitial spaces in the fabric, sufficient to accommodate a needle or to facilitate yarn redistribution, are necessary if mechanical damage is to be avoided. If the fabric (woven or knitted) has a high cover, yarn relocation is impeded. With needle penetration of a fabric with maximum cover, the fabric tends to buckle from the horizontal plane. As a result warp/weft yarns in a woven fabric are pushed together, and where the fabric is near its maximum cover, this is possible only if the fabric buckles from the horizontal plane. The action of the presser foot and the throat plate tend to inhibit buckling, and yarn breakages may result. Damage to knitted fabrics needs also to be considered as a three dimensional phenomenon. Yarns are stretched in the plane of the fabric with yarn robbing, and also stretched in the direction of the needle axis with deformation of the fabric by the needle point. In some fabrics, the length of yarn in a knitted loop is less than the circumference of the needle, so yarn robbing from adjacent loops occurs. The case with which any one loop can draw from adjacent loops is critical. Two separate frictional forces are in operation: that generated between the needle surface and the yarn around it: and that generated as the yarns in the fabric move over each other changing their configuration to accommodate the needle. As yarns are forced out of the path of the needle, they are stretched. If the extension forces are equal to or exceed the breaking load of the yarn, it will break and result in a hole in the fabric at the seam (Laing and Webster, 1998).

Fabric is controlled in a horizontal plane during stitching by the feed dog and the spring-loaded presser foot. The needle acts vertically on the cloth and filaments/yarns may form a glove around the needle point. Yarns may be forced into the throat plate hole. With knitted fabric, only those loops or parts of loops not held by the presser foot are free to move. Cotton interlock, which is often badly damaged in machine sewing, is not damaged at all if the fabric is held freely in the hand and hand stitched, thus illustrating the detrimental effect of clamping.

16.5 Control of fusing and pressing operations, storage and packaging

The following section discusses the process control issues relating to fusing and pressing operations, storage and packaging.

16.5.1 Fusing operations

Using fusible interlinings can prevent or reduce the problems of seam pucker. A fusible interlining is defined as a base fabric having a deposit of thermoplastic adhesive resin that can be bonded to another fabric by the application of heat and pressure (Cooklin, 1990). Fusible interlining enabled the apparel industry to make great strides in efficiency and quality in the past several decades, but the matter of finding the right interlining for a specific shell fabric continues to present challenges. If fusing is not handled properly, the shell fabric can change in colour or surface appearance, shrink, stretch or stiffen (Fan and Hunter, 2009).

To ensure good appearance and drape of the finished garment, the interlining should be compatible with the shell fabric. However, fusible interlinings vary in terms of base fabric composition, base fabric structure, and type of adhesive and finish. Fusible interlining suppliers are numerous. It is not an easy task to select the most suitable fusible interlining for a specific shell fabric or a specific range of shell fabrics. In general, most companies have guidelines on how to select fusible interlinings. The traditional method of choosing a suitable interlining fabric is by trial and error and/or by past experience. In general, many problems in fusible interlining selection and use can be avoided if manufacturers collaborate with their interlining suppliers in choosing the right fusible and conduct pre-production testing to ensure consistent performance.

Table 16.9 Causes and solutions for strike-through

Causes	Solution
Too long fusing time	Reduce fusing time. Shorten fusing time for thinner fabrics. Carry out a fusing test before bulk production.
Too high fusing temperature	Reduce fusing temperature. Carry out a fusing test before bulk production. Use a temperature measurement strip to find the optimum fusing temperature.
Too high fusing pressure	Fusing test should be carried out to check for the bond, surface and handle of fabrics before bulk production. Pressure setting should follow machine manufacturer's instructions. Reduce the pressure for lightweight and loosely constructed outer fabrics.
Incompatible fusible interlining (too much adhesive on interlining)	Avoid using interlining with excessive adhesive. Reduce size of the adhesive dots on the interlining. Reduce the weight of the adhesive dots on the interlining. Carry out a fusing test before bulk production.
Inappropriate fusing method for thin and loosely constructed fabrics	Use frame fusing for thin and loosely constructed fabrics. Carry out a fusing test before bulk production.

Strike-through and strike-back

During fusing, three parameters must be considered: the fusing method; the fusing machine; and the condition of the fusing (viz. temperature, time pressure and cooling). If the parameters are not properly selected and/or adjusted fusing problems (e.g. strike-through and strike-back, colour changes and fusing distortion) can occur. Strike-through means that the adhesive resin appears on the outside of the fabric being fused. Table 16.9 summarises the causes and solutions for strike-through. Strike-back means that adhesive resin appears on the non-adhesive side of the fusible interlining after fusing. Table 16.10 lists the causes and solution of strike-back (Anon, 2010; Fan and Hunter, 2009; Galuszynski and Robinson 1986).

Colour changes during fusing

Inappropriate fusing can cause temporary or permanent colour changes (shine/glazing and/or discolouration) in fabrics. Colour changes during

Table 16.10 Causes and solutions for strike-back

Causes	Solutions
In appropriate fusible interlining	Avoid using too light substrate materials. Avoid using interlining with excessive adhesive. Reduce size of the adhesive dots on the interlining. Carry out a fusing test before bulk production.
Too long fusing time	Reduce fusing time. Carry out a fusing test before bulk production.
Too high fusing temperature	Reduce fusing temperature to avoid adhesive becoming too fluid. Carry out a fusing test before bulk production. Use a temperature measurement strip to find the optimum fusing temperature.
Too high fusing pressure	Fusing test should be carried out to check the bond, surface and handle of fabrics before bulk production. Pressure setting should follow machine manufacturers' instructions. Reduce the pressure for lightweight and loosely constructed outer fabrics

fusing are caused by the action of heat on certain dyes. The causes and solutions for this phenomenon are listed in Table 16.11 (Anon, 2010; Fan and Hunter, 2009; Galuszynski and Robinson 1986).

Fusing distortion

Fusing distortion means that panels are distorted during the fusing process. This problem should be prevented, as such distorted panels cannot be corrected and must be discarded as waste. Fusing distortion may be caused by inadequate fusing conditions, improper handling of fusing panels, poor stability of the fabric and wrong fusing directions. The cause and solutions of fusing distortion are summarised in Table 16.12 (Anon, 2010; Fan and Hunter, 2009; Galuszynski and Robinson 1986).

16.5.2 Pressing operations

Pressing often represents the final opportunity to enhance apparel shape, and to smooth or flatten creases in the apparel. Good pressing can also help to preserve the shape and fit of apparel. Pressing performance can govern the appearance of a garment, as presented to a potential customer. Well-pressed apparel makes a good impression on potential customers, and can command a higher price. Good pressing appearance gives an overall smooth and undisturbed appearance without any shine, scorching, melting, clamp marks or water marks. There should be no unplanned creases or pleats,

Table 16.11 Causes and solutions for colour changes during fusing

Causes	Solutions
Some synthetic fabrics change colour temporarily after heat press/iron heat	Regulate the pressing temperature, pressure and time to avoid overheating and over pressing. Allow fused fabrics to cool down for a while to regain their colour.
Excessive temperature, pressure colour during the fusing process	Put a cloth (same material or cotton) to cover the fusing material during heat pressing. Select multivariable adhesive (MVA) that requires lower temperature, pressure and time for fusing.

Table 16.12 Causes and solutions for fusing distortion

Causes	Solutions
Poor fusing machine setting and conditions, e.g. speed of the continuous fusing press, temperature, pressure and setting time	Reduce the speed of the fusing press. Reduce the pressure of the fusing press. Establish optimum fusing machine setting by consulting the interlining supplier. Allow time for the fused parts to cool before handling. Test the fusing conditions before bulk fusing.
Bad orientation of fusing panels	Change orientation of panels in order to avoid the corner of the garment panels being caught first by the pressure rollers of the fusing press. Pre-set the fusing panels by hand-ironing before passing the fusing panels to the fusing press. Use flatbed fusing press, if available.
Unstable fabric/substrate properties	Use stable fabrics-avoid fabrics which have low shear rigidity. Take extra care in fabric handling during the fusing process.
Wrong fusing direction	Interlining and outer fabrics should be fused in the same direction, with similar dimensional changes.

pressed wrinkles, or hems pressed on the wrong side. Seams and side seams of pants and shorts should be pressed seam-on-seam with a smooth centre crease (Brown and Rice, 2001; Fan and Hunter, 2009). Fabrics should be engineered to have optimum pressing performance, and apparel should be pressed under the proper pressing conditions.

Preparation and handling

Before pressing, remove all the tacking threads and pins so as to avoid any marks on the garment. In order to prevent scorching, melting, hardening or shrivelling, first test the heat (temperature) of the iron prior to pressing so as to ensure that it is suitable for the garment to be pressed. Test the heat on a spare piece of the actual fabric before applying the iron to the apparel, so that a suitable heat can be found. When placing a garment onto a pressing machine, make sure the grain of the fabric is not pulled out of line, so as to ensure that it is not unnecessarily creased. Keep all seams in a straight line. To avoid shine or iron marking on the right side, press on the wrong side of the apparel. If necessary, slip a strip of paper underneath seam turnings to prevent their marking the right side of the fabric or apparel (Butler 1972; Fan and Hunter, 2009).

Pressing conditions

In practice, pressing can be conducted either with industrial or domestic irons (with or without steam) or else with a heavy press. Whatever the pressing method chosen, pressing consists of five components namely, heat (temperature), moisture (usually as steam), pressure, time and equipment. When pressing low-volume, high-fashion or highly structured apparel, an industrial flat iron is usually used, similar to those used by the consumers at home. For pressing performed on high-volume, ready-to-wear apparel, presses or pressing machines, such as flatbed, buck presses, form presses and upright presses are normally used, these being more efficient than hand pressing. Generally, the apparel is slipped over the machine; steam and pressure are applied automatically and then a vacuum extracts any excess moisture. Ideal pressing conditions vary for different types of fabrics and apparel (Brown and Rice, 2001; Fan and Hunter, 2009; Crowley, 1978).

Effect of fabric properties on pressing performance

Different fabrics have different properties, which require different pressing conditions. For example, microfibre apparel has poor resistance to heat and can be hard to press and finish without any shine, marks or glazes. In order to solve this problem, it is recommended that apparel comprising microfibres or extremely light fabrics be pressed from inside using foam finishes. If they need to be pressed from the 'right side' of the fabric, the temperature has to be very low, whilst the steam has to be perfectly dry since synthetic cannot take any moisture. The press duration and ambient relative humidity for both preconditioning and post conditioning of the fabrics also has a significant influence on the pressing performance (Fan and Hunter, 2009).

16.5.3 Storage and packaging

After pressing, the garment should be free from wrinkle and creases, and have a good shape. Apparel needs to be stored and packed for delivery to the potential customers. Nowadays, there are manual or semi-automatic packaging machines available in the industry, but careless or inappropriate storage and packaging will cause a deterioration in the appearance of the final product.

The pressed apparel should be stored in a cool, dry place. The storage should have sufficient room to accommodate the apparel without being too closely packed, which could not only cause wrinkles in the apparel, but also block the air circulation (causing a moisture build-up leading to mildew on the apparels), or result in excessive heat causing the plastic bags to bond to the fabric. Also, the storage area should be as clean as possible so that dust and dirt, the acid pressure in atmospheric pollution, as well as the presence of moths or other pests, do not cause apparel to deteriorate during storage. Apparel in storage should not be subjected to any strain or movement that could cause the fibres to become weak and break. Avoid pressing in sharp folds. Different types of apparel should have different conditions of storage and packaging. Knitted or stretched apparel should preferably be folded rather than hung, and should be stored in a plastic bag. Nylon hosiery should be stored and packed in plastic bags to prevent snagging. Sweaters should also be stored in plastic bags to keep them from becoming contaminated by lint. Clothing containing wool should be stored in a moth-free or mothproofed place (Finch and Putnam, 1977; Fan and Hunter, 2009).

Vacuum packing is recommended for tailored jackets; this not only reduces the storage space required for the packed apparel, but also prevents reintroducing creases during storage and transportation. This packing is also reusable, which can reduce environmental pollution. Vacuum packaging is now commonly used in households, and available in different sizes for different kinds of apparel. Vacuum packing also saves space, and protects the apparel against dirt, moisture and insects (because it is waterproof and airtight) (Fan and Hunter, 2009).

16.6 Quality evaluation of apparel: testing for tailorability

The following section discusses in general the relationship between tailorability and low stress mechanical properties. Evaluation of low stress mechanical properties using KES (Kawabata Evaluation System) and FAST (Fabric assurance by simple testing) instruments are also discussed.

16.6.1 Tailorability and low-stress mechanical properties

Low-stress mechanical properties of fabrics have established themselves as an objective measure of quality and performance. There are two major reasons why they are important in tailoring:

- Fabrics are more extensible in the low-load region than in the higher-load region.
- The property in this region is closely related to the tailoring process and the comfort of a wearer.

In the tailoring process, an initially flat fabric is formed into stable, complex, three dimensional apparel shapes. The conformation of a flat fabric to any three dimensional surface requires complex mechanical deformation of the fabric, such as bending, extension, longitudinal compression and shearing in the fabric plane at very low loads. Apparel patterns often require different lengths of fabrics to be sewn together by overfeeding the longer of the two fabrics to form an intermediate length as a means of imparting a three dimensional character to the apparel. The seam is inserted either in the warp or weft direction or at some angle. Higher fabric extensibility in the initial region causes difficulty in the handling of fabrics during the cutting and sewing processes. Thus the fabric tensile, longitudinal compressional and shear properties are the main mechanical properties relevant to tailoring (Kothari, 1999a).

Fabric mechanical properties are measured under low loads so that conditions similar to the actual fabric deformation in production and use of apparels are taken. The hysteresis behaviour in tensile, shear, bending and compression is measured to determine the fabric resilience. The fabric surface properties are also measured to detect roughness by human senses. Test systems such as KES (Kawabata Evaluation System) and FAST (fabric assurance by simple testing) were developed for conducting tests on fabric samples in the low-stress region.

16.6.2 Kawabata evaluation system

Kawabata, in order to establish the relationship between subjectively assessed hand values and properties of fabrics, have identified certain properties such as tensile and shear, bending, compressional and surface frictional characteristics. These properties were considered as important from the point of view of fabric handle. In each group certain characteristic values have been identified that can represent the property and can be used for establishing interrelationships between fabric hand and these properties. A total of 16 such parameters given in Table 16.13 have been identified. Kawabata Evaluation System of Fabric (KES-F) has four modules for measuring low stress and surface characteristic of fabrics (Apurba Das and Alagirusamy, 2010; Kawabatta, 1980).

KES-F1 (tensile and shear tester)

The principle of KESF-1 system is shown in Fig 16.16. The fabric specimen is clamped between two jaws (one attached with drum for tensile force

Table 16.13 The sixteen parameters describing fabric properties

Parameter	Symbol	Description
Tensile	LT	Linearity of load extension curve
	WT	Tensile energy (gf.cm/cm^2)
	RT	Tensile resilience (%)
Shear	G	Shear rigidity (gf. cm/degree)
	2HG	Hysteresis of shear force at 0.5 degree shear angle
	2HG5	Hysteresis of shear force at 5 degree shear angle
Bending	B	Bending rigidity (gf.cm^2/cm)
	2HB	Hysteresis of bending moment (gf.cm/cm)
Compression	LC	Linearity of compression-thickness curve
	WC	Compressional energy (gf.cm/cm^2)
	RC	Compressional resilience (%)
Surface	MIU	Coefficient of friction
	MMD	Mean deviation of MIU
	SMD	Geometrical roughness (μm)
Fabric construction	W	Fabric weight per unit area (mg/cm^2)
	T	Fabric thickness (mm)

application, and the other attached with slide for shear force application) and subjected to a constant tension of 10 gf/cm by a weight attached to the drum on which one jaw is mounted. Tensile force is applied by allowing the drum to rotate freely. The tensile force is measured by the tensile force detector by measuring the torque and the tensile strain is measured by tensile strain detector from the data of rotational angle of the drum. The shear force is measured by a transducer connected to the other jaw (attached with slide) which moves sideways to apply shear deformation. The shear force is measured by the shear force detector by measuring the force required to slide, and the shear strain is measured by shear strain detector from the resulting displacement of the slide.

KES-F2 (bending tester)

The principle of the KESF-2 system is shown in Fig. 16.17. The fabric specimen is gripped by two jaws. One jaw is attached with the bending arrangement, which moves in a circular direction to apply bending force. The other jaw is connected with the torque sensor, which detects the torque value of steel wire during bending of specimen. The curvature of bending is obtained from the drive to the bending arrangement. The fabric specimen is bent with the help of bending arrangement between the curvatures of -2.5 cm^{-1} and $+2.5$ cm^{-1}.

KES-F3 (compression tester)

The working principle of KESF-3 system is shown in Fig. 16.18. The fabric specimen is compressed between two plates, that is anvil and pressure foot. The

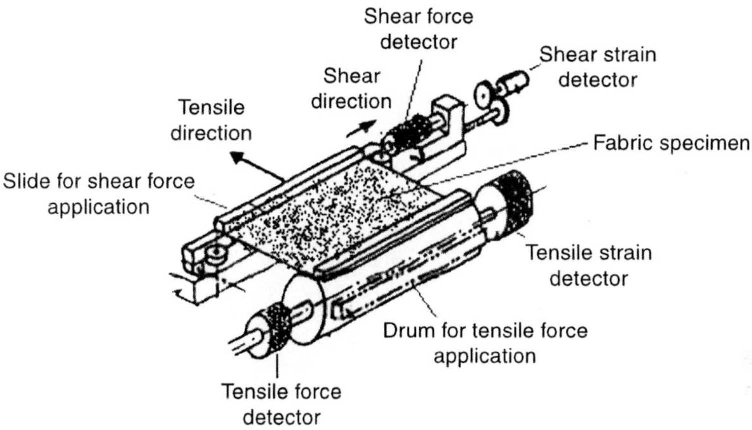

16.16 Working principle of KESF-1 system.

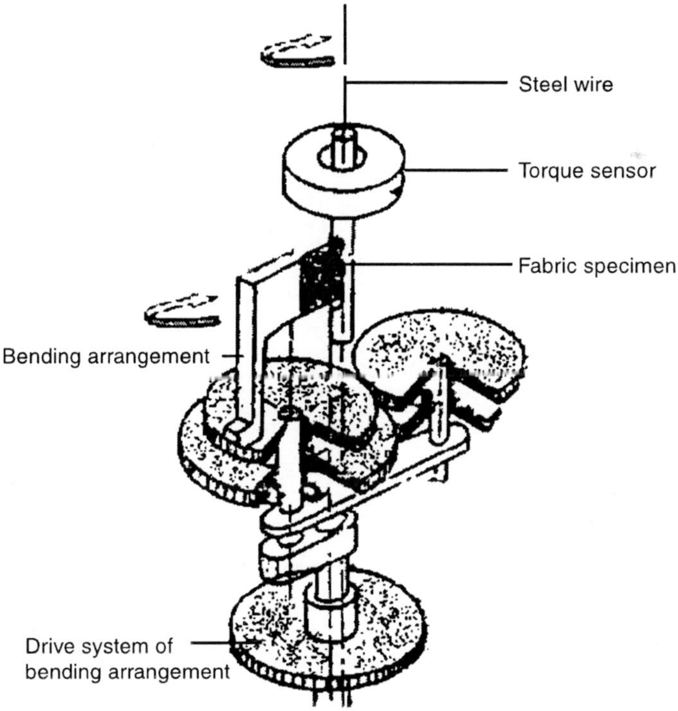

16.17 Working principle of KESF-2 system.

fabric specimen is placed on the anvil and pressure is increased with the help of the pressure foot, while continuously monitoring the sample thickness with a thickness detector. The compressive pressure is detected by the compressive force detector. The pressure foot is driven from the drive arrangement.

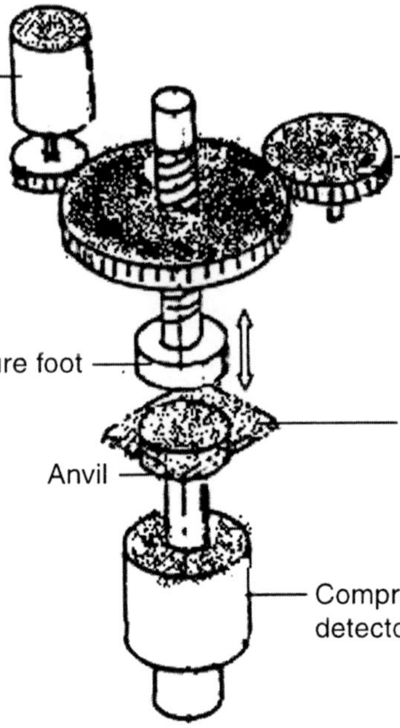

16.18 Working principle of KESF-3 system.

KES-F4 (surface tester)

Friction and surface roughness characteristics of fabrics are measured by KESF-4 system. The working principle is shown in Fig. 16.19, where the surface roughness and surface friction of the fabric specimen are being measured at the same time. The fabric specimen, kept at constant tension by a hanging dead weight, gets to-and-fro motion from a drum that rotates intermittently clockwise and anticlockwise. The frictional force between the fabric specimen and the friction surface at the friction point is detected by a frictional force detector. The surface roughness of the fabric is measured by a detection system that is basically a displacement sensor. The probe of the displacement sensor is in touch with the fabric surface. When the fabric moves in the horizontal plane, due to the surface roughness the probe is deflected vertically. This vertical deflection of the probe measures of the surface roughness of fabric.

From all the above parameters, which are obtained from KESF instruments, the total handle value (THV) can be calculated, which is the indicator of fabric handle behaviour of fabric.

16.6.3 Fabric assurance by simple testing (FAST)

The FAST system has been developed by CSIRO (Australia) primarily for quality control and assurance of fabrics. It measures properties that are

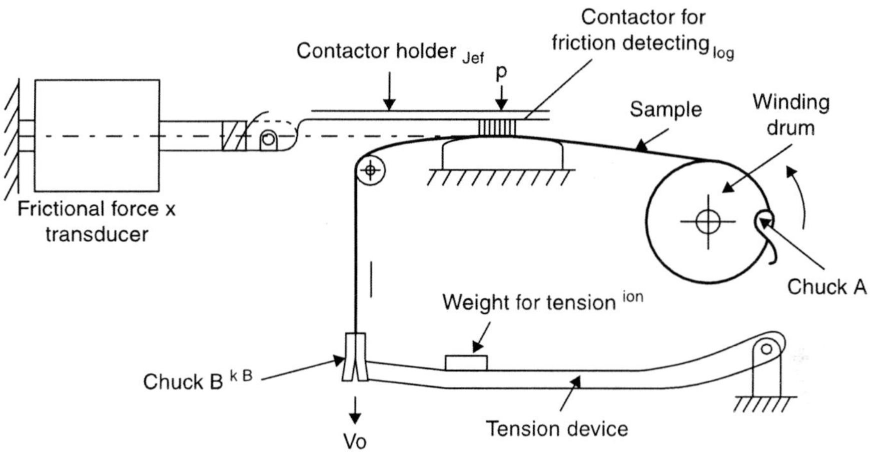

16.19 Working principle of KESF-4 system.

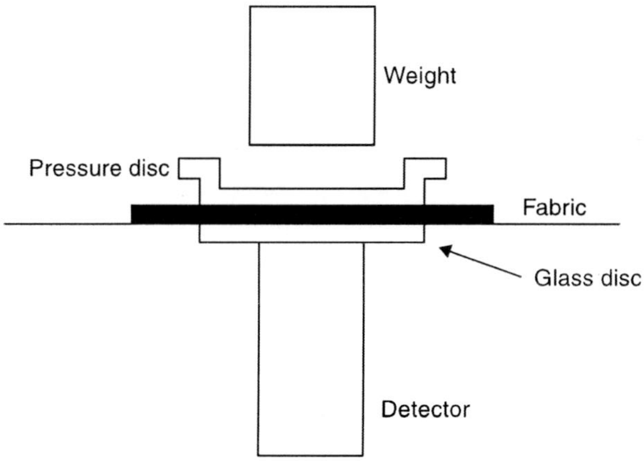

16.20 Working principle of FAST-1 system.

closely related to the ease of apparel manufacturing, handle characteristics and the durability of surface finishing. Unlike the KES system, FAST measures only the resistance of fabric to deformation and not its recovery from deformation. However, the FAST system is inexpensive, simple to use and robust in construction. It consists of a series of three instruments and a test method (Apurba Das and Alagirusamy, 2010; Kothari 1999c).

FAST-1 (compression meter)

A schematic diagram of FAST-1 system is shown in Fig. 16.20. It measures the fabric thickness over a range of loads, the variability and the durability of the thickness of the fabric surface layer. It can measure fabric thickness to

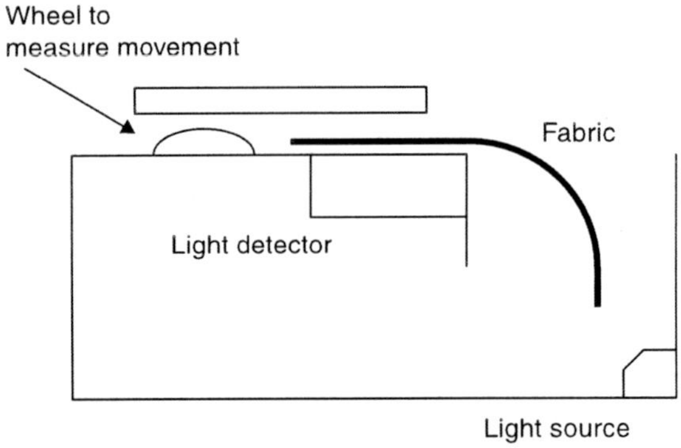

16.21 Working principle of FAST-2 system.

micrometre resolution at two predetermined loads, and thereby enables the accurate measurement of surface layer thickness. The fabric thickness (T) is measured at a pressure of 2 gf/cm^2. Surface thickness (ST) is the difference in thickness of a fabric measured at pressures of 2 and 100 gf/cm^2. This gives information about the hairiness or surface bulk of the fabric (closely related to surface treatment like brushing, singeing, finishing etc). Released surface thickness (STR) is the measure of the ST after the fabric is exposed to steam or water. The increase in fabric ST obtained by this steaming process simulates the actual change in surface characteristics that occur during actual use of the apparel.

FAST-2 (bending meter)

A schematic diagram of the FAST-2 system is shown in Fig. 16.21. FAST-2 measures the bending length (BL) and bending rigidity (B) of fabric. The fabric BL simulates the draping behaviour of fabric and B is related to the quality of stiffness when a fabric is handled. B is particularly crucial in the tailoring of lightweight fabrics as a very flexible fabric (low B) may cause seam puckering while a high B fabric can be more manageable in sewing and so produce a flat seam. Operator error in aligning the sample is eliminated with the use of an optical sensor. BL is displayed automatically, so chances of the error due to the operator's judgement are eliminated.

FAST-3 (extension meter)

A schematic diagram of the FAST-3 system is shown in Fig. 16.22. It measures extensibility of fabric at various loads as well as its shear rigidity. It is capable of measuring the fabric extensibility in warp, weft and bias directions over a range of loads, with direct reading of extension as a percentage of the initial gauge length. The fabric extension is displayed as a percentage

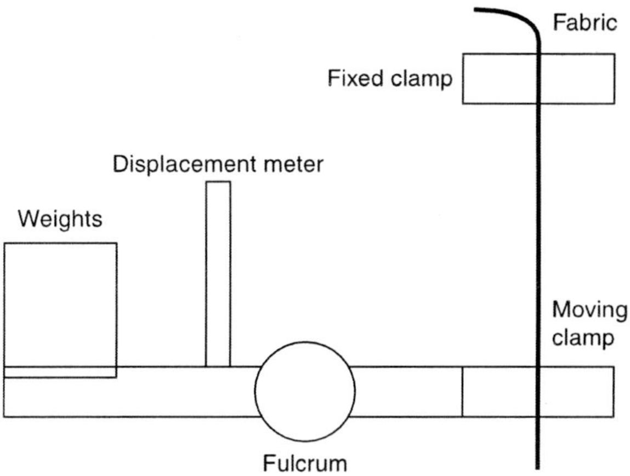

16.22 Working principle of FAST-3 system.

with a 0.1% resolution. Extensibility is measured at three loads 5 gf/cm (E5), 20 gf/cm (E20) and 100 gf/cm (E100). The difference between fabric extensibilities at between E5 and E20 is used to calculate fabric formability (F), which is a parameter related to the incidence of seam pucker. Fabric extensibility is combined with bending rigidity to calculate F, which is a measure of the ability of a fabric to absorb compression in its own plane without buckling. E100 is used in FAST control chart (FAST fabric fingerprint) as the measure of fabric extensibility. If the value is below approximately 2%, then the fabric will be difficult to extend during seam overfeed. Apart from the fabric extensibility, another important fabric handle related parameter, shear rigidity (G), is measured in the FAST-3 system. The bias extension is converted to G, which is directly related to fabric looseness. For G below 30 N/m, the fabric deforms so easily that it may give problems in handling, laying up and sewing. Conversely if it is above 80 N/m then the fabric can be difficult to overfeed, mould, etc.

FAST-4 (dimensional stability test method)

Poor dimensional stability of fabrics is one of the main causes of poor appearance in apparel. This is a test method for measurement of relaxation shrinkage (RS) and hygral expansion (HE), which can test these parameters in less than an hour as compared to the conventional one-day test. The drying is done in a forced convection oven. A template and a ruler are the only equipment required to do the test. The results from this method simulates the change in fabric dimensions that may occur during the actual wear as the fabric is subjected to washing and changing humidity conditions. RS is mainly due to the recovery of fabric structure that became strained during manufacturing, while HE (and hygral contraction) is caused by the swelling

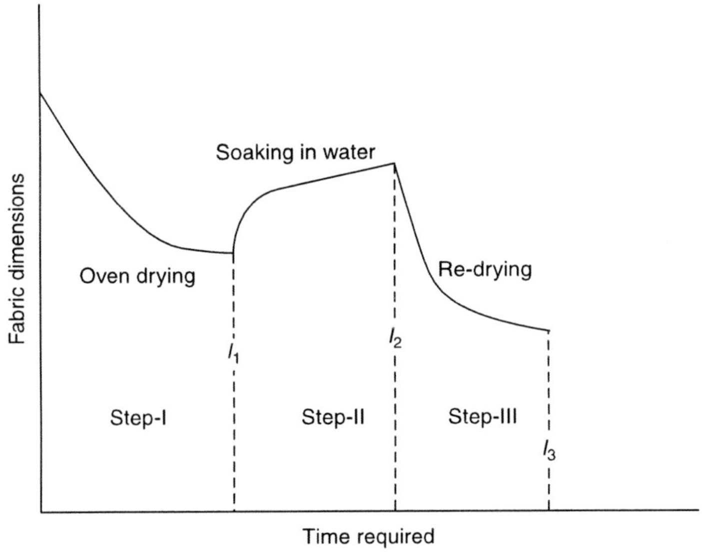

16.23 Steps to test the dimensional stability in FAST-4 system.

or de-swelling of hygroscopic fibres. Very high RS results in the problem of change of size, puckering, etc. Similarly, higher HE may result in seam pucker, fabric waviness, buckling and overall poor appearance.

The testing is completed in following three difference steps (Fig. 16.23):

- Step-I: The fabric specimen is first over dried (0% moisture) to measure its dry dimensions (l_1).
- Step-II: It is then soaked in water to measure its wet relaxed dimensions (l_2).
- Step-III: The specimen is dried again to measure its final dry dimensions (l_3).

RS and HE are calculated from the above dimensions, using following relationships:

$$RS = [(l_1 - l_2) \times 100] / l_1$$

$$HE = [(l_2 - l_3) \times 100] / l_3$$

The FAST data analysis software, included as part of the FAST system package, automatically plots the appropriate values and joins the various plotted points together to form a fabric 'FAST control chart', which is unique to each particular fabric. Figure 16.24 shows a typical FAST control

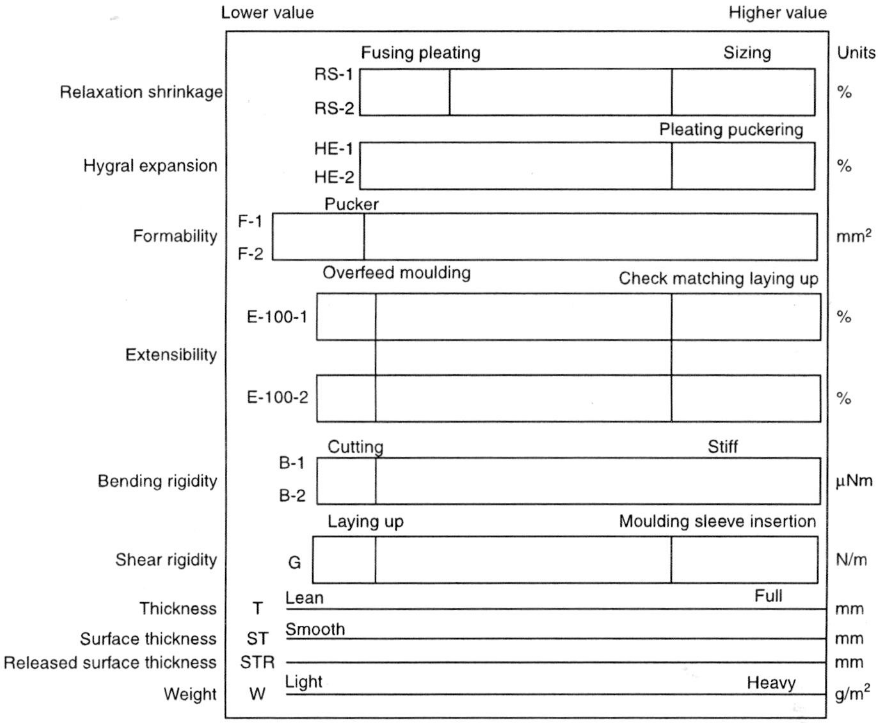

16.24 Typical FAST control chart.

chart. Each value has a separate scale showing a graphical representation of the range of values in the appropriate units that we expect for each of the various measurements. For example relaxation shrinkage 'RS-1' represents the warp value and 'RS-2' that of the weft. In addition to the values for the various measurements, each scale contains one or more shaded zones. If the fingerprint falls into one of these zones, a potential problem with the particular aspect(s) of fabric performance influenced by that property is indicated.

16.7 Quality evaluation of apparel: testing for sewability

Sewability is defined as the ability and ease with which fabric components can be quantitatively and qualitatively seamed together to convert a two dimensional fabric into three dimensional apparel. The quality and serviceability of a garment not only depends on the quality of the fabric, but also on the quality of the seam. Quality characteristics of the seam can be measured by seam parameters such as seam strength, seam slippage, seam damage or needle cutting and seam appearance (Kothari, 1999a).

16.7.1 Seam strength

Seam failure in apparel can occur because of either the failure of the sewing thread, leaving the fabric intact, or fabric rupture, leaving the seam intact or both breaking at the same time. Seam strength in woven and knit fabrics is tested in almost the same manner as fabric breaking and bursting strength respectively. A specimen 11×10 cm^2 is placed between two sets of jaws, 7.6 cm apart in such a way that the seam is in the middle of the 7.6 cm distance. These jaws are then pulled away from each other creating a tensile force on the specimen, ultimately resulting in a break. The force, elongation and nature of the break (seam break or fabric rupture) are noted [ASTM D1683]. In the case of a knitted specimen, the specimen is clamped in the ball burst tester in such a way that the seam is in the middle of the clamps. The force is exerted against the specimen by the steel ball, until rupture occurs.

The strength of a seam or stitching should equal that of the material it joins in order to have a balanced construction that will withstand the forces encountered in the use of the apparel of which the seam is a part. The stitch type, thread strength, stitches per inch, thread tension, seam type, and seam efficiency of the material affect the seam strength.

Generally, a seam made with chain stitch will be stronger than a seam made using lock stitch. Obviously, stronger the sewing thread, the stronger the seam. A higher number of stitches per centimetre, up to a point, will give higher seam strength, but too many stitches will weaken the fabric, so that the seam may stay intact but the fabric may rupture resulting in seam failure. Higher thread tension will give higher seam strength but too high a thread tension will result in seam puckering. Lap felled seams will be stronger than lapped seams. Fabric with higher seam efficiency will provide stronger seams than fabric with lower seam efficiency. Seam efficiency is seam strength expressed as a percentage of fabric breaking strength. The elasticity of a seam or stitching should be slightly greater than that of the material that it joins, so that the material will support its share of the forces encountered in the end use of the apparel. The elasticity of a seam or stitching depends on the stitch type and thread elasticity.

16.7.2 Seam slippage

In some apparel, before seam failure occurs, enough yarn slippage (filling yarns shifting over warp yarns or vice versa) develops to render the apparel unusable, because such failure is not readily repairable by seaming. Seam slippage is measured by subtracting the elongation of the fabric from the elongation of the fabric with a seam in it. The difference is indicated as seam slippage, which can be considered one form of failure of seam assembly [ASTM D434 and D1683].

16.7.3 Seam damage or needle cutting

Seam damage or needle cutting in the fabric is objectionable because it may result in reduced seam strength or poor appearance or both due to frayed yarns. To find the needle cutting index, sewing threads are removed from specimens. The count of the number of fabric yarns and the count of the number of severed and fused fabric yarns in the direction nearly perpendicular to the direction of sewing thread are used.

$$\text{Needle cutting index (\%)} = \frac{\text{No. of yarns cut / unit length}}{\text{No. of yarns in fabric / unit length}} \times 100$$

Seam damage or needle cutting occurs due to stiffness of the yarns in fabric and their lack of mobility. Instead of moving and deforming when the needle penetrates the fabric structure, the yarns remain taut and are ruptured or burned. Also some damage may occur due to excessive heat generated due to friction of the sewing needle and fabric.

16.7.4 Seam appearance

The overall seam performance can be best judged by examining surface look of the seam line with a microscope. Although this method is quite subjective and needs expertise to predict the performance, it can however identify various aspects of seam appearance. The microscopic study helps in examining seam and surface damage.

16.8 Quality evaluation of accessories

Accessories add value to the apparel, hence the quality evaluation of accessories is very important. The following section discusses the quality evaluation of accessories such as zippers, sewing threads, buttons and snap fasteners.

16.8.1 Zippers

Zippers can be tested using any one or more of the following test methods (Kothari, 1999b; Pradip and Satish, 1998).

Durability of finish of zippers to laundering

The durability of the finish of zippers to laundering is evaluated by laundering the test specimen in a Launderometer (AATCC test method 61).

The effect of the test on the zipper coating is evaluated by noting the loss of coating on the zipper chain or components or both (ASTM D2051).

Colourfastness of zippers to dry-cleaning, light and crocking

The colourfastness of a zipper to dry-cleaning is tested by subjecting the zipper stringer (tape) to commercial dry-cleaning with a multifibre fabric. The dry-cleaned specimen is compared with the original specimen and any change in the colour of the specimen or staining of the multifibre fabric is then assessed using the AATCC gray scale for colour change or the chromatic transference scale. The colourfastness of zippers to light and crocking is tested in the same way as the colourfastness of fabrics to light and crocking (ASTM D2052, ASTM D2053 and ASTM D2054).

Colourfastness of zippers to laundering

The colourfastness of zippers to laundering is tested by subjecting the zipper with a multifibre test fabric to home laundering according to the intended care instructions for the apparel on which this particular zipper would be used. The alteration in shade of the zipper stringer (tape) and the degree of staining of the multifibre test fabric are evaluated by the AATCC gray scale for staining and colour change or the chromatic colour transference scale (ASTM D2057).

Durability of finish of zippers to dry-cleaning

The durability of the finish of zippers to dry-cleaning is tested by subjecting the zipper to dry-cleaning, as in AATCC test method 86, but the zipper is air-dried rather than hot-pressed. The specimen is then evaluated visually for any exposed base metal compared to a new zipper or compared to a sample illustrating an acceptable degree of coating loss (ASTMD2058).

Resistance of zippers to salt spray

Sometimes, due to corrosion, a zipper will not operate smoothly and its crosswise strength may be reduced. Such deterioration in a zipper can be evaluated by subjecting the zipper to a salt spray test. Of course, plastic/nylon zippers do not corrode, and therefore, this test applies only to metal zippers. In this test method (ASTM D2059), specimens are subjected to salt spray (5% salt solution at 33°C to 36°C) for 24 h continuously. The exposed specimens are then visually evaluated for any sign of corrosion and tested for case of operation and crosswise strength, and the results are compared with the case of operation and crosswise strength of the original specimens (ASTM D2059).

Operability of zippers

The operability of zippers is tested by pulling the slider with a force indicator (such as a pull gauge) along the zipper chain alternately in the opening and closing directions and recording the force required to maintain each movement. This force is a measure of the case with which the zipper will operate in end-use applications (ASTM D2062).

Strength tests of zippers

The usefulness of a zipper in service can be evaluated by the following strength tests. No test determines the suitability of a zipper for a specific end use. Since the tests are interrelated, more than one may be needed for a complete evaluation. Zipper strength is usually tested in the following areas (ASTM D2061):

- *Cross wise strength:* The ability of a zipper chain to withstand lateral stress is measured by loading to destruction a 2.5 cm section of the specimen in a tensile testing machine.
- *Scoop pull-off:* The gripping strength of a scoop (tooth) around the bead is determined by pulling a single scoop from the bead at right angles to the stringer using a tensile testing machine with a specially designed fixture.
- *Holding strength of stops:* The ability of stops to perform their intended purpose is determined through the use of five different methods that simulate the major stresses encountered in the end use of the zippers.
- *Scoop slippage:* The ability of a scoop to resist longitudinal movement along the bead of the tape is determined with a tensile testing machine fitted with a specially designed fixture.
- *Resistance to cushioned compression of sliders:* The lower plateau of a compression tester is cushioned with a neoprene pad. The specimen is laid on the pad and a load is applied. Then, the operability of the zipper is tested and compared to the operability of the original zipper.
- *Slide deflection and recovery:* There are two procedures for determining the resistance of zipper slider planes to an opening or spreading force. In one procedure, the force is applied to the mouth of the slider. In the other, an alternative method, the force is applied through the slider pull and backplane of the slider.
- *Resistance to twist of pull and slider:* In this method, the twist resistance of a pull and slider assembly against a torsional force applied to the pull of the zipper is evaluated. A fixture is used with a torque wrench to apply a specified twisting force to a slider pull. The amount of permanent twist

imparted to the slider pull or other permanent damage or deformation are noted. The specimen is also examined for any other effects, such as breaking or deformation of the lug or any other part of the assembly.

- *Resistance to pull-off of slider pull:* In this test, with a special fixture, tensile load is applied to the slider pull to determine how much force is required to pull off the slider pull.

16.8.2 Sewing threads

It may be necessary sometimes to test sewing thread for any one or more of the following characteristics (Kothari, 1999b; Pradip and Satish, 1998).

Yarn diameter

Knowledge of thread diameter is important because it can affect sewing performance and seam appearance. Sewing performance can be influenced because thread is required to pass through restrictions, such as a needle's eye and tension discs. Seam appearance can be adversely affected when the diameter of a thread is large enough to displace fabric yarn and results in a puckered seam. Sewing thread diameter is also a consideration when selecting sewing threads for embroidery, contrast stitching, or other decorative applications. The diameter of a thread is determined either with a thickness gauge (preferred method) or optical method (alternative method).

The procedure (ASTM D204) for measuring sewing-thread diameter by a thickness gauge is given below. Draw the thread from the side of the sewing thread holder, taking care not to disturb the twists. Place four strands of the thread side by side on the anvil and approximately midway between the sides of the pressure foot of the thickness gauge. Measure the thickness to the nearest 0.0025 cm under 240 gm/cm^2 at 10 m points along the thread and calculate the average as the diameter of the sewing thread The optical method for measuring sewing thread diameter is not recommended because it is difficult to determine the exact boundaries of threads having hairy fibres on the surface.

Yarn number of a sewing thread

Yarn number is a measure of the fineness or size of a yarn, expressed either as mass per unit length or length per unit mass. There are two systems of expressing yarn number or yarn count: a direct system and an indirect system. Under the direct yarn-numbering system, yarn number is expressed in terms of mass per unit length. The most frequently used units of the direct system are denier (weight of 9000 m of yarn in grams) and tex (weight of

1000 m of yarn in grams). In an indirect yarn-numbering system, yarn number is expressed in terms of length per unit mass. The most widely used unit for expressing the yarn number of a sewing thread in the indirect system is cotton count, which is the number of 840-yard lengths of yarn per pound, generally used for yarns spun on a cotton system.

Yarn number based on short-length specimens can be determined by (ASTM D1059). In this method, a short length (25–50 cm) of yarn is taken from a spool, conditioned, and weighed. The yarn number is then calculated from the weight and the measured length of the yarn. This is a quick method for the determination of the approximate yarn number. Because any error present in the reported length of the yarn specimen is multiplied many times when calculating the theoretical yarn number, it is extremely important that the length be measured as precisely as possible. Another method (ASTM D1907) for determining yarn number uses a much longer yarn length. Under this method, specified lengths of yarn are wound on a reel as skeins and weighed. The yarn number is calculated from the mass and length of the yarn in the skein. For all practical purposes, the calculation of the yarn number of a sewing thread based on short-length specimen is adequate.

Yarn ply

For checking the number of plies in a sewing thread, simply cut a short length of the sewing thread from the spool, hold two ends in each hand and untwist the sewing thread to a point that one can count the number of yarns plied together to make that sewing thread unless it is single ply (Kothari, 1999b; Pradip and Satish, 1998).

Strength and elongation

The strength and elongation of sewing thread must be adequate for good sewing performance as well as good seam strength. Strength and elongation of sewing threads are determined by breaking a single sewing thread on a tensile testing machine and recording the force and elongation at break (ASTM D2256). Sometimes, it is necessary to measure loop strength and loop elongation of a sewing thread, as these properties are also very relevant to the performance. The loop strength and elongation of a sewing thread are a measure of the thread's ability to contribute to seam performance. The loop strength of a thread bears a direct relationship to stitch strength and hence to seam strength. Loop elongation is an indication of the degree to which a seam, under stress, can be stretched without a thread breaking. Besides loop elongation, the ultimate elongation of a seam is dependent on the material stitched, the stitch and seam type, and number of stitches per inch. In a loop strength and elongation test, each specimen consists of two pieces of yarn taken from one package or end. Both ends of one piece are secured in one clamp of the testing machine so that the length of the loop is about one-half the gauge length. One end of the second loop is passed though the loop formed by the first piece of the sewing thread, and both ends of the second piece are clamped in the other clamp of the tensile testing machine. The machine is started, and the force and elongation, when the loop breaks, are observed and noted, which are the values of loop strength and elongation.

Twist balance and number of twists

The determination of twist balance of a sewing thread is important in predicting the snarling tendency of the thread during actual sewing operations. In this method, about a metre of conditioned thread from a spool is withdrawn, in the same manner as that in which it is delivered to the sewing machine, and formed into a loop, positioning the ends of the threads so that they are 10 cm apart at the top of the loop. The twist balance is reported in terms of the complete rotations that the loop makes.

The number of twists per centimetre is measured by counting the number of turns of twist in a known length of sewing thread as they are being removed by rotating one end of the specimen while the other end remains fixed until the elements of the thread being tested are parallel and free from twist (ASTM D1423).

Shrinkage

Good shrinkage performance of sewing thread is important because shrinkage can cause puckering of a seam, thus adversely affecting seam appearance. A conditioned single end thread is measured under a prescribed tensioning force before and after exposure to boiling water for ½ h or dry heat 152 ±3°C for 1 h. The change in length is expressed as a percentage of the length before exposure (ASTM D204).

Length per thread holder

The length of sewing thread on a thread holder is measured in metres or yards while being removed from the thread holder (ASTM D204 and D3693).

16.8.3 Buttons

Durability of buttons can be tested by an impact test. Individual buttons are placed on a surface centred under a tube, through which a preselected mass falls from a preselected height. After the mass impacts the button, the button is removed from the testing device and visually examined using a 5x magnifying glass for breakage, cracking, or chipping. This practice is used for acceptance testing of buttons. The buttons are classified as class A, B1 through B3 depending upon the impact resistance, the class A buttons being the most durable (ASTM D5171).

16.8.4 Snap fasteners

The quality of snap fasteners is generally judged by how easily they can be fastened or unfastened and their holding power. The resistance to unsnapping of snap fasteners can be determined by testing snap fasteners mounted on a strip of material near the end are tested on standard tensile testing machines equipped for testing the strength of textile fabrics and having sensitivity for accurate low force levels. Test are made on snap fasteners before laundering and after a predetermined number of launderings or dry-cleanings (ASTM D4846).

16.9 References

AATCC Test Method 61, Colorfastness to Laundering, Home and Commercial: Accelerated, Technical Manual of the AATCC.

Anon, Technical know-how in garment processing, www.apparelkey.com

Apurba Das and Alagirusamy R (2010), *Science in clothing comfort*, New Delhi, Woodhead publishing India.

ASTM D204, Standard Test Methods for Sewing Threads. West Conshohocken, ASTM International.

ASTM D1059, Standard Test Method for Yarn Number Based on Short-Length Specimens, West Conshohocken, ASTM International.

ASTM D1423, Standard Test Method for Twists in Yarns by Direct-Counting method, West Conshohocken, ASTM International.

ASTM D1907, Standard Test Method for Yarn Number by the Skein Method, West Conshohocken, ASTM International.

ASTM D2051, Standard Test Method for Durability of Finish of Zippers to Laundering, West Conshohocken, ASTM International.

ASTM D2052, Standard Test Method for Colorfastness of Zippers to Dry-cleaning, West Conshohocken, ASTM International.

ASTM D2053, Standard Test Method for Colourfastness of Zippers to light, West Conshohocken, ASTM International.

ASTM D2054, Standard Test Method for Colorfastness of Zipper Tapes to Crocking, West Conshohocken, ASTM International.

ASTM D2057, Standard Test Method for Colorfastness of Zippers to Laundering, West Conshohocken, ASTM International.

ASTM D2058, Standard Test Method for Durability of Finish of Zippers to Dry-cleaning, West Conshohocken, ASTM International.

ASTM D2059, Standard Test Method for Resistance of Zippers to Salt Spray (Fog). West Conshohocken, ASTM International.

ASTM D2060, Standard Practice for Measuring Zipper Dimensions, West Conshohocken, ASTM International.

ASTM D2061, Standard Test Methods for Strength Tests for Zippers, West Conshohocken, ASTM International.

ASTM D2062, Standard Test Methods for Operability of Zippers, West Conshohocken, ASTM International.

ASTM D2256, Standard Test Method for Tensile Properties of Yarns by the Single-Strand Method, West Conshohocken, ASTM International.

ASTM D 3693, Specification for Labeled Length per Holder of Sewing Thread, West Conshohocken, ASTM International.

ASTM D4910, Standard Tables of Body Measurements for, Infants, Ages 0 to 18 Months, West Conshohocken, ASTM International.

ASTM D5171, Standard Test Method for Impact Resistance of Plastic Sew-Through Flange Buttons, West Conshohocken, ASTM International.

Brown P and Rice J (2001), *Ready-to-wear apparel analysis*, New Jersey, Prentice Hall.

Butler M G (1972), *Clothes their choosing making and care*, London, B T Batsford Ltd.

Carr H and Latham B (1994), *The technology of clothing manufacture, Second edition*, United Kingdom, Blackwell science.

Carr H and Latham B (2000), *The technology of clothing manufacture, Third edition*, United Kingdom, Blackwell science.

Cooklin C (1990), *Fusing technology*, Manchester, The Textile Institute.

Crowley C H (1978), *All about clothes*, USA, McGraw-Hill Book Company.

Crum R J (1983), *Methods of joining fabrics: A technical literature survey*, Manchester, Shirley Institute.

Dorkin C M C and Chamberlain N H (1961), Seam pucker its cause and prevention: Report No. 10, London, Clothing Institute.

Dorkin C M C and Chamberlain N H (1963), The facts about needle heating: Report No. 13, London, Clothing Institute.

Fan J and Hunter L (2009), 'Engineering apparel fabrics and garments' in Fan J, Lau L and Hunter L (eds.), *Appearance issues in garment processing*, Cambridge, Woodhead, 130–60.

Finch K and Putnam G (1977), *Caring for textiles*, New York, Watson-Guptill publications

Galuszynski S and Robinson G A (1986), Fusing of polyester/viscose fabrics, SAWTRI Technical Report, No. 581.

Hati S and Das B R (2011), Seam pucker in apparels, Asian network for scientific information.

Hurt F N and Tyler D J (1971), An investigation of needle heating and associated problems in machine sewing-I, HATRA Research Report No.19.

Hurt F N and Tyler D J (1972a), An investigation of needle heating and associated problems in machine sewing-II, HATRA Research Report No.20.

Hurt F N and Tyler D J (1972b), An investigation of needle heating and associated problems in machine sewing-III, HATRA Research Report No.21.

Kawabatta S (1980), The standardisation and analysis of hand evaluation, Japan, Text mach soc.

Kothari V K (1999a), 'Testing and Quality management' in Behera B K (ed.), *Fabric quality and apparel industry*, New Delhi, IAFL Publications, 386–402.

Kothari V K (1999b), 'Testing and Quality management' in Pradip V M (ed.), *Quality control in apparel production*, New Delhi, IAFL Publications, 403–93.

Kothari V K (1999c), 'Testing and Quality management' in Kothari V K (ed.), *Physical testing of fabrics*, New Delhi, IAFL Publications, 326–59.

Laing R M and Webster J (1998), *Stitches and seams*, Manchester, The Textile Institute

Pradip V M and Satish K B (1998), *Managing quality in the apparel industry*, New Delhi, New Age International publishers.

Solinger J (1980), *Apparel manufacturing handbook*, New York, Nostrand Reinhold.

Solinger J (1988), *Apparel manufacturing handbook, Second edition*, Columbia, Bobbin publications.

Sondheim W S (1953), 'Causes of seaming damage: Needle heating', *J Text Inst*, **44**, 580–5.

Worthington A P (1975), Effects of sewing machine needle in clothing manufacture: A technical literature survey, Manchester, Shirley Institute.

Index